Energy in Electromagnetism
by H. G. Booker

Errata

We apologise for the following errors that have occurred in the typesetting of this book:

Page 36, Problem, **1.9**. In B_θ all three terms should be positive
Page 60. In eqn. 2.71 replace κ_L by $1 + \kappa_L$
    In eqn. 2.75 replace $P_b \ (3\epsilon_0)$ by $P_b/(3\epsilon_0)$
Page 72, Problem 2.16. In line 14 replace 2.15 by 2.14, but in line 15 retain 2.15
Page 100, Section 4.3. In line 14 replace Dds by DdS
Page 244. In the second of eqns 11.24 increase the phase angle by π
Page 335, Fig. 14.7. The markings on the abscissa should read, in order,

$$0, \tfrac{1}{4}\lambda_g, \tfrac{1}{2}\lambda_g, \tfrac{3}{4}\lambda_g, \lambda_g$$

Pages 342 and 343. Interchange the two pages.

IEE ELECTROMAGNETIC WAVES SERIES 13
SERIES EDITORS: PROFESSOR P. J. B. CLARRICOATS,
G. MILLINGTON
E. D. R. SHEARMAN
AND J. R. WAIT

ENERGY IN ELECTROMAGNETISM

H.G. BOOKER

ENERGY IN ELECTROMAGNETISM

H.G. BOOKER

Henry G. Booker
University of California
San Diego

PETER PEREGRINUS LTD.
On behalf of the
Institution of Electrical Engineers

197992

Published by: The Institution of Electrical Engineers, London
and New York
Peter Peregrinus Ltd., Stevenage, UK, and New York

© 1982: Peter Peregrinus Ltd.

British Library Cataloguing in Publication Data

Booker, Henry G.
Energy in electromagnetism. — (IEE electromagnetic
waves series; 13)
1. Electromagnetism
I. Title   II. Series
537   QC760

ISBN 0-906048-59-1

Typeset at the Alden Press Oxford London and Northampton
Printed in England by Short Run Press Ltd., Exeter

Contents

Preface

Is the teaching of electricity and magnetism in need of change? One might imagine that any change needed was accomplished with the development of the *Berkeley Physics Course* (McGraw Hill, New York, 1963–5) and with the appearance of *The Feynman Lectures in Physics* (Addison-Wesley, New York, 1964). Not so. Feynman himself says in his preface: 'In the second year I was not so satisfied. In the first part of the course, dealing with electricity and magnetism, I could not think of any really unique or different way of doing it – of any way that would be particularly more exciting than the usual way of presenting it. So I don't think I did very much in the lectures on electricity and magnetism.' The preface to the *Berkeley Physics Course* says: 'Our specific objectives were to introduce coherently into an elementary curriculum the ideas of special relativity, of quantum mechanics, and of statistical physics.' There is no mention of comprehensive rethinking of the presentation of electricity and magnetism.

The preface to Volume II of the *Berkeley Physics Course* dealing with electricity and magnetism says: 'The sequence of topics is, in rough outline, not unusual. . . . . The difference is most conspicuous in Chapters 5 and 6 where, building on the work of Vol. I, we treat electric and magnetic fields of moving charges as manifestations of relativity and the invariance of electric charge.' The book certainly does make it easier for a student to understand particle accelerators, but not ordinary electrical machines that do not involve relativistic velocities. I doubt if a student who has studied the book could explain in terms of simple physics not involving corrections of order $(v/c)^2$ how, for the same machine performing the same task, different charge distributions are acquired as seen by two different observers, one fixed relative to the field-winding and the other fixed relative is the armature-winding.

The preface to Volume II of the *Berkeley Physics Course* goes on to say: 'This approach focuses attention on some fundamental questions, such as: charge conservation, charge invariance, the meaning of field.' Yet the meaning of field is obscured in the book precisely where charge conservation is concerned. The Maxwell equations divide into two groups. There are two equations involving E and B that are directly concerned with the exercise of force in accordance with the expression $E + v \times B$; they correspond to the kinetic equations in mechanics.

The other two Maxwell equations are relations between the quantities H, D, J, ρ that are concerned with current and charge geometry; the unit of H is current per unit distance, that of D is charge per unit area, that of J is current per unit area, and that of ρ is charge per unit volume. These two Maxwell equations may be likened to the equations of kinematics in mechanics; they incorporate the principle of conservation of charge embodied in the equation of continuity. I know of no book that adequately brings out the quite different physical bases that exist for the E, B Maxwell equations and for the H, D, J, ρ Maxwell equations.

These considerations led me to question the conventional presentation of electricity and magnetism. In consequence, the entire basis of electricity and magnetism, as well as the methods used for teaching it, have been subjected to the kind of original thinking customarily reserved for research. In some areas this has confirmed the wisdom of the traditional approach, but in other areas improvements have come to light. In some cases, the fresh thought has actually resurrected in novel form ideas that people once discarded. But wherever the electric-flux-density vector D is concerned, the changes triggered by the fresh thinking have turned out to be substantial. Even in this respect, however, an interesting link exists to the early thinking of Faraday.

Radiation from antennas has always been treated by dissecting the current into elements and regarding each current element, together with the associated charges at its ends, as an electric dipole. It is possible, and frequently desirable, to treat any flow of free electric current as a distribution of free electric moment per unit volume P_f. As in the case of bound current and charge associated with bound electric moment per unit volume P_b, the free current density is the time-derivative of P_f and the free charge density is the negative divergence of P_f. Propagation of electromagnetic waves in a plasma has been treated for half a century by handling the free electric moment per unit volume P_f in a plasma in the same way as is done for the bound electric moment per unit volume P_b in a dielectric. This means that the electric-flux-density vector used in connection with the complex dielectric constant (or tensor) of a plasma is not identical with the vector D introduced in most elementary textbooks in connection with dielectric insulators.

Care is necessary to distinguish between three different versions of the electric-flux-density vector in materials. For an oscillatory field of angular frequency ω in material of dielectric constant ϵ/ϵ_0 and conductivity σ, the three versions of the complex electric-flux-density vector at a point where the complex electric vector is E are $\epsilon_0 E$, ϵE, and $(\epsilon - j\sigma/\omega)E$. The first of these is the electric-flux-density vector to be used if one simply regards the material as creating an additional distribution of current and charge in free space. On the other hand, ϵE is an electric-flux-density vector that suppresses in Maxwell's equations the currents and charges associated with the bound electric moment per unit volume P_b; this is convenient when discussing the design and operation of equipment where special attention needs to be paid to the free currents and charges on good conductors. But $(\epsilon - j\sigma/\omega)E$ is an electric-flux-density vector that suppresses in Maxwell's equations the currents and charges associated with both the bound electric moment per unit volume P_b and the free electric moment per unit volume P_f; it is usually the convenient one to use when the wave character of the electromagnetic field under study is dominant. For any time-varying electromagnetic field in any material, including non-linear and non-isotropic materials, the three versions of the electric-flux-density vector are, respectively, $\epsilon_0 E$, $\epsilon_0 E + P_b$, and $\epsilon_0 E + P_b + P_f$.

The upshot of the thinking outlined in this preface has been a presentation of electricity and magnetism in three parts. These do not correspond to electrostatics, magnetostatics, and electromagnetism. Instead they deal with:

(1) electromagnetic fields in free space,
(2) electromagnetic fields in materials, and
(3) energy in electromagnetism.

Consistent with not requiring the reader to take leaps in the dark, Part 1 follows a direct route to the calculation of the electromagnetic field in free space caused by any known distribution of time-varying electric current and its associated distribution of time-varying electric charge. Dielectric and magnetic materials do not appear explicitly in Part 1; conductors appear only as locations for electric currents and charges.

Materials constitute, from an electromagnetic standpoint, distributions of current and charge in free space additional to the primary sources of the field. But usually these additional currents and charges are not immediately known. The local dipole moments are not known because they depend on the local electromagnetic field, which is just what we are trying to calculate. This is the fundamental complication created in electromagnetic theory by the presence of materials. It is this complication that is the subject of Part 2.

The magnetic vector potential at a point in an electromagnetic field is presented as the electromagnetic momentum per unit charge of a test charge located at the point, the total momentum of a particle being the vector sum of its mechanical momentum and its electromagnetic momentum. Currents round circuits are then seen as electronic fly-wheels for which the effect of charge is normally much greater than that of mass. Also, in Part 3, the reader is encouraged to think of the Poynting vector as representing a flow of energy in space that can be intercepted and measured in substantially the way in which flow of energy along a power line can be intercepted and measured. The vector field $E \times H$ is not then a by-product of Poynting's theorem plagued by an indeterminacy involving closed flow of energy. On the contrary, the Poynting vector is something as vivid as the product VI on the basis of which consumers of electric power pay their bills.

In the division of subject matter described above, the present volume constitutes Part 3. Chapters 1 and 2 summarise the contents of Parts 1 and 2 concerning electromagnetic fields in free space and in materials. They therefore deal in outline form with a number of the approaches developed in Parts 1 and 2. A few problems are provided, not as exercises, but to incorporate in suscinct form relevant additional material.

A number of the ideas employed in this text have been under development for a considerable period of time. Some originated in the Bell Telephone Laboratories before World War II, some came to the fore in connection with radar research, some appeared in lectures given by the author at the University of Cambridge, and some in a text (Henry G. Booker, An Approach to Electrical Science, McGraw Hill, New York, 1959) written at Cornell University. But it is a sequence of courses presented in recent years at the University of California, San Diego, that has provided the opportunity to assemble this book. The work was supported by an Instructional Improvement Grant from the University of California. My thanks are due to Pat Norvell for converting almost illegible manuscript into readable typescript, to Jerry Ferguson and Reza MajidiAhi for performing the computations needed for some of the diagrams, and to Hari Vats for reading the proof.

Henry G. Booker
January 1981

Chapter 1

Electromagnetic fields in free space

1.1 Introduction

Before we can study energy in electromagnetism, it is necessary to understand electromagnetic fields. In this chapter, we shall summarise the facts concerning the electromagnetic fields created by known time-varying distributions of electric charge and current in otherwise free space.

Material consists of an aggregate of particles in space many of which are charged and are in motion. Such a distribution of moving charged particles constitutes electric charge and current in space. However, the electric charge and current densities in a material near a point 0 depend on the electromagnetic field in the neighbourhood of 0, that is, on the very electromagnetic field that we seek to calculate. This causes a complication that we shall not address until the following chapter.

Generally, we shall suppose in this chapter that the distributions of electric charge and current exist in otherwise free space, and that they are known functions of position and time. The problem is to calculate the electromagnetic field that they create. However, it will sometimes be instructive to regard the electromagnetic field as known in space and time, and to enquire what distributions of electric charge and current are needed to create it.

1.2 Force in an electromagnetic field

The exercise of force in an electromagnetic field is described by means of the electric field and the field of magnetic flux density. At a point in an electromagnetic field where the electric vector is E and the magnetic-flux-density vector is B, the force per unit charge that would act on a test charge at the point is

$$E + v \times B \tag{1.1}$$

where v is the velocity of the test charge relative to the observer. Different observers in motion relative to each other observe different values of v, but they also observe different values of E and B. When all observers have evaluated $E + v \times B$

they are in agreement with each other on the force exerted. This agreement constitutes equality when the velocities are small compared with the velocity c of light in free space.

The fields described by the electric vector E and the magnetic-flux-density vector B are related by Faraday's law of induction. If ds is a vector element of length of any closed curve C in any electromagnetic field, and if E is the electric vector at ds, then (see Fig. 1.1)

$$\int_C E \cdot ds \tag{1.2}$$

is the circulation of E round C in the direction of ds. If S is a surface spanning C and dS is a vector element of area of S at a point where the magnetic-flux-density vector is B, then

$$\int_S B \cdot dS \tag{1.3}$$

is the magnetic flux crossing S in the direction of dS. In terms of these concepts, Faraday's law of induction states that, for any closed curve C fixed relative to the observer and spanned by a surface S,

$$\int_C E \cdot ds = -\frac{\partial}{\partial t} \int_S B \cdot dS \tag{1.4}$$

where t denotes time. Faraday's law of induction expresses the fact that the circulation of the electric vector round any fixed closed curve C in any electromagnetic field is equal to the time-rate of decrease of the magnetic flux crossing a surface S spanning C. The law implies the adoption of a sign convention, and we employ the right-hand-screw rule. We regard the vector product in expr. 1.1 as defined with the aid of the right-hand-screw rule, and we also regard the direction of the vector element of area dS of S in Fig. 1.1 as related by the right-hand-screw rule to the direction of the vector element of length ds of the rim C.

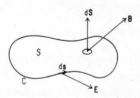

Fig. 1.1   *Illustrating calculation of the circulation of the electric vector round a closed curve C and calculation of the magnetic flux crossing a spanning surface S*

Eqn. 1.4 applies in particular to the rim of every fixed element of area dS in space, no matter what its location or what its orientation. The curl of a vector field E (see Appendix A) is a derived vector field such that, for any vector element of area dS at a location where the derived vector field is $\operatorname{curl} E$, the right-hand related circulation of E round the rim of dS is $(\operatorname{curl} E) \cdot dS$. Hence, by definition of curl, eqn. 1.4 implies that, at all locations at all times,

$$\operatorname{curl} E = -\frac{\partial B}{\partial t} \tag{1.5}$$

This is Maxwell's electric curl equation. It is a relation between the two vector fields (E, B) involved in describing how force is exerted in an electromagnetic field. It is true in every electromagnetic field at every point of space where the derivatives involved in the equation exist.

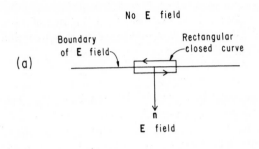

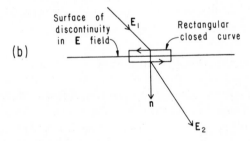

Fig. 1.2 *Illustrating vanishing of the surface curl of E*
a At a fixed boundary of an E field
b At a fixed surface of discontinuity in an E field
Note that the rectangular loop, the vector E_1 and the vector E_2 are not in general coplanar

At a boundary of an E field, the space derivatives of E involved in eqn. 1.5 do not in general exist. However, at a boundary fixed relative to the observer, we can apply the integral version of eqn. 1.5, given in eqn. 1.4, to the perimeter of a narrow rectangular element of area for which the long edges are on opposite sides of the boundary (see Fig. 1.2a). As the length of the narrow edges of the rectangle tends to zero, the contribution from the right-hand side of eqn. 1.4 also tends to zero. Moreover, the circulation of E round the rim of the rectangle reduces to the

the product of one of the long edges and the component of E along it. Since all orientations of this edge in the boundary are possible, we deduce that the tangential component of E must always vanish at a fixed boundary of an electric field. This may be expressed by saying that, if n is a unit normal to the boundary directed into the field at a point where the adjacent electric vector is E, then (see Fig. 1.2a)

$$n \times E = 0 \tag{1.6}$$

The boundary of an electric field is a special case of a surface of discontinuity. At any fixed surface of discontinuity of the E field we may apply eqn. 1.4 to the perimeter of a narrow rectangular element of area for which the long edges are on opposite sides of the surface (see Fig. 1.2b). We deduce that there is no discontinuity in the tangential component of E even though there is a discontinuity in the normal component. This may be expressed by saying that, if n is a unit normal to a fixed surface of discontinuity at a point where $E \mid$ is the discontinuous increase in E experienced on crossing the surface in the direction of n, then

$$n \times E \mid = 0 \tag{1.7}$$

Eqn. 1.6 is what eqn. 1.7 becomes when E vanishes on one side of the surface of discontinuity. For moving surfaces of discontinuity, see Table 2.4.

The expression $n \times E$ appearing in eqn. 1.6 is (c.f. $\nabla \times E$) the surface curl of E at the boundary of the field. Likewise, the expression $n \times E \mid$ appearing in eqn. 1.7 is the surface curl of E at any surface of discontinuity. We may say that the counterpart of Maxwell's electric curl equation (1.5) at a fixed boundary of an E field, or at any fixed surface of discontinuity in an E field, is expressed by the vanishing of the surface curl of E.

1.3 The vector field of magnetic flux density B

The vector field B may be described by means of curves, or lines, of magnetic flux such that, at each point P of the field, B is directed along the line through P. The aggregate of lines of magnetic flux drawn through a specified closed curve constitutes a tube of magnetic flux. The entire magnetic field may by analysed into thin tubes of magnetic flux.

A thin tube of magnetic flux has the property that, if dS is the vector area of a cross-section of the tube (not necessarily a normal cross-section) at a point where the magnetic-flux-density vector is B then, at each instant of time, the quantity

$$B \cdot dS \tag{1.8}$$

has the same value for all locations along the tube and for all cross-sections of the tube. The quantity (1.8) is known as the strength of the tube. Tubes of magnetic flux have the property that they have no beginning and no end. They frequently take the form of closed tubes. Tubes of magnetic flux are therefore endless tubes of constant strength.

Let dS be a vector element of area of an unclosed surface S, and let B be the magnetic-flux-density vector at dS at time t. Then, in accordance with expr. 1.8, the tube of magnetic flux formed by lines of magnetic flux through the rim of dS at time t has strength $B \cdot dS$, and the sum of the strengths of the tubes crossing S at time t is

$$\int_S B \cdot dS. \tag{1.9}$$

This is the magnetic flux crossing S at time t in the direction defined by dS.

If C is the rim of the surface S, the integral (1.9) is also described as the magnetic flux threading C at time t. Because $B \cdot dS$ does not vary along a tube of magnetic flux, the sum of the strengths of the tubes threading a closed curve C may be calculated from expr. 1.9 using any surface S spanning C. The magnetic flux threading a closed curve C is independent of the spanning surface S used in the calculation. This property of the B field is relevant to Faraday's law of induction. The value of the surface integral on the right-hand side of eqn. 1.4 is independent of the surface S used to span the closed curve C round which the circulation of E is evaluated.

The fact that each thin tube of magnetic flux not only has the same value of $B \cdot dS$ at all points, but is also endless, has the following consequence. For any surface Σ enclosing a volume V in space, each tube that conveys magnetic flux into V across Σ also conveys the same amount of flux across Σ out of V (see Fig. 1.3). Hence, for any closed surface Σ in any electromagnetic field,

$$\int_\Sigma B \cdot dS = 0 \tag{1.10}$$

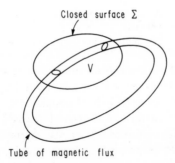

Closed surface Σ

V

Tube of magnetic flux

Fig. 1.3   *Illustrating the vanishing of the magnetic flux out of a closed surface Σ*

Eqn. 1.10 applies in particular to the surfaces of every element of volume in space. The divergence of a vector field B (see Appendix A) is a derived scalar field such that, for any element of volume $d\tau$ at a location where the derived scalar field is $\mathrm{div} B$, the flux of B out of the element of volume is $(\mathrm{div} B) d\tau$. Hence, by definition of divergence, eqn. 1.10 implies that, at all locations at all times,

$$\mathrm{div} B = 0 \tag{1.11}$$

This result is Maxwell's magnetic divergence equation. It is an expression of the fact that tubes of magnetic flux are endless tubes of constant strength. It is true in very electromagnetic field at every point of space where the derivatives involved in the definition of div B exist.

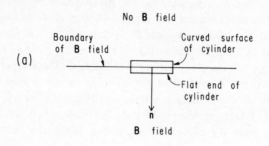

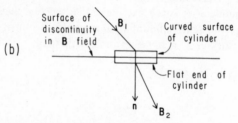

Fig. 1.4   *Illustrating the vanishing of the surface divergence of B*
a At a boundary of a B field
b At a surface of discontinuity in a B field

At a boundary of a B field, the space derivatives of B involved in eqn. 1.11 do not in general exist. However, we can then apply the integral version of eqn. 1.11, given in eqn. 1.10, to the surface of a short cylindrical element of volume for which the flat ends are on opposite sides of the boundary (see Fig. 1.4a). As the length of the cylinder tends to zero, so also does the magnetic flux out of the curved surface. Moreover, the flux out of the two flat ends of the cylinder reduces to the product of the area of one end and the component of B normal to this end. We deduce that the normal component of B must vanish at a boundary of a magnetic field. This may be be expressed by saying that, if n is a unit normal to the boundary directed into the field at a point where the adjacent magnetic-flux-density vector is B, then

$$n \cdot B = 0 \tag{1.12}$$

The boundary of a magnetic field is a special case of a surface of discontinuity. At any surface of discontinuity of the B field we may apply eqn. 1.10 to a short cylindrical element of volume for which the flat ends are on opposite side of the surface (see Fig. 1.4*b*). We deduce that there is no discontinuity in the normal component of *B* even though there is a discontinuity in the tangential component. This may be expressed by saying that, if *n* is a unit normal to the surface of discontinuity at a point where *B*| is the discontinuous increase in *B* experienced on crossing the surface in the direction of *n*, then

$$n{\cdot}B| \,=\, 0 \tag{1.13}$$

Eqn. 1.12 is what eqn. 1.13 becomes when *B* vanishes as one side of the surface of discontinuity.

The expression *n*·*B* appearing in eqn. 1.12 is (c.f. $\nabla{\cdot}B$) the surface divergence of *B* at the boundary of the field. Likewise the expression *n*·*B*| appearing in eqn. 1.13 is the surface divergence of *B* at any surface of discontinuity. We may say that the counterpart of Maxwell's magnetic divergence eqn. 1.11 at a boundary of a *B* field, or at any surface of discontinuity in a *B* field, is expressed by the vanishing of the surface divergence of *B*.

1.4 Electric charge and current

Electric charge (for which the unit is the coulomb) and electric current (measured in coulomb second^{-1}, known as an ampere) may be distributed continuously in three-dimensional space, or they may be localised on a surface (such as the surface of a good conductor), or on a curve (such as a wire). If distributed in three-dimensional space, electric charge is specified by means of its volume density ρ measured in units of charge per unit volume (coulomb meter^{-3}). Electric current in three-dimensional space is specified by means of the field of electric current density *J* for which the direction gives the direction of flow and the magnitude gives the charge per unit time crossing unit area at right angles to the flow (ampere meter^{-2}).

Both ρ and *J* are in general functions of time and position, but they are not independent. Electric current involves movement of electric charge from one place to another in such a way that charge is neither created nor destroyed. The time rate of decrease of the charge contained in any volume is the direct consequence of the rate at which charge is departing from the volume due to current across its bounding surface. If we apply this principle to each element of volume in space we obtain, by definition of divergence,

$$-\frac{\partial \rho}{\partial t} \,=\, \mathrm{div}\,J \tag{1.14}$$

This is the equation of continuity. It is an expression of the principle of conservation of charge, and it is true in every electromagnetic field at every point of space where the derivatives involved in the equation exist.

If we suppose that any system of charges and currents in which we are interested at time t has been created from a neutral configuration at $t = -\infty$, we may integrate eqn. (1.14) to obtain

$$\rho = -\int_{-\infty}^{t} \operatorname{div} J \, dt \tag{1.15}$$

This equation expresses the distribution of charge existing at time t in terms of the currents that have been used to create the system from a state of neutrality. The distribution of charge is deducible from the distribution of current provided that we know the time-history of the currents.

The concept of surface density of charge plays an important role in electromagnetic theory. For charge distributed over a surface, statements about volume distributions of charge need to be rephrased. Consider a situation in which there is a flow of electric charge in a volume V which is bounded by a fixed closed surface Σ; there is no flow exterior to Σ. At a point P of Σ, let n be a unit normal directed into the field of current occupying V. If J is the current density in V close to P, then the rate at which charge is leaving Σ per unit area at P is $n \cdot J$. If q is the charge per unit area on Σ at P, then the principle of conservation of charge requires that

$$-\frac{\partial q}{\partial t} = n \cdot J \tag{1.16}$$

Likewise, at any fixed surface where there is a discontinuity in the current density J, the principle of conservation of charge requires that

$$-\frac{\partial q}{\partial t} = n \cdot J| \tag{1.17}$$

where $J|$ is the discontinuous increase in the current density experienced on crossing the surface in the direction of the unit normal n.

The quantity $n \cdot J$ (c.f. $\nabla \cdot J$) in eqn. 1.16 is the surface divergence of J at the boundary of the current flow, while $n \cdot J|$ in eqn. 1.17 is the surface divergence of J at any surface of discontinuity in the current flow. Eqns. 1.16 and 1.17 may be summarised by saying that, at a fixed surface of discontinuity, the time rate of decrease of the surface density of charge is equal to the surface divergence of the field of current density. This statement is the surface counterpart of eqn. 1.14, which states that the time rate of decrease of the volume density of charge in space is equal to the volume divergence of the field of current density.

1.5 The vector field of electric flux density D

In any electromagnetic field in free space there exists a vector field of electric flux density D that describes a way of instantaneously discharging the system and rendering it electrically neutral. At a particular time, the direction of the vector D at each point gives the direction of the instantaneous flux of discharge at that point, and the magnitude of D gives the amount of charge that would be displaced per unit area across a surface perpendicular to D. The total charge that would cross a surface S in the process of instantaneous discharge is therefore

$$\int_S D \cdot dS \tag{1.18}$$

where *dS* is a vector element of area of S pointing in the direction in which it is desired to measure the flux of discharge. The integral (1.18) is known as the electric flux across the surface S in the direction of *dS*. Electric flux is measured in units of charge (coulomb), while the magnitude of the electric-flux-density vector *D* at a point is measured in units of charge per unit area (coulomb meter^{-2}).

For a closed surface Σ for which *dS* is an outward-pointing vector element of area, the integral

$$\int_{\Sigma} D \cdot dS \tag{1.19}$$

is the total flux of charge out of Σ that would be involved in instantaneously discharging the system. By the principle of conservation of charge, this must be equal to the charge within Σ immediately before the instantaneous discharge, and this is (see Fig. 1.5)

$$\int_{V} \rho \, d\tau \tag{1.20}$$

where ρ is the charge per unit volume at the element of volume $d\tau$, and V is the volume contained within Σ. Hence, for any closed surface Σ containing a volume V in any electromagnetic field

$$\int_{\Sigma} D \cdot dS = \int_{V} \rho \, d\tau \tag{1.21}$$

This is Gauss' law. It is an expression of the principle of conservation of charge.

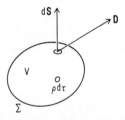

Fig. 1.5   *Illustrating the equality of (i) the flux of D out of a surface Σ enclosing a volume V, and (ii) the included charge $\int_V \rho \, d\tau$*

Eqn. 1.21 may be applied to every element of volume in space. Hence, if ρ is the volume density of charge at a point where the electric-flux-density vector is D, eqn. 1.21 becomes, by definition of divergence,

$$\text{div} \, D = \rho \tag{1.22}$$

This is Maxwell's electric divergence equation. It is an expression of the principle of conservation of charge, and it is true in every electromagnetic field at every point of space where the mathematical expressions involved in the equation exist.

The significance of eqn. 1.22 may be expressed by saying that the distribution of charge required to create a field of electric flux density D is given by

$$\rho = \text{div}\,D \tag{1.23}$$

However, this result requires restatement at the boundary of a D field. If n is a unit normal to the boundary directed into the field, the charge per unit area that would leave the surface in the process of instantaneous discharge described by the D field is $n \cdot D$. Hence the charge per unit area on the boundary is

$$q = n \cdot D \tag{1.24}$$

(a)

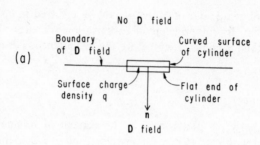

(b)

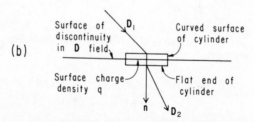

Fig. 1.6   *Illustrating the equality of the surface divergence of D to the surface density of charge*
a At a boundary of a D field
b At a surface of discontinuity in a D field

This may be verified by applying Gauss' law to the surface of the cylindrical element of volume indicated in Fig. 1.6a (c.f. Fig. 1.4a). Likewise, at a surface of discontinuity in the D field, the charge density on the surface is (see Fig. 1.6b)

$$q = n \cdot D| \tag{1.25}$$

where $D|$ is the discontinuous increase in the electric-flux-density vector experienced on crossing the surface in the direction of the unit normal n. We may summarise eqns. 1.23, 1.24 and 1.25 by saying that the charge density required to maintain a given D field is the divergence of D, where we are to use volume density and volume divergence in space but surface density and surface divergence at a surface of discontinuity.

It should be noted that the flux of discharge described by the field of electric flux density D is not usually one that actually takes place. It is one that we could imagine taking place instantaneously at any moment along tubes of electric flux. Ensuring that this flux of electric charge would always instantaneously discharge the system and render it electrically neutral is a way to ensure that the principle of conservation of charge is satisfied at all times.

Because the flux of charge described by the vector field of electric flux density D would discharge the system, it follows that

$$\frac{\partial D}{\partial t} \tag{1.26}$$

is a field of current density that would keep the system discharged, that is, one that would discharge the system as the actual current density J charges it. Since $\partial D/\partial t$ is a field of current density that would keep the system discharged, it follows that a possible field of current density for charging the system is

$$-\frac{\partial D}{\partial t} \tag{1.27}$$

However, in most electromagnetic systems, this is not the current-density actually used for charging. The actual current density is J, and it exceeds the current density $-\partial D/\partial t$ by

$$J + \frac{\partial D}{\partial t} \tag{1.28}$$

This is a closed flow in which J effects the charging process and $\partial D/\partial t$, if it existed, would nullify it. That the flow represented by expr. 1.28 is closed may be verified by taking its divergence, and then using eqns. 1.14 and 1.23 to deduce that this divergence vanishes.

The current density $\partial D/\partial t$ in free space is Maxwell's displacement current density. It describes a current that does not flow. Combined with the current density J that does flow, it gives the closed flow represented by expr. 1.28. It is the fact that the current density $\partial D/\partial t$ does not flow that results in the actual current density J being an unclosed flow that leads to the accumulation of charge as described by eqn. 1.15.

1.6 The magnetic vector H

In the previous section we saw that an electromagnetic system could be charged by means of the current density $-\partial D/\partial t$, but that it is usually charged by means of some different current density J, involving an additional closed flow. It is this additional closed flow that leads to the existence of magnetic field. No magnetic field is caused if the charging current density is simply $-\partial D/\partial t$. But if the charging current density is in fact J, then there is an additional close flow of current whose density is given by expr. 1.28. It is this closed flow that causes magnetic field. The closed flow may be analysed into closed solenoidal currents that flow over the surfaces of the tubes of magnetic flux.

Not only are tubes of magnetic flux endless tubes of constant strength as described in Section 1.3, but each tube can exist by itself in space if there is a suitable solenoidal current over its surface. To describe this current it is convenient to introduce, at each point of space, a vector *H* known as the magnetic vector. At a point P of a selected thin tube of magnetic flux in free space, the magnetic vector *H* points along the tube in the direction of the magnetic flux. If the tube is not in motion relative to the observer, the magnitude *H* of the magnetic vector is the solenoidal current per unit length that must exist round the tube at P if the tube is to exist by itself (see Fig. 1.7). The direction of this current is related by the right-hand-screw rule to the direction of the magnetic flux in the tube.

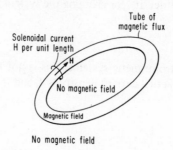

Fig. 1.7   *Illustrating how the magnetic vector H describes the solenoidal current needed round a tube of magnetic flux in free space to keep the tube in existence by itself*

If *n* is a unit inward normal to the surface of a thin tube of magnetic flux at a location where the magnetic vector is *H*, then the solenoidal current per unit length needed round the tube to keep it in existence by itself is given in magnitude and direction as

$$i = n \times H \tag{1.29}$$

The quantity *i* in this equation is the surface density of current on the surface of the tube. Its direction gives the direction of flow on the surface. Its magnitude gives the charge per unit time crossing a unit length drawn in the surface at right angles to the flow. Both *i* and *H* are measured in units of current per unit length (ampere meter^{-1}). The magnitude of the magnetic vector *H* is the magnetic field strength. The quantity *n* × *H* on the right-hand side of eqn. 1.29 is described (c.f. $\nabla \times H$) as the surface curl of *H* at the boundary of the tube, thought of as existing by itself.

In any magnetic field, each thin endless tube of magnetic flux may be regarded as created individually by solenoidal current over its surface, and the entire magnetic field may be formed by packing the individual tubes together. If this is done, there is a large amount of cancellation of current on the common surfaces of adjacent tubes, and we need to know the net current required to maintain the magnetic field after this cancelation has been taken into account. It is in order to perform this operation in the limit when the cross-sectional areas of the tubes tend to zero that the curl operator is designed. If the solenoidal currents round the individual tubes are described by means of the magnetic vector *H* in accordance with eqn. 1.29, then the net flow for the aggregate of tubes is described by the current density vector

$$\text{curl}\, H \tag{1.30}$$

At a fixed boundary of the magnetic field, the surface density of current is the surface curl of *H*, namely

$$i = n \times H \tag{1.31}$$

where *n* is a unit normal to the boundary directed into the field (c.f. eqn. 1.29). At a fixed surface where there is a discontinuity in the magnetic vector, the surface density of current is

$$i = n \times H| \tag{1.32}$$

where *H*| is the discontinuous increase in the magnetic vector experienced on crossing the surface in the direction of the unit normal *n*. The net current involved in an aggregate of thin tubes of magnetic flux, each maintained by solenoidal currents described by the field *H*, is the curl of *H*, where we are to use the volume curl in space and the surface curl at a fixed surface of discontinuity. For moving surfaces of discontinuity, see Table 2.4.

Maxwell's magnetic curl equation states that the closed flow described by curl *H* in expr. 1.30 is the closed flow that we have identified as the cause of magnetic field (expr. 1.28). This means that

$$\text{curl}\, H = J + \frac{\partial D}{\partial t} \tag{1.33}$$

This statement is true in every electromagnetic field at every point of space where the mathematical expressions involved in the equation exist. If the result is written in the integral form

$$\int_C H \cdot ds = \int_S \left(J + \frac{\partial D}{\partial t} \right) \cdot dS \tag{1.34}$$

it incorporates not only eqn. 1.33 but also the corresponding statements appropriate to the boundary of a magnetic field and to a surface of discontinuity in a magnetic field. In eqn. 1.34, S is a surface spanning any closed curve C fixed relative to the observer, and the vector elements of area *dS*, and of length *ds*, have directions that are related by the right-hand-screw rule. If we apply eqn. 1.34 to the perimeter of a narrow rectangular element of area for which the long sides are on opposite sides of a fixed boundary of a magnetic field (see Fig. 1.8*a*), the contribution from the term involving $\partial D/\partial t$ tends to zero as the length of the narrow sides of the rectangle tends to zero (c.f. the term involving $\partial B/\partial t$ in eqn. 1.5 and Fig. 1.2a); consequently, we derive eqn. 1.31 (c.f. eqn. 1.6). Likewise, at a fixed surface of discontinuity in a magnetic field (see Fig. 1.8b), we derive eqn. 1.32 (c.f. eqn. 1.7).

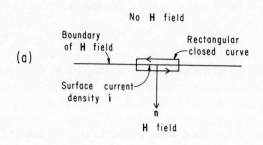

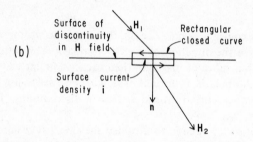

Fig. 1.8   *Illustrating the equality of the surface curl of H to the surface density i of current*
a At a fixed boundary of an *H* field
b At a fixed surface of discontinuity in an *H* field
Note that the rectangular loop, the vector H_1 and the vector H_2 are not in general coplanar

Eqn. 1.34 is Ampere's law. In any electromagnetic field in space the law applies to any closed curve C fixed relative to the observer. It states that the circulation of the magnetic vector H round C is equal to a right-hand-related current threading C that is calculated by vector addition of:

(i) the current density J that actually flows in free space, and
(ii) the Maxwell displacement current density $\partial D/\partial t$ that does not flow in free space.

1.7 Electromagnetic fields as aggregates of elementary inductors and capacitors

In the previous section we have seen that each tube of magnetic flux in an electromagnetic field in space can exist by itself. Such an isolated tube is an inductor.

Moreover, every section of the tube is an inductor for which the rest of the tube constitutes a guard inductor. At any instant of time the entire magnetic field of any electromagnetic system in space may be dissected into cylindrical elements of volume each of which is an ideal inductor consisting of a short section of a thin tube of magnetic flux. For the local elementary inductor at a point where the magnetic vector is *H*, the inductor is aligned along *H* and the current per unit length round it is *H*, right-hand-related to *H*. The significance of the magnetic vector *H* at a point 0 in an electromagnetic field may therefore be visualized in terms of the curent per unit length round a local elementary inductor at 0.

In a similar manner the significance of the electric flux-density vector *D* at a point 0 in an electromagnetic field in space may be visualised in terms of the charges per unit area on the plates of a local elementary capacitor at 0. Although unclosed, each thin tube of electric flux may be dissected into short sections, each of which constitutes a cylindrical element of volume. The electric field in each element of volume may be regarded as created by capacitor plates coinciding with the flat ends of the element. In accordance with eqn. 1.24, the charges per unit area on the plates associated with this element of volume are ± *D*, and the direction of *D* is from the positive plate to the negative plate. No fringing needs to be taken into account because the field of each local elementary capacitor is appropriately guarded by that in the surrounding elementary capacitors. Simultaneous discharge of each elementary capacitor would give the flux of charge that is capable of instantaneously discharging the system at each moment of time; this flux of discharge is described by the field of electric flux density *D*.

By slicing tubes of electric and magnetic flux into cylindrical elements of volume, every electromagnetic field in space may be dissected into elementary capacitors and inductors. At each point in the field at each instant of time, the vector *D* gives the information concerning the orientation of the local elementary capacitor and its charges per unit area, while the vector *H* gives the information concerning the orientation of the local elementary inductor and its current per unit length. When, in accordance with the principle of conservation of charge, we have taken account of the cancellation of the currents round adjacent inductors and the cancellation of the charges associated with adjacent capacitors, and have allowed the sizes of the elementary inductors and capacitors to tend to zero, then the net current and charge in space have densities (eqns. 1.33 and 1.23)

$$J = \operatorname{curl} H - \frac{\partial D}{\partial t}, \quad \rho = \operatorname{div} D \tag{1.35}$$

At a fixed surface of discontinuity in the electromagnetic field where *H*| and *D*| are the discontinuous increases in *H* and *D* experienced on crossing the surface in the direction of the unit normal *n*, the net surface current and charge have surface densities (eqns. 1.32 and 1.25)

$$i = n \times H|, \quad q = n \cdot D| \tag{1.36}$$

For moving surfaces of discontinuity, see Table 2.4.

Eqns. 1.35 summarise the facts concerning the geometry of current and charge distributions in any electromagnetic field in space. They incorporate the principle of conservation of charge; by substitution, we may verify that the current density J and the charge density ρ in eqns. 1.35 satisfy the equation of continuity (1.14).

1.8 The relations of the (E, B) fields to the (D, H) fields in free space

As explained in Sections 1.2 and 1.3, the (E, B) fields are useful for describing how force is exerted in an electromagnetic field. The force per unit charge on a particle moving with velocity v relative to the observer is $E + v \times B$, and the (E, B) fields satisfy Maxwell's electric curl equation and Maxwell's magnetic divergence equation. These are eqns. 1.5 and 1.11, which take the form of eqns. 1.7 and 1.13 at a fixed surface of discontinuity. They express as differential equations and boundary conditions the following integral relations:

(i) For any fixed closed curve C (having element of length ds) spanned by a surface S (having right-hand-related element of area dS) in space we have (eqn. 1.4)

$$\int_C E \cdot ds + \frac{\partial}{\partial t} \int_S B \cdot dS = 0 \tag{1.37}$$

(ii) For any closed surface Σ (having element of area dS) in space we have (eqn. 1.10)

$$\int_\Sigma B \cdot dS = 0 \tag{1.38}$$

As explained in the previous section, the (D, H) fields are useful for describing how an electromagnetic field may be visualised as a volume distribution of elementary capacitors and inductors. The relations of the (D, H) fields to the net charge and current distributions needed to create the electromagnetic system are given by eqns. 1.35, which take the form of eqns. 1.36 at a fixed surface of discontinuity. These are the volume and surface versions of Maxwell's magnetic curl equation and Maxwell's electric divergence equation. They express as differential equations and boundary conditions the following integral relations:

(i) For any fixed closed curve C (having element of length ds) spanned by a surface S (having right-hand-related element of area dS) in space we have (eqn. 1.34)

$$\int_C H \cdot ds - \frac{\partial}{\partial t} \int_S D \cdot dS = \text{threading current} \tag{1.39}$$

(ii) For any closed surface Σ (having outward directed element of area dS) in space we have (eqn. 1.21)

$$\int_\Sigma D \cdot dS = \text{included charged} \tag{1.40}$$

Assembling the differential equations and boundary conditions, we see that the electromagnetic field (E, B, D, H) in space satisfies the Maxwell equations

$$\begin{cases} \operatorname{curl} E + \dfrac{\partial B}{\partial t} = 0, & \operatorname{div} B = 0 \end{cases} \tag{1.41}$$

$$\begin{cases} \operatorname{curl} H - \dfrac{\partial D}{\partial t} = J, & \operatorname{div} D = \rho \end{cases} \tag{1.42}$$

of which the counterparts at a fixed surface of discontinuity are the boundary conditions

$$\begin{cases} n \times E| = 0, & n \cdot B| = 0 \tag{1.43} \\ n \times H| = i, & n \cdot D| = q \tag{1.44} \end{cases}$$

where $(E|, B|, D|, H|)$ are the discontinuous increases in (E, B, D, H) experienced on crossing the surface in the direction of the unit normal n. Eqns. 1.41 and 1.43 are concerned with the exercise of force, whereas eqns. 1.42 and 1.44 are concerned with current and charge geometry.

If desired, the continuity equation (1.14) may be derived from the Maxwell eqns. 1.42 by taking the divergence of the curl equation and the time-derivative of the divergence equation. Alternatively, we may say that, given the truth of the equation of continuity, there is some redundance in the Maxwell eqns. 1.42. By taking the divergence of the curl equation and using the equation of continuity, we obtain (identity 5, Appendix B)

$$\frac{\partial}{\partial t}(\operatorname{div} D - \rho) = 0 \tag{1.45}$$

We may therefore say that, if $\operatorname{div} D - \rho$ vanishes at any particular instant of time, then it vanishes at all other times. There is also some redundance in the Maxwell eqns. 1.41. By taking the divergence of the curl equation, we deduce that

$$\frac{\partial}{\partial t}(\operatorname{div} B) = 0 \tag{1.46}$$

Hence the Maxwell curl equations ensure that, if the Maxwell divergence equations are postulated to be true at only a single instant of time, then they are automatically true at all other times. In particular, if it is assumed that the system started at $t = -\infty$ from an unchanged condition involving no charge flow and no electromagnetic field, then the truth of Maxwell's divergence equations follows from the truth of Maxwell's curl equations.

Maxwell's equations are insufficient to determine the electromagnetic field (E, B, D, H). We need to know in addition what the relations are between the (E, B) fields associated with the exercise of force and the (D, H) fields associated with the geometry of the charge and current distributions in otherwise free space. These connection relations are

$$D = \epsilon_0 E, \quad H = \frac{1}{\mu_0} B \tag{1.47}$$

where ϵ_0 and μ_0 are constants that depend on the units employed.

Solution of the electromagnetic eqns. (1.41, 1.42 and 1.47) shows that electromagnetic waves are propagated in free space with the velocity

$$c = (\mu_0 \epsilon_0)^{-1/2} \tag{1.48}$$

which is known as the velocity of light in free space and which is measured to be

$$c = 2 \cdot 998 \times 10^8 \, \text{m/s} \tag{1.49}$$

The international system of units (meter, kilogram, second, coulomb) is chosen so that

$$\mu_0 = 4\pi \times 10^{-7} \, \text{H/m} \tag{1.50}$$

and it then follows from eqns, 1.48 and 1.49 that

$$\epsilon_0 = 8 \cdot 854 \times 10^{-12} \, \text{F/m} \tag{1.51}$$

If a pair of opposite faces of a unit cube in free space are occupied by conducting plates, then the resulting capacitor has capacitance ϵ_0, provided that guard-plates are employed to avoid non-uniformity of the electric field in the cube. If, instead, the other four faces of the unit cube are occupied by a conducting loop, then the resulting inductor has inductance μ_0, provided that a guard-inductor is employed to avoid non-uniformity of the magnetic field in the cube. The quantities ϵ_0 and μ_0 may be called capacitivity and inductivity of free space, although they are usually called the absolute permitivity and absolute permeability.

1.9 The electromagnetic potentials

Any electromagnetic field (E, B) in space may be explored by means of a test particle of mass m and charge e moving with velocity v relative to the observer. The charge e should be small enough not to upset the electromagnetic field under study. The test particle possesses mechanical momentum mv on account of its mass. But it also possesses electromagnetic momentum eA on account of its charge, so that the total momentum of the particle is

$$p = mv + eA \tag{1.52}$$

The electromagnetic momentum per unit charge of a test charge at a point is known as the magnetic vector potential A of the electromagnetic field at the point. Furthermore, the electric potential energy per unit charge of the test charge at the point is known as the electric scalar potential ϕ of the electromagnetic field at the point (see Problem 1.5).

In an electromagnetic field created by a time-varying distribution of electric current of density J in space, and an associated distribution of electric charge of density ρ, the electromagnetic potentials satisfy the wave equations

$$\left.\begin{aligned}
\nabla^2 A - \frac{1}{c^2}\frac{\partial^2 A}{\partial t^2} &= -\mu_0 J \\[2mm]
\nabla^2 \phi - \frac{1}{c^2}\frac{\partial^2 \phi}{\partial t^2} &= -\frac{1}{\epsilon_0}\rho
\end{aligned}\right\}$$

$$(1.53)$$

where c is given by eqn. 1.48. Moreover, if ϕ and A are defined so that they vanish in the complete absence of an electromagnetic field, the electromagnetic potentials satisfy the Lorentz relation

$$-\frac{1}{c^2}\frac{\partial \phi}{\partial t} = \operatorname{div} A \tag{1.54}$$

In terms of the electromagnetic potentials, the electric vector E and the magnetic-flux-density vector B are given by

$$E = -\operatorname{grad}\phi - \frac{\partial A}{\partial t}, \quad B = \operatorname{curl} A \tag{1.55}$$

If the current and charge densities J and ρ are prescribed as functions of position and time, eqns. 1.53 may be solved for A and ϕ, and the electromagnetic field deduced from eqns. 1.55.

The unit of electric scalar potential is the volt, so that the unit of electric field strength is the volt meter^{-1}. The unit of magnetic flux is the weber, so that the unit of magnetic flux density is the weber meter^{-2} and that of magnetic vector potential is the weber meter^{-1}. The first of eqns. 1.55 shows that a volt is identical with a weber second^{-1}. Faraday's law of induction implies that, when the magnetic flux threading a fixed closed curve is decreasing at the rate 1 weber second^{-1}, then the right-hand-related circulation of the electric vector round C is 1 volt.

Suppose that a test particle of mass m and charge e is seen by the observer to be at rest at a particular location in an electromagnetic field. The particle then possesses no mechanical momentum. But it does possess electromagnetic momentum equal, per unit charge, to the magnetic vector potential A at the location. Moreover, the electromagnetic momentum of the stationary test particle is, in general, changing with time. The first of eqns. 1.55 states that the time rate of increase of this electromagnetic momentum per unit charge is equal to the local excess of the negative gradient of the electric scalar potential over the electric vector. In the absence of such an excess, the electromagnetic momentum remains static.

The second of eqns. 1.55 shows that the magnetic field may, if desired, be completely described with the aid of the magnetic vector potential A. In particular, instead of expressing the magnetic flux Φ crossing a surface S as an integral over S (expr. 1.3), we can use the second of eqns. 1.55 to express it as an integral round the rim C of S. If dS is a vector element of area of S pointing in the direction of the desired magnetic flux, and if ds is a vector element of length of C pointing in the direction related to dS by the right-hand-screw rule, we obtain

$$\Phi = \int_S B \cdot dS = \int_C A \cdot ds \tag{1.56}$$

The equality of these two expressions for Φ follows by multiplying the second of eqns. 1.55 scalarly by dS, integrating over S, and then applying the curl theorem (Identity 18, Appendix B). Moreover, using eqns. 1.56, we may write Faraday's law of induction (eqn. 1.37) in the form

$$\oint_C \left(E + \frac{\partial A}{\partial t} \right) \cdot ds = 0 \tag{1.57}$$

where C is any closed curve fixed relative to the observer. Eqn. 1.57 is therefore an alternative version of Faraday's law of induction. It states that, in any electromagnetic field, the circulation of the vector field $E + \partial A/\partial t$ round any closed curve fixed relative to the observer vanishes. This means that the vector field $E + \partial A/\partial t$ may be expressed as the gradient of a scalar function of position (c.f. electrostatics), and this is what is stated in the first of eqns. 1.55.

The expressions for E and B in eqns. 1.55 satisfy eqns. 1.41 identically. But, for them also to satisfy the equations obtained by substituting for D and H from eqns. 1.47 into eqns. 1.42, it is necessary for ϕ and A to be solutions of eqns. 1.53.

As discussed in Section 1.4, J and ρ on the right-hand sides of eqns. 1.53 are not independent, being related by the continuity relation (eqn. 1.14). Moreover, A and ϕ in eqns. 1.53 are not independent, being related by the Lorentz relation (eqn. 1.54), which is similar in form to eqn. 1.14. These facts permit us to express both A and ϕ in terms of a single vector potential known as the electric Hertzian potential Π. To do this we express both the current and the charge densities J and ρ in terms of a single vector field known as the electric moment per unit volume P.

1.10 Electric moment per unit volume

In discussing materials one encounters the concept of electric moment per unit volume in a dielectric. However, this concept also provides a valuable way of handling all charges and currents whether or not they are bound in atoms and molecules.

Any flow of current described by a field of current density J may be dissected into tubes of flow. Each tube of flow may then be dissected into current elements. Let a typical cylindrical current element have a small normal cross-sectional area S and a small length l, so that it constitutes a volume element Sl carrying a current JS. If the volume element is considered by itself, charges $\pm Q$ are required at its ends in accordance with the principle of conservation of charge. The relation between J and Q is

$$JS = \frac{\partial Q}{\partial t} \tag{1.58}$$

The current element occupying the volume S constitutes an electric dipole of moment Ql. The resulting electric moment per unit volume is

$$P = (Ql)/(Sl)$$

and, in terms of P, eqn. 1.58 may be rewritten

$$J = \frac{\partial P}{\partial t}$$

Moreover, if P is the vector electric moment per unit volume, then

$$J = \frac{\partial P}{\partial t} \tag{1.59}$$

Any distribution of current occupying a volume V may be regarded as an ensemble of current elements, and hence may be described by means of its distribution of electric moment per unit volume P. The current density J is deduced from P by means of eqn. 1.59. By integrating this equation with regard to time and assuming that the system started from an electrically neutral configuration at $t = -\infty$ with no current flowing, we may define the electric moment per unit volume for the current distribution as

$$P = \int_{-\infty}^{t} J \, dt \tag{1.60}$$

From P we are not only able to derive the current distribution in accordance with eqn. 1.59 but we can also derive the associated charge distribution that is mandated by the principle of conservation of charge. By taking the volume divergence of eqn. 1.60 and using the continuity equation (1.14), we see that, in the volume V occupied by the current distribution,

$$\operatorname{div} P = -\rho \tag{1.61}$$

Similarly, for the surface divergence at a fixed boundary S of V, we multiply eqn. 1.60 scalarly by a unit normal n directed into V and use eqn. 1.16, thereby obtaining

$$n \cdot P = -q \tag{1.62}$$

Eqns. 1.61 and 1.62 show that a distribution of electric moment per unit volume P occupying a volume V bounded by a fixed closed surface S involves a net charge density ρ in V, and q on S, given by

$$\rho = -\operatorname{div} P, \quad q = -n \cdot P \tag{1.63}$$

where n is a unit normal to S directed into V.

By dissecting the current flow into current elements, each with its appropriate terminal charges, we ensure that the principle of conservation of charge is satisfied. The resulting concept of a distribution of electric moment per unit volume P in space provides, for many purposes, the simplest way of describing a current distribution. In accordance with eqns. 1.59 and 1.63, the volume densities of current and charge are derived from the electric moment per unit volume P by means of the equations

$$J = \frac{\partial P}{\partial t}, \quad \rho = -\operatorname{div} P \tag{1.64}$$

These expressions for J and ρ automatically satisfy the continuity equation (1.14).

1.11 The electric Hertzian potential

The technique employed in the previous section of using the equation of continuity to express J and ρ in terms of a single vector field P may also be employed to express the electromagnetic potentials A and ϕ in terms of a single vector field Π by making use of eqn. 1.54.

Let us assume that the electromagnetic potentials vanish at $t = -\infty$, and let us define the electric Hertzian potential as (c.f. eqn. 1.60)

$$\Pi = c^2 \int_{-\infty}^{t} A \, dt \tag{1.65}$$

Then it follows that (c.f. eqns. 1.64)

$$A = \frac{1}{c^2} \frac{\partial \Pi}{\partial t}, \quad \phi = -\operatorname{div} \Pi \tag{1.66}$$

The first of eqns. 1.66 follows from eqn. 1.65 by differentiation, and the second may be derived from eqn. 1.54 by integrating with respect to time and assuming that ϕ vanishes at $t = -\infty$.

Moreover, if we take Π to satisfy the wave-equation

$$\nabla^2 \Pi - \frac{1}{c^2} \frac{\partial^2 \Pi}{\partial t^2} = -\frac{1}{\epsilon_0} P \tag{1.67}$$

then the time-derivative of this equation gives the first of eqns. 1.53 and the divergence gives the second. Furthermore, substitution from eqns. 1.66 into eqns. 1.55 shows that

$$E = \operatorname{grad} \operatorname{div} \Pi - \frac{1}{c^2} \frac{\partial^2 \Pi}{\partial t^2}, \quad B = \frac{1}{c^2} \operatorname{curl} \frac{\partial \Pi}{\partial t} \tag{1.68}$$

For any given current and charge distribution described by its distribution of electric moment per unit volume P in space and time, we can solve the wave-equation (1.67) for the electric Hertzian vector Π, and then derive the electromagnetic field from eqns. 1.68.

The solution of eqn. 1.67 when there is a single elementary electric dipole of moment $p(t)$ at the origin of spherical polar coordinates (r, θ, ψ) in free space is

$$\Pi = \frac{1}{4\pi\epsilon_0} \frac{p(t - r/c)}{r} \tag{1.69}$$

showing that the electric Hertzian potential at distance r from the dipole at time t

depends on the moment of the dipole at the earlier time $t - r/c$. The interval of time r/c is that required to transmit information from the dipole to the field-point with the velocity of light c. The electromagnetic field is obtained by substituting from eqn. 1.69 into eqns. 1.68; see Problems 1.6 and 1.7.

By integrating eqn. 1.69 with respect to volume, it follows that, for any given time-varying distribution of current and charge for which the electric moment per unit volume at an element of volume $d\tau$ is $P(t)$, the solution of eqn. 1.67 is

$$\Pi = \frac{1}{4\pi\epsilon_0} \int_V \frac{P(t - r/c)}{r} d\tau \tag{1.70}$$

where r is the distance of the element of volume $d\tau$ from the location at which Π is evaluated, and the integral is taken througout the volume V occupied by current and charge. The corresponding electromagnetic field is obtained by substituting from eqn. 1.70 into eqns. 1.68.

Eqns. 1.68 and 1.70 constitute a complete solution of the electromagnetic equations for any situation in which the distribution of current (and therefore of charge and of electric moment per unit volume) is known as a function time and position. However, this is not the situation encountered in materials. Materials possess conduction, dielectric and magnetic properties that result in the local electric current depending on the local electromagnetic field that it is desired to calculate. This complication will be addressed in the next chapter.

1.12 Magnetic moment per unit volume

Although the method described in the previous section may be used to calculate the electromagnetic field of any distribution of charge and current that is prescribed in space and time, this method does not always constitute the most convenient procedure. In particular, an alternative method is often simpler when dealing with situations involving what is known as magnetic moment. In discussing materials, one encounters the concept of magnetic moment per unit volume in magnets. However, this concept is also relevant for solenoidal flow of current in space.

Consider a thin solenoid of length l and small cross-sectional area S. Let I be the current flowing uniformly round the curved surface, so that the current per unit length is

$$M = \frac{I}{l} \tag{1.71}$$

In the limit when the cross-sectional area of the solenoid tends to zero, the magnetic field strength inside the solenoid is also

$$H = \frac{I}{l} \tag{1.72}$$

Hence the magnetic flux density inside the solenoid is $\mu_0 I/l$ and the magnetic flux threading the solenoid is

$$\Phi = \frac{\mu_0 S}{l} I \tag{1.73}$$

This magnetic flux emerges from one end of the thin solenoid and returns to the other end, completing closed tubes of magnetic flux by threading through the solenoid (see Fig. 1.9). The exterior magnetic field of the solenoid may be regarded as that due to point sources of magnetic flux $\pm \Phi$ at the ends of the solenoid. However, the tubes of magnetic flux do not start from one end of the solenoid and stop at the other; they are piped through the solenoid to complete closed tubes of magnetic flux. Nevertheless, calculations about that part of the magnetic field exterior to the thin solenoid are most simply performed by thinking of two point sources of magnetic flux $\pm \Phi$ at a distance l apart.

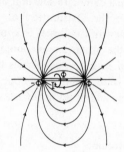

Fig. 1.9   *Illustrating magnetic flux Φ threading through a thin solenoid carrying current uniformly distributed along its length*

The product IS is called the magnetic moment of the solenoid. In accordance with eqn. 1.73, magnetic moment differs from the product Φl by a factor $1/\mu_0$. If desired, this factor $1/\mu_0$ may be absorbed by defining magnetic pole strength not as Φ but as Φ/μ_0. The effect of omitting the factor $1/\mu_0$ in the definition of magnetic moment will be noticeable in various equations in this section and the next; compare, for example, eqns. 1.85 with eqns. 1.64.

Let S be the cross-sectional area of the solenoid, expressed as a vector related by the right-hand-screw rule to the positive direction of the current round the solenoid. Then the vector magnetic moment of the solenoid is

$$m = IS \tag{1.74}$$

A thin solenoid may be regarded as a magnetic dipole for which the field of magnetic flux may be calculated by a procedure similar to that for an electric dipole, provided that we remember to add the internal field consisting of a magnetic flux Φ threading through the solenoid (see Problems 1.1 and 1.9).

The magnetic moment per unit volume M is obtained by dividing the magnetic moment in eqn. 1.74 by the volume lS of the solenoid. We obtain (c.f. eqn. 1.71)

$$M = \frac{I}{l}\hat{S} \tag{1.75}$$

where $\hat{S}$ is a unit vector in the direction of S. We see that the magnitude of the magnetic moment per unit volume is the current per unit length, and that the direction of M is related by the right-hand-screw rule to the direction of the solenoidal current. The vector M describes the current round a thin solenoid of finite length in the same way that the vector H describes the solenoidal current round an endless tube of magnetic flux that exists by itself (Fig. 1.7). The magnetic vector and the magnetic-flux-density vector inside the thin solenoid are (c.f. eqns. 1.71 and 1.72)

$$H = M, \quad B = \mu_0 M \tag{1.76}$$

Let us now consider a distribution of magnetic moment per unit volume M occupying a volume V bounded by a fixed surface S. When cancellation of the currents on the common surfaces of adjacent solenoids has been taken into account, we have a net volume current density J in V given by (c.f. expr. 1.30) the volume curl of M and a net surface current density i on S given by (c.f. eqn. 1.31) the surface curl of M. Hence

$$J = \operatorname{curl} M, \quad i = n \times M \tag{1.77}$$

where n is a unit normal to S directed into V.

The electromagnetic field created by the volume distribution of magnetic moment per unit volume M may be calculated from the net flow of electric current described in eqns. 1.77. As an alternative, however, we may employ a procedure similar to that used for a volume distribution of electric moment per unit volume. In this method we regard each elementary solenoid as a pair of equal and opposite sources of magnetic flux, neglecting initially the magnetic flux piped through the solenoid. Let us suppose that the resulting electromagnetic field at a point 0 is (E', B', D', H'). If 0 is outside the volume V occupied by the volume distribution of elementary solenoids, no correction is required for the internal field of any solenoid, and (E', B', D', H') is the actual electromagnetic field. But if 0 is inside the volume distribution of magnetic moment, it is also within the local elementary solenoid whose internal magnetic field is given by eqns. 1.76. Hence the actual electromagnetic field (E, B, D, H) is calculated from (E', B', D', H') by the relations

$$\begin{cases} E = E', \quad B = B' + \mu_0 M & (1.78) \\ D = D', \quad H = H' + M & (1.79) \end{cases}$$

1.13 The magnetic Hertzian potenial

The field (E', B') appearing in eqns. 1.78 and associated with a distribution of magnetic moment per unit volume M in a volume V bounded by a surface S may be calculated by a process that is the dual of the calculation of the electromagnetic field of a distribution of electric moment per unit volume. The process involves appropriately interchanging electric and magnetic quantities. Use is made of a magnetic Hertzian potential Π' that satisfies the wave equation (c.f. eqn. 1.67)

$$\nabla^2 \Pi' - \frac{1}{c^2} \frac{\partial^2 \Pi'}{\partial t^2} = -\mu_0 M \tag{1.80}$$

and one derives $(E', B',)$ from Π' by means of the relations (c.f. eqns. 1.68)

$$E' = -\text{curl}\, \frac{\partial \Pi'}{\partial t}, \quad B' = \text{grad div}\, \Pi' - \frac{1}{c^2} \frac{\partial^2 \Pi'}{\partial t^2} \tag{1.81}$$

The solution of eqn. 1.80 for a single elementary magnetic dipole of moment $m(t)$ at the origin of spherical polar coordinates (r, θ, ψ) is (c.f. eqn. 1.69)

$$\Pi' = \frac{\mu_0}{4\pi} \frac{m(t - r/c)}{r} \tag{1.82}$$

and the solution for a distribution of magnetic moment per unit volume M occupying a volume V is (c.f. eqn. 1.70)

$$\Pi' = \frac{\mu_0}{4\pi} \int_V \frac{M(t - r/c)}{r} d\tau \tag{1.83}$$

We substitute for Π' into eqns. 1.81 to obtain the $(E', B',)$ field, and then into eqns. 1.78 to obtain the actual electromagnetic field (E, B).

If desired, one can make use of an electric vector potential A' and a magnetic scalar potential ϕ' defined as (c.f. eqns. 1.66)

$$A' = \epsilon_0 \frac{\partial \Pi'}{\partial t}, \quad \phi' = -\frac{1}{\mu_0} \text{div}\, \Pi'. \tag{1.84}$$

Moreover, one can regard the (E', B') field as arising from a distribution of magnetic current of volume density J', and a distribution of magnetic charge of volume density ρ', defined by (c.f. eqns. 1.64)

$$J' = \mu_0 \frac{\partial M}{\partial t}, \quad \rho' = -\mu_0 \text{div}\, M. \tag{1.85}$$

If this is done then, at a fixed boundary of a volume V occupied by magnetic moment, account must be taken of a surface density of magnetic charge defined by (c.f. the second of eqns. 1.63)

$$q' = -\mu_0 n \cdot M \tag{1.86}$$

where n is a unit normal to the boundary directed into V.

1.14 The degree of utility of the concepts of magnetic charge and current

Although magnetic charge and current do not exist, these concepts nevertheless have a theoretical utility for describing certain portions of electromagnetic fields that do exist. One example is the external field of a thin solenoid carrying a magnetic flux Φ (Fig. 1.9). Another example is the fringing field of a thin parallel plate-capacitor of voltage V; the complete field is obtained by adding to a uniform internal electric field within the capacitor the field of a magnetic current of strength V round the rim of the capacitor (Problem 1.15).

If we were to imagine that there is not merely a volume distribution of electric current and charge (J, ρ) but also a volume distribution of magnetic current and charge (J', ρ'), Maxwell's equation would become (c.f. eqns. 1.41 and 1.42)

$$\left\{ \operatorname{curl} E + \frac{\partial B}{\partial t} = -J', \quad \operatorname{div} B = \rho' \right. \tag{1.87}$$

$$\left. \operatorname{curl} H - \frac{\partial D}{\partial t} = J, \qquad \operatorname{div} D = \rho \right. \tag{1.88}$$

The threading of magnetic current by electric flux would be opposite-handed to the threading of electric current by magnetic flux.

The counterparts of eqns. 1.87 and 1.88 at a fixed surface of discontinuity in the electromagnetic field would be the boundary conditions (c.f. eqns. 1.43 and 1.44)

$$\left\{ n \times E| = -i', \quad n \cdot B| = q' \right. \tag{1.89}$$

$$\left. n \times H| = i, \qquad n \cdot D| = q \right. \tag{1.90}$$

where (i, q) are the surface densities of electric current and charge, (i', q') are the surface densities of magnetic current and charge, and $(E|, B|, D|, H|)$ are the discontinuous increases of (E, B, D, H) experienced on crossing the surface in the direction of the unit normal n.

Eqns. 1.89 and 1.90 imply that any electromagnetic field could be dissected along any fixed surface, and the field on one side abolished, provided that we could introduce, on the dissecting surface, electric and magnetic currents and charges whose surface densities are

$$\left\{ i' = -n \times E \quad q' = n \cdot B \right. \tag{1.91}$$

$$\left. i = n \times H \qquad q = n \cdot D \right. \tag{1.92}$$

where n is a unit normal directed into the field that is retained. Because magnetic currents and charges are not available, it follows from eqns. 1.91 that dissection of electromagnetic fields by fixed surfaces is only possible for surfaces at which the tangential component of E and the normal component of B vanish.

Although magnetic currents and charges do not exist, nevertheless the concept is useful for handling certain aspects of the electromagnetic properties of materials. A magnet is a volume distribution of magnetic moment per unit volume M, and we saw in the two preceding sections that one way of handling this is by first calculating a field (E', B', D', H') which is the electromagnetic field of an associated distribution of magnetic current and charge given by eqns. 1.85 and 1.86, and then using eqns. 1.78 and 1.79.

Even in handling a dielectric, which is a volume distribution of electric moment per unit volume P, it transpires that it is convenient to give consideration to an associated distribution of magnetic current. Consider a thin solenoid for which the solenoidal current is not electric but magnetic. For such an arrangement, electric flux would thread through the solenoid in the direction related by the left-hand-screw rule to the solenoidal magnetic current. The electric flux would emerge from one end and return to the other, forming an external field identical with that of an electric dipole. If this were done for a volume distribution of electric moment per unit volume P, we would have a volume distribution of solenoidal magnetic current. At a particular point 0, let (E', B', D', H') be the electromagnetic field due to the volume distribution of solenoidal magnetic current, and let (E, B, D, H) be that due to the actual electric moment per unit volume P. Then (c.f. eqns. 1.78 and 1.79)

$$\begin{cases} E = E' - \dfrac{1}{\epsilon_0}P, & B = B' & (1.93) \\[2mm] D = D' - P, & H = H' & (1.94) \end{cases}$$

For a volume distribution of electric moment in dielectric material, it will emerge in Section 2.4 that, although the actual electromagnetic field is (E, B, D, H), nevertheless the field (E', B', D', H') is of interest for identifying what is known as the free electric charge and current in the system.

In general it is possible to consider a volume distribution of time-varying electric moment per unit volume P and a volume distribution of time-varying magnetic moment per unit volume M, which are jointly creating an electromagnetic field (E, B, D, H). If each elementary electric dipole were replaced by a corresponding short thin solenoid of magnetic current, and each elementary magnetic dipole composed of a short thin solenoid of electric current were replaced by a corresponding pair of equal and opposite magnetic charges, then the electromagnetic field would be changed to (E', B', D', H'). The relation of this field to the actual electromagnetic field (E, B, D, H) is (c.f. eqns. 1.78, 1.79, 1.93 and 1.94)

$$\begin{cases} E = E' - \dfrac{1}{\epsilon_0}P & B = B' + \mu_0 M & (1.95) \\[2mm] D = D' - P & H = H' + M & (1.96) \end{cases}$$

The distribution of magnetic current and charge that would create the electromagnetic field (E', B', D', H') may be called the magnetic conjugate of the distribution of electric current and charge that creates the electromagnetic field (E, B, D, H). The two distributions are sometimes said to be equivalent, but the field (E', B', D', H') is in fact only identical to the field (E, B, D, H) where P and M vanish, that is, outside the volume distributions of electric and magnetic moment.

1.15 Relative motion in electromagnetism

Relative motion in electromagnetism involves more complicated considerations than in mechanics even if the differential speed v of the observer is such that $(v/c)^2$ can be neglected. For non-relativistic mechanical systems the magnitude of c is sufficiently irrelevant that it may be regarded as infinite. The situation is different, however, for electromagnetic systems. The finite value of $2 \cdot 998 \times 10^8$ m/s for c is essential to the description of all electromagnetic systems that vary with time. In non-relativistic mechanics, terms of order $(v/c)^2$ are neglected and terms of order v/c are unimportant. In electromagnetism, even if terms of order $(v/c)^2$ can be neglected, terms of order v/c appear that are of great importance.

Let us suppose that a pair of observers are in relative motion, and that the relative velocity is constant in magnitude and direction. Let observer 0 be arbitrarily described as being 'at rest', and let the observer in motion relative to 0 with steady velocity v be described as the 'moving' observer $\vec{0}$. All quantities measured by the moving observer, whether scalar or vector, will be denoted by a superscript arrow; alternatively a superscript bar may be used if preferred. The absence of a superscript arrow implies a quantity associated with the observer that is regarded as at rest.

If each observer uses a co-ordinate system fixed relative to himself, the co-ordinates that the two observers attach to a particular point in the electromagnetic system are different. At a point that the fixed observer regards as having a position vector r (cartesian components x, y, z) the moving observer regards as having a position vector $\vec{r}$ (cartesian components $\vec{x}, \vec{y}, \vec{z}$). Moreover, the two observers do not have the same concept of time, especially when comparing times at two radically different points in space. If time for the fixed and moving observer is denoted by t and $\vec{t}$, respectively, then t is not in general the same as $\vec{t}$. It is only at the same point of space that we can put $t = \vec{t}$, and then only when the differential speed v of the two observers is such that $(v/c)^2$ can be neglected.

The difference in the concept of time for the two observers arises from the fact that each regards the velocity of an electromagnetic wave in space as c in spite of the fact that they are in motion relative to each other. Let the frame $(\vec{x}, \vec{y}, \vec{z})$ fixed relative to observer $\vec{0}$ be in motion with velocity v along the z axis in the frame (x, y, z) which is fixed relative to the observer 0. At time zero for both observers let the two frames coincide, and let an impulse of light be radiated from the plane $z = \vec{z} = 0$ in the direction in which both z and $\vec{z}$ increase. For observer 0 the impulse reaches position z at time t, where

$$z = ct \tag{1.97}$$

For observer $\vec{0}$ the impulse reaches position $\vec{z}$ at time $\vec{t}$, where

$$\vec{z} = c\vec{t} \tag{1.98}$$

Meanwhile, the frames of the two observers have separated. According to $\vec{0}$ the distance of separation is $v\vec{t}$ and according to 0 it is vt. Hence,

$$\begin{cases} z = \vec{z} + v\vec{t} & (1.99) \\ \vec{z} = z - vt & (1.100) \end{cases}$$

From eqns. 1.97, 1.98 and 1.99 it follows that

$$t = \{1 + (v/c)\}\vec{t} \qquad (1.101)$$

and from eqns. 1.97, 1.98 and 1.100 it follows that

$$\vec{t} = \{1 - (v/c)\}t \qquad (1.102)$$

By multiplying eqns. 1.101 and 1.102 together we see that they are consistent provided that we neglect $(v/c)^2$.

It is important to notice that the times t and $\vec{t}$ in eqns. 1.101 and 1.102 refer to planes that are separated by a distance that is equal to z (eqn. 1.97) according to observer 0 and to $\vec{z}$ (eqn. 1.98) according to observer $\vec{0}$. It follows from eqns. 1.98 and 1.101 that

$$t = \vec{t} + c^{-2}v\vec{z}$$

and from eqns. 1.97 and 1.102 that

$$\vec{t} = t - c^{-2}vz$$

Hence the complete expressions for (x, y, z, t) in terms of $(\vec{x}, \vec{y}, \vec{z}, \vec{t})$ for a differential speed v such that $(v/c)^2$ can be neglected are

$$x = \vec{x}, \quad y = \vec{y}, \quad z = \vec{z} + v\vec{t}, \quad t = \vec{t} + c^{-2}v\vec{z} \qquad (1.103)$$

Alternatively, the expressions for $(\vec{x}, \vec{y}, \vec{z}, \vec{t})$ in terms of (x, y, z, t) are

$$\vec{x} = x, \quad \vec{y} = y, \quad \vec{z} = z - vt, \quad \vec{t} = t - c^{-2}vz \qquad (1.104)$$

Eqns. 1.104 constitute the algebraic solution of eqns. 1.103 for $(\vec{x}, \vec{y}, \vec{z}, \vec{t})$ neglecting $(v/c)^2$. However, eqns. 1.104 would also follow from eqns. 1.103 if one regarded observer $\vec{0}$ as being 'at rest' and observer $\vec{0}$ as being in motion relative to 0 with steady velocity $(0, 0, -v)$.

The fourth of eqns. 1.103 and the fourth of eqns. 1.104 show that the gradient of $c(t - \vec{t})$ in the direction of the relative velocity v is v/c as perceived by either observer. Hence, although for a pair of observers moving with speed v relative to each other $t - \vec{t}$ vanishes at the same location in space, nevertheless the gradient of $t - \vec{t}$ in the direction of relative motion is of order v/c and is not negligible even if $(v/c)^2$ is being neglected. When considering time-separations of the order of t for points in space that are separated by distances of the order of ct, the fractional difference between t and $\vec{t}$ is of the order of v/c and therefore needs to be taken into account even if $(v/c)^2$ is being neglected. It is only when considering time-separations of the order of t for points in space that are separated by distances of the order of vt or less that the fractional difference between t and $\vec{t}$ is of the order of $(v/c)^2$.

It may be noticed however that, although the distinction between t and $\vec{t}$ is essential in the fourth of eqns. 1.103 and the fourth of eqn. 1.104, it is not essential in the third of these equations. Substitution for $\vec{t}$ from the fourth of eqns. 1.104 into the third of eqns. 1.103 gives

$$z = \vec{z} + vt - (v/c)^2 z$$

and the last term in this equation must be dropped if we are neglecting quantities of order $(v/c)^2$. Hence, to order (v/c), the third of eqns. 1.103 would still be true if $\vec{t}$ were replaced by t.

The distinction between t and $\vec{t}$ is required in order that both observers may regard the velocity of light in space as c, and both observers may regard Maxwell's equations as satisfied. In the absence of the distinction between t and $\vec{t}$, the velocity of light for the observer regarded as 'moving' would be $c - v$ in the direction of his motion and $c + v$ in the opposite direction as happens in the case of sound waves in air. For electromagnetic waves in free space, however, it is impossible to decide which observer is 'moving' and which is 'at rest'. Both regard the velocity of light in free space as having the same value c, and this mandates the distinction between the concepts of time for the two observers.

Eqns. 1.103 and 1.104 may be expressed in terms of the vector velocity v of the moving observer $\vec{0}$ relative to the fixed observer 0, and of the position vectors r and $\vec{r}$ of a point in space relative to origins attached to observers 0 and $\vec{0}$, respectively. Neglecting $(v/c)^2$ we obtain

$$\left.\begin{aligned} r &= \vec{r} + v\vec{t} \quad t = \vec{t} + c^{-2}v\cdot\vec{r} \\ \vec{r} &= r - vt \quad\;\; \vec{t} = t - c^{-2}v\cdot r \end{aligned}\right\} \tag{1.105}$$

These equations can be written more symmetrically as

$$\left.\begin{aligned} r &= \vec{r} + (v/c)(c\vec{t}) \quad ct = c\vec{t} + (v/c)\cdot r \\ \vec{r} &= r - (v/c)(ct) \quad\;\; c\vec{t} = ct - (v/c)\cdot r \end{aligned}\right\} \tag{1.106}$$

Eqns. 1.105, and particularly the difference between the concepts of time for the two observers, lead to differences of the greatest importance in the electromagnetic fields perceived by the two observers even when their differential speed v is such that $(v/c)^2$ is negligible. These differences constitute the low-speed Lorentz transformation summarised in Table 1.1. The equations on the right of the Table are the solution of those on the left neglecting $(v/c)^2$. Each half is obtained from the other by interchanging the quantities associated with the two observers and changing the sign of v.

If an observer 0 sees a field of electric current density J (and associated electric charge density ρ) that causes an electromagnetic field (E, B, D, H) at position r at time t, then an observer $\vec{0}$ moving relative to 0 with steady velocity v [$(v/c)^2$ neglected] sees the electromagnetic system as possessing a field of current density $\vec{J}$ (and associated charge density $\vec{\rho}$) that causes an electromagnetic field $(\vec{E}, \vec{B}, \vec{D}, \vec{H})$ at position $\vec{r}$ at time $\vec{t}$, where the quantities are related as shown in lines 1–4 of Table

Table 1.1 *The low-speed Lorentz transformation*

1a	$\vec{r} = r - tv$	$\vec{t} = t - c^{-2}v\cdot r$	$r = \vec{r} + \vec{t}v$	$t = \vec{t} + c^{-2}v\cdot\vec{r}$
1b	$\vec{\nabla} = \nabla + c^{-2}v\partial/\partial t$	$\partial/\partial\vec{t} = \partial/\partial t + v\cdot\nabla$	$\nabla = \vec{\nabla} - c^{-2}v\partial/\partial\vec{t}$	$\partial/\partial t = \partial/\partial\vec{t} - v\cdot\vec{\nabla}$
2	$\vec{E} = E + v\times B$	$\vec{B} = B - c^{-2}v\times E$	$E = \vec{E} - v\times\vec{B}$	$B = \vec{B} + c^{-2}v\times\vec{E}$
3	$\vec{H} = H - v\times D$	$\vec{D} = D + c^{-2}v\times H$	$H = \vec{H} + v\times\vec{D}$	$D = \vec{D} - c^{-2}v\times\vec{H}$
4	$\vec{J} = J - \rho v$	$\vec{\rho} = \rho - c^{-2}v\cdot J$	$J = \vec{J} + \vec{\rho}v$	$\rho = \vec{\rho} + c^{-2}v\cdot\vec{J}$
5	$\vec{A} = A - c^{-2}\phi v$	$\vec{\phi} = \phi - v\cdot A$	$A = \vec{A} + c^{-2}\vec{\phi}v$	$\phi = \vec{\phi} + v\cdot\vec{A}$
6	$\vec{P} = P - c^{-2}v\times M$	$\vec{M} = M + v\times P$	$P = \vec{P} + c^{-2}v\times\vec{M}$	$M = \vec{M} - v\times\vec{P}$
7	$\vec{\Pi} = \Pi - v\times\Pi'$	$\vec{\Pi'} = \Pi' + c^{-2}v\times\Pi$	$\Pi = \vec{\Pi} + v\times\vec{\Pi'}$	$\Pi' = \vec{\Pi'} - c^{-2}v\times\vec{\Pi}$

A superscript arrow indicates a quantity appropriate to the 'moving' observer

1.1 (see Problem 1.10). It is assumed that the origins for the two frames of reference coincide at time zero for both observers.

If the electromagnetic field is described with the aid of the magnetic vector
potential and the electric scalar potential (Section 1.9), the relation between the
potentials experienced by the two observers is as shown on line 5 (see Problem
1.11). If the electric currents and charges are described with the aid of the electric
and magnetic moments per unit volume (Sections 1.10 and 1.12), the relation
between the moments seen by the two observers is as shown on line 6, and the relation between the electric and magnetic Hertzian potentials experience by the two
observers is as shown on line 7 (see Problem 1.12).

If the relative speed v of the two observers is such that $(v/c)^2$ cannot be neglected, the concept of the Lorentz contraction must be introduced. When observer 0
looks at the moving frame of observer $\vec{0}$, he sees distances in that frame parallel to
the direction of motion as contracted by a fraction γ^{-1}; but distances perpendicular
to the direction of motion are unaffected. When observer $\vec{0}$ looks at the frame of
observer 0, he sees it moving backwards with speed v. Observer $\vec{0}$ likewise sees distances in the frame of observer 0 parallel to the direction of motion as contracted
by the same fraction γ^{-1}; but distances perpendicular to the direction of motion are
unaffected. Each observer sees the other's frame as contracted parallel to the direction of relative motion by the same fraction γ^{-1}. This means that eqns. 1.99 and
1.100 become

$$\left. \begin{array}{l} z = \gamma(\vec{z} + v\vec{t}) \\ \vec{z} = \gamma(z - vt) \end{array} \right\} \tag{1.107}$$

In consequence, eqns. 1.101 and 1.102 become

$$\left. \begin{array}{l} t = \gamma\{1 + (v/c)\}\vec{t} \\ \vec{t} = \gamma\{1 - (v/c)\}t \end{array} \right\} \tag{1.108}$$

and multiplication of this pair of equations now gives

$$\gamma = \{1 - (v/c)^2\}^{-1/2} \tag{1.109}$$

As a result, eqns. 1.103 and 1.104 become

$$\left. \begin{array}{llll} x = \vec{x}, & y = \vec{y}, & z = \gamma(\vec{z} + v\vec{t}), & t = \gamma(\vec{t} + c^{-2}v\vec{z}) \\ \vec{x} = x, & \vec{y} = y, & \vec{z} = \gamma(z - vt), & \vec{t} = \gamma(t - c^{-2}vz) \end{array} \right\} \tag{1.110}$$

It follows from eqns. 1.110 that multiplying factors γ must be appropriately included in Table 1.1 if $(v/c)^2$ cannot be neglected. The multiplying factor γ has to be
included in the scalar quantities (time, charge and electric potential) but only in
one component of the vector quantities, either that parallel to the direction of
relative motion or that perpendicular to it. For position vector, current density and
vector potential the factor γ appears in the parallel component, but for all other
vectors it appears in the perpendicular component. The consequences of eqns.
1.110 in mechanics are summarised in Appendix C.

Problems

1.1 A small static electric dipole of moment p is located at the origin of spherical polar coordinates (r, θ, ψ) in free space and points along the axis $\theta = 0$. At a point whose distance from the dipole is large compared with the linear dimensions of the dipole, show that the electric scalar potential is

$$\phi = \frac{p}{4\pi\epsilon_0} \frac{\cos\theta}{r^2}$$

Deduce that the electric field has spherical polar components

$$E = \frac{p}{4\pi\epsilon_0} \frac{1}{r^3}(2\cos\theta, \sin\theta, 0)$$

Show that a similarly located small static magnetic dipole of moment m has magnetic scalar potential

$$\phi' = \frac{m}{4\pi} \frac{\cos\theta}{r^2}$$

and a field of magnetic flux density whose spherical polar components are

$$B = \frac{\mu_0 m}{4\pi} \frac{1}{r^3}(2\cos\theta, \sin\theta, 0)$$

1.2 A steady current I flows round a loop C in free space in the presence of a non-uniform external magnetostatic field. With respect to a point 0 in the field, r is the position vector of a vector element of length ds of C that points in the direction of I. If B is the external magnetic-flux-density vector at ds, show that the resultant force on the loop is

$$F = I \oint_C ds \times B$$

and that the resultant torque about 0 is

$$T = I \oint_C r \times (ds \times B)$$

Show further that, if the loop suffers a small non-uniform displacement in which the element ds is displaced ds', then the work done by the forces acting on the loop is $Id\Phi$, where $d\Phi$ is the resulting increment in the external magnetic flux Φ threading C in the direction related by the right-hand-screw rule to the current I.

1.3 In the previous problem, the loop is replaced by a long thin solenoid conveying a steady magnetic flux Φ from outside the external magnetic field to a point P at which the external magnetic vector is H. Use the principle of work to show that the resultant of all the forces acting on the solenoid is a single force ΦH acting through P.

1.4 In the previous problem, the external field is produced by a loop C round which flows a steady current I. With respect to the point P where the magnetic flux Φ emerges from the thin solenoid, r is the position vector of an element of length ds

of C that points in the direction of the current I, and $\hat{r}$ is a unit vector in the same direction. By first calculating the force exerted on the loop C by the magnetic flux emerging from the solenoid, show that the resultant of all forces exerted on the solenoid by the current I round C is a force

$$\frac{I\Phi}{4\pi} \oint_C \frac{\hat{r} \times ds}{r^2}$$

acting through P. Deduce the Biot–Savart law stating that the magnetic vector produced at P by the current I round C is

$$H = \frac{I}{4\pi} \oint_C \frac{\hat{r} \times ds}{r^2}$$

1.5   A particle of charge e and mass m is in motion with velocity v in a prescribed time-varying electromagnetic field (E, B) for which the electromagnetic potentials are (A, ϕ). Taking the motional energy of the particle as

$$W_M = \tfrac{1}{2}mv^2 + ev\cdot A$$

verify that the momentum of the particle is

$$\frac{\partial W_M}{\partial v} = mv + eA$$

Taking the potential energy of the particle as

$$W_P = e\phi$$

write down Lagrange's equations of motion for the particle, and show that they reduce to

$$\frac{d}{dt}(mv) = e(E + v \times B)$$

1.6   A time-varying electric Hertzian potential Π at the point (r, θ, ψ) in spherical polar co-ordinates in free space is such that its direction is everywhere parallel to the axis $\theta = 0$ and its magnitude Π is a function of the radial distance r and time t only. Show that the electric scalar potential at the point (r, θ, ψ) is $-\cos\theta \; \partial\Pi/\partial r$, and that the r, θ, and ψ components of the electromagnetic field are

$$E_r = \frac{\partial^2\Pi}{\partial r^2}\cos\theta - \frac{1}{c^2}\frac{\partial^2\Pi}{\partial t^2}\cos\theta$$

$$E_\theta = -\frac{1}{r}\frac{\partial\Pi}{\partial r}\sin\theta + \frac{1}{c^2}\frac{\partial^2\Pi}{\partial t^2}\sin\theta$$

$$B_\psi = -\frac{1}{c^2}\frac{\partial^2\Pi}{\partial r\partial t}\sin\theta$$

$$E_\psi = B_r = B_\theta = 0$$

where c is the velocity of light in free space.

1.7   A small electric dipole having time-varying moment $p(t)$ exists in free space at the origin of spherical polar coordinates (r, θ, ψ), the positive direction of the dipole being along the axis $\theta = 0$. Write down the electric Hertzian potential Π at any point (r, θ, ψ) at time t. Use the results of the previous problem to show that the r, θ, and ψ components of the electromagnetic field at the point (r, θ, ψ) at time t are

$$E_r = \frac{1}{4\pi\epsilon_0} \frac{p'(t-r/c)}{cr^2} 2\cos\theta + \frac{1}{4\pi\epsilon_0} \frac{p(t-r/c)}{r^3} 2\cos\theta$$

$$E_\theta = \frac{1}{4\pi\epsilon_0} \frac{p''(t-r/c)}{c^2 r} \sin\theta + \frac{1}{4\pi\epsilon_0} \frac{p'(t-r/c)}{cr^2} \sin\theta + \frac{1}{4\pi\epsilon_0} \frac{p(t-r/c)}{r^3} \sin\theta$$

$$B_\psi = \frac{1}{4\pi\epsilon_0} \frac{p''(t-r/c)}{c^3 r} \sin\theta + \frac{1}{4\pi\epsilon_0} \frac{p'(t-r/c)}{c^2 r^2} \sin\theta$$

$$E_\psi = B_r = B_\theta = 0$$

where c is the velocity of light in free space and a prime denotes differentiation.

1.8 If the electric dipole in the previous problem consists of a stationary proton at the origin together with an electron that vibrates along the axis $\theta = 0$ near the origin, justify the statement that the local field depends on the displacement of the electron from the origin whereas the distant radiated field depends on the acceleration of the electron. Verify that the dipole does not radiate along its axis.

1.9 A small magnetic dipole is located in free space at the origin of spherical polar co-ordinates (r, θ, ψ). Its moment is directed along the axis $\theta = 0$ and it has the value $m(t)$ at time t. From the magnetic Hertzian potential deduce that the electromagnetic field at the point (r, θ, ψ) at time t has spherical polar components

$$B_r = \frac{\mu_0}{4\pi} \frac{m'(t-r/c)}{cr^2} 2\cos\theta + \frac{\mu_0}{4\pi} \frac{m(t-r/c)}{r^3} 2\cos\theta$$

$$B_\theta = \frac{\mu_0}{4\pi} \frac{m''(t-r/c)}{c^2 r} \sin\theta - \frac{\mu_0}{4\pi} \frac{m'(t-r/c)}{cr^2} \sin\theta + \frac{\mu_0}{4\pi} \frac{m(t-r/c)}{r^3} \sin\theta$$

$$E_\psi = -\frac{\mu_0}{4\pi} \frac{m''(t-r/c)}{cr} \sin\theta - \frac{\mu_0}{4\pi} \frac{m'(t-r/c)}{r^2} \sin\theta$$

$$B_\psi = E_r = E_\theta = 0$$

where c is the velocity of light in free space and a prime denotes differentiation. Verify that the dipole does not radiate along its axis.

1.10 An electromagnetic field in free space is viewed by two observers 0 and $\vec{0}$, the latter being in steady motion with velocity v relative to the former $[(v/c)^2$ neglected]. The frames of reference for the two observers may be assumed to coincide at time zero for both observers. Observer 0 sees an electromagnetic field (E, B, D, H) at position r at time t. Observer $\vec{0}$ sees the same electromagnetic field as

$(\vec{E}, \vec{B}, \vec{D}, \vec{H})$ at position $\vec{r}$ at time $\vec{t}$. Use Identities 9 and 15 in Appendix B to show that, neglecting terms of order $(v/c)^2$,

$$\vec{\nabla} \times \vec{E} + \partial\vec{B}/\partial\vec{t} = (\nabla \times E + \partial B/\partial t) + (\nabla \cdot B)v$$

$$\vec{\nabla} \cdot \vec{B} = \nabla \cdot B + c^{-2}v \cdot (\nabla \times E + \partial B/\partial t)$$

$$\vec{\nabla} \times \vec{H} - \partial\vec{D}/\partial\vec{t} = (\nabla \times H - \partial D/\partial t) - (\nabla \cdot D)v$$

$$\vec{\nabla} \cdot \vec{D} = \nabla \cdot D - c^{-2}v \cdot (\nabla \times H - \partial D/\partial t)$$

From the first pair of equations show that, if Maxwell's electric curl and magnetic divergence equations are satisfied for observer 0, then they are also satisfied for observer $\vec{0}$. From the second pair of equations show that the same is true for Maxwell's magnetic curl and electric divergence equations provided that the densities of electric current and charge perceived by the two observers are related as shown in line 4 of Table 1.1.

1.11 An electromagnetic field in free space is viewed by two observers 0 and $\vec{0}$, the latter being in steady motion with velocity v relative to the former $[(v/c)^2$ neglected]. The frames of reference for the two observers may be assumed to coincide at time zero for both observers. Observer 0 sees an electromagnetic field caused by densities of electric current and charge (J, ρ) and described by a magnetic vector potential A and an electric scalar potential ϕ at position r at time t. Observer $\vec{0}$, veiwing the same electromagnetic field, experiences densities of electric current and charge $(\vec{J}, \vec{\rho})$, a magnetic vector potential $\vec{A}$ and an electric scalar potential $\vec{\phi}$ at position $\vec{r}$ at time $\vec{t}$. If $(\vec{A}, \vec{\phi})$ are related to (A, ϕ) in accordance with line 5 of Table 1.1 show that, neglecting terms of order $(v/c)^2$,

$$\vec{\nabla}^2\vec{A} - c^{-2}\partial^2\vec{A}/\partial\vec{t}^2 + \mu_0\vec{J} = (\nabla^2 A - c^{-2}\partial^2 A/\partial t^2 + \mu_0 J) -$$
$$- c^{-2}v(\nabla^2\phi - c^{-2}\partial^2\phi/\partial t^2 + \rho/\epsilon_0)$$

$$\vec{\nabla}^2\vec{\phi} - c^{-2}\partial^2\vec{\phi}/\partial\vec{t}^2 + \vec{\rho}/\epsilon_0 = (\nabla^2\phi - c^{-2}\partial^2\phi/\partial t^2 + \rho/\epsilon_0) -$$
$$- v \cdot (\nabla^2 A - c^{-2}\partial^2 A/\partial t^2 + \mu_0 J)$$

Deduce that, if (A, ϕ) satisfy the differential equations for magnetic vector potential and electric scalar potential for observer 0, then $(\vec{A}, \vec{\phi})$ satisfy the corresponding equations for observer $\vec{0}$.

1.12 An electromagnetic field in free space is viewed by two observers 0 and $\vec{0}$, the latter being in steady motion with velocity v relative to the former $[(v/c)^2$ neglected]. The frames of reference for the two observers may be assumed to coincide at time zero for both observers. Observer 0 sees an electromagnetic field caused by electric and magnetic moments per unit volume (P, M) and described by electric and magnetic Hertzian potentials (Π, Π') at position r at time t. Observer $\vec{0}$, viewing the same electromagnetic field, experiences electric and magnetic moments per unit volume $(\vec{P}, \vec{M})$ and electric and magnetic Hertzian potentials $(\vec{\Pi}, \vec{\Pi}')$ at position $\vec{r}$ at time $\vec{t}$. If $(\vec{\Pi}, \vec{\Pi}')$ are related to (Π, Π') in accordance with line 7 of Table 1.1 show that, neglecting terms of order $(v/c)^2$,

$$\vec{\nabla}^2 \vec{\Pi} - c^{-2}\partial^2 \vec{\Pi}/\partial \vec{t}^2 + \vec{P}/\epsilon_0 = (\nabla^2 \Pi - c^{-2}\partial^2 \Pi/\partial t^2 + P/\epsilon_0) -$$
$$- v \times (\nabla^2 \Pi' - c^{-2}\partial^2 \Pi'/\partial t^2 + \mu_0 M)$$

$$\vec{\nabla}^2 \vec{\Pi}' - c^{-2}\partial^2 \vec{\Pi}'/\partial \vec{t}^2 + \mu_0 \vec{M} = (\nabla^2 \Pi' - c^{-2}\partial^2 \Pi'/\partial t^2 + \mu_0 M) +$$
$$+ c^{-2}v \times (\nabla^2 \Pi - c^2 \partial^2 \Pi/\partial t^2 + P/\epsilon_0)$$

Deduce that, if (Π, Π') satisfy the differential equations for electric and magnetic Hertzian potentials for observer 0, then $(\vec{\Pi}, \vec{\Pi}')$ satisfy the corresponding equations for observer $\vec{0}$.

1.13 The unit of electric charge is changed numerically and dimensionally so as to make $(\mu_0/\epsilon_0)^{1/2}$ dimensionless and numerically equal to unity. Show that Maxwell's equations and the corresponding boundary conditions as given in eqns. 1.41–1.44 are unchanged, but that the connection relations in eqns. 1.47 become $D = c^{-1}E, H = cB$. Show that the force per unit charge on a test charge moving with velocity v relative to the observer may then be written $E + c^{-1}v \times H$, and that eqns. 1.41–1.44, if expressed in terms of E and H, become

$$\begin{cases} \operatorname{curl} E + c^{-1}\partial H/\partial t = 0, & \operatorname{div} H = 0 \\ \operatorname{curl} H - c^{-1}\partial E/\partial t = J, & c^{-1} \operatorname{div} E = \rho \end{cases}$$

$$\begin{cases} n \times E| = 0, & n \cdot H| = 0 \\ n \times H| = i, & c^{-1}n \cdot E| = q \end{cases}$$

1.14 Show that the unit of magnetic scalar potential is the ampere. A fixed tube of magnetic flux is created in free space by steady electric current flowing solenoidally over its surface. Show that the total solenoidal current between any two surfaces of constant magnetic scalar potential is equal to the difference of magnetic scalar potential between the two surfaces.

1.15 Show that the unit of magnetic current is the volt. In free space a flat parallel-plate capacitor has thin closely-spaced plates, and the rim of the capacitor is denoted by C. If the capacitor is charged to a steady voltage V, show that the fringing electric field is that of a magnetic current of strength V round C. Show also that the direction of the magnetic current is related by the right-hand-screw rule to the direction of the internal electric field in the capacitor.

Chapter 2

Electromagnetic fields in materials

2.1 Introduction

The methods summarised in the previous chapter permit us to calculate the electromagnetic field for distributions of electric charge and current that are specified as functions of position and time in otherwise free space. But the presence of materials causes a difficulty.

It is true that the distribution and motion of the charged particles in a material merely create additional distributions of electric charge and current in space. But they are not ones that are specified directly as functions of position and time. Near a point 0 in material, they are specified instead as functions of the electromagnetic field (E, B) that exists in the neighbourhood of 0. In this chapter we summarize the procedures for handling this complication.

2.2 The constitutive functions for materials

A conductor is a material which contains free charges (often electrons) that are capable of drifting through the material to form what is known as a free electric current. An electric field E at a point 0 in a conductor causes the free charges near 0 to drift, thereby creating a free electric current. In simple conducting material the electric current is in the direction of E, but presence of a magnetic field of flux density B at 0 can modify this. In all cases we may say that the free electric current density in the neighbourhood of any point in a conductor is a function of E and B in the neighbourhood of that point, and we may write it as $J_f(E, B)$.

As described in Section 1.10, free electric current may be dissected into tubes of flow, and each tube may then be dissected into current elements, each of which is an electric dipole. If this is done, use may be made of the concept of free electric moment per unit volume P_f. In terms of P_f, the volume densities of free electric current and free electric charge are given by (eqns. 1.64)

$$J_f = \frac{\partial P_f}{\partial t}, \qquad \rho_f = - \operatorname{div} P_f \tag{2.1}$$

The values of P_f depend on the time-history of E and B, but only the contemporary values of E and B are involved in the values of J_f and ρ_f.

A dielectric is an insulating material that develops an electric moment per unit volume if placed in an electric field. This induced electric moment per unit volume is associated with electrons that are bound in the atoms and molecules, and it is called the bound electric moment per unit volume P_b. The bound electric moment per unit volume in the neighbourhood of a point 0 of a material depends on the electric vector E in the neighbourhood of 0. It also depends slightly on the magnetic-flux-density vector B in the neighbourhood of 0, and this dependence can become important if the magnetic field is strong. In all cases we may say that the bound electric moment per unit volume in the neighbourhood of any point in a dielectric is a function of E and B in the neighbourhood of that point, and we may write it as $P_b(E, B)$.

Vibration of the aggregate of bound electrons in an insulating dielectric creates an electric current in the material and, in accordance with the principle of conservation of charge, the electric current generates a distribution of electric charge. These electric currents and charges in an insulating dielectric are known as bound currents and charges. Their volume densities are (c.f. eqns. 2.1)

$$\frac{\partial}{\partial t} P_b(E, B), \qquad - \operatorname{div} P_b(E, B) \tag{2.2}$$

Materials not normally considered to be magnets develop some magnetic moments per unit volume if placed in a magnetic field. This is called induced magnetic moment per unit volume, and it is associated mainly with bound charges in atoms and molecules. The induced magnetic moment per unit volume in the neighbourhood of a point 0 due to bound charges depends on the electromagnetic field (E, B) in the neighbourhood of 0, and we write it $M_b(E, B)$. The associated volume density of electric current is, according to the first of eqns. 1.77,

$$\operatorname{curl} M_b(E, B) \tag{2.3}$$

The divergence of this current density vanishes in accordance with Identity 5 of Appendix B. It follows from the equation of continuity (eqn. 1.14) that this current generates no density of electric charge.

In a conductor under the influence of a strong magnetic field, free charges can spiral round the field and thereby create some magnetic moment per unit volume. This is an example of free magnetic momentum per unit volume, and we write it $M_f(E, B)$. The corresponding volume density of electric current is (c.f. expr. 2.3)

$$\operatorname{curl} M_f(E, B) \tag{2.4}$$

and there is no associated density of electric charge.

A particular piece of material usually has conduction, dielectric and magnetic properties. In general, the total volume densities of electric current J and of electric charge ρ in a material are found by summing the various contributions. We obtain

$$
\left.
\begin{aligned}
J &= J_f(E, B) + \frac{\partial}{\partial t} P_b(E, B) + \operatorname{curl} M_b(E, B) + \operatorname{curl} M_f(E, B) \\
\rho &= \rho_f(E, B) - \operatorname{div} P_b(E, B)
\end{aligned}
\right\}
\tag{2.5}
$$

If use is made of the free electric moment per unit volume $P_f(E, B)$ then, in accordance with eqns. 2.1, we may write eqns. 2.5 as

$$
\left.
\begin{aligned}
J &= \frac{\partial}{\partial t} P_f(E, B) + \frac{\partial}{\partial t} P_b(E, B) + \operatorname{curl} M_b(E, B) + \operatorname{curl} M_f(E, B) \\
\rho &= -\operatorname{div} P_f(E, B) - \operatorname{div} P_b(E, B)
\end{aligned}
\right\}
\tag{2.6}
$$

In addition to the volume densities of electric current and charge there are also surface densities of electric current i and of charge q that can be described with the aid of the concepts of surface curl and surface divergence appearing in eqns. 1.77 and 1.63. If the boundary of the material is at rest relative to the observer and if n is a unit normal to the boundary directed into the material, we have

$$
\left.
\begin{aligned}
i &= n \times M_f(E, B) + n \times M_b(E, B) \\
q &= -n \cdot P_f(E, B) - n \cdot P_b(E, B)
\end{aligned}
\right\}
\tag{2.7}
$$

At a fixed interface between two different materials the surface densities of electric current and charge are

$$
\left.
\begin{aligned}
i &= n \times M_f(E, B)| + n \times M_b(E, B)| \\
q &= -n \cdot P_f(E, B)| - n \cdot P_b(E, B)|
\end{aligned}
\right\}
\tag{2.8}
$$

where a vertical line implies the discontinuous increment that takes place in the preceding quantity as the interface is crossed in the direction of the unit normal n.

The various functions of E and B appearing in this section are constitutive functions for the material under discussion. They permit us to specify the electric current and charge that exist in the material when it is known what electromagnetic field exists in the material. The constitutive functions may be calculated on the basis of a model of the material as outlined in Section 2.12. Such models usually involve parameters whose values must be determined experimentally.

2.3 Version 1 of the electromagnetic equations for materials

If the distribution of electric current and charge were specified as functions of position and time, we would calculate the total electric and magnetic moments per unit volume from the equations

$$
P = P_f + P_b, \qquad M = M_f + M_b
\tag{2.9}
$$

We would then substitute into eqns. 1.70 and 1.83 to calculate the electric and magnetic Hertzian potentials, and hence deduce the electromagnetic field (E, B) as described in Sections 1.11–1.13. In practice, the electric currents and charges are usually specified as functions of position and time only in limited locations known as sources. They are not so specified in the surrounding materials. As described in the preceding section, the electric current and charge in the neighbourhood of a point in material is specified instead of terms of the electromagnetic field (E, B) in the neighbourhood of the point, that is, in terms of the very electromagnetic field that we are seeking to calculate. This is the complication that has to be faced when materials are present in an electromagnetic field.

Maxwell's equations are (eqns. 1.41 and 1.42)

$$\begin{cases} \operatorname{curl} E + \dfrac{\partial B}{\partial t} = 0, & \operatorname{div} B = 0 & (2.10) \\[2em] \operatorname{curl} H - \dfrac{\partial D}{\partial t} = J & \operatorname{div} D = \rho & (2.11) \end{cases}$$

and the relations connecting (D, H) to (E, B) are (eqns. 1.47)

$$D = \epsilon_0 E, \qquad H = \frac{1}{\mu_0} B \qquad (2.12)$$

But, in material, J and ρ on the right-hand sides of eqns. 2.11 are functions of (E, B), and these functions are specified either by eqns. 2.5 or by eqns. 2.6. This complicates the solution of eqns. 2.10, 2.11 and 2.12 for (E, B). If we substitute for J and ρ from eqns. 2.5 into eqns. 2.11, we obtain

$$\operatorname{curl} H - \frac{\partial D}{\partial t} = J_f(E, B) + \frac{\partial}{\partial t} P_b(E, B) + \operatorname{curl} M_b(E, B) + \operatorname{curl} M_f(E, B)$$

$$\operatorname{div} D = \rho_f(E, B) - \operatorname{div} P_b(E, B)$$

$$(2.13)$$

In material, the equations to be solved for (E, B) are eqns. 2.10, 2.13 and 2.12. These constitute the most fundamental version of the electromagnetic equations in material. They simply recognize the fact that material consists of elementary particles in a vacuum, and that the resulting electric current and charge in the neighbourhood of each point of the material are determined by the local electromagnetic field (E, B).

2.4 Version 2 of the electromagnetic equations for materials

The fact that electric current and charge in material depends on the local electromagnetic field is a complication that we seek to alleviate by means of a change of variables. In place of the fields (D, H) we introduce modified fields (D', H'), which are in fact the fields already encountered in eqns. 1.96. The change of variables will have the effect of simplifying eqns. 2.13, which are differential equations, at the expense of complicating eqns. 2.12, which are algebraic equations. The modified electric-flux-density vector D' and the modified magnetic vector H' are defined by

$$D' = D + P_b(E, B), \qquad H' = H - M_b(E, B) \tag{2.14}$$

Eqns. 2.13 may be rearranged as

$$\left. \begin{aligned} \operatorname{curl} \{H - M_b(E, B)\} - \frac{\partial}{\partial t} \{D + P_b(E, B)\} &= J_f(E, B) + \operatorname{curl} M_f(E, B) \\ \operatorname{div} \{D + P_b(E, B)\} &= \rho_f(E, B) \end{aligned} \right\}$$

and they then become, with the aid of eqns. 2.14,

$$\left. \begin{aligned} \operatorname{curl} H' - \frac{\partial D'}{\partial t} &= J_f + \operatorname{curl} M_f(E, B) \\ \operatorname{div} D' &= \rho_f(E, B) \end{aligned} \right\} \tag{2.15}$$

These equations have the same form as eqns. 2.11 but:

(i) only the free electric current and charge appear on the right-hand sides of eqns. 2.15, and

(ii) the electric-flux-density vector and the magnetic vector appearing in eqns. 2.15 are the modified vector fields (D', H') defined in eqns. 2.14.

Electric currents and charges associated with bound electrons in material appear in eqns. 2.15 only by virtue of the fact that (D', H') are used instead of (D, H). We have effected a simplification in the magnetic curl and electric divergence equations in material.

However, if we are to use the modified electric-flux-density vector and the modified magnetic vector in the magnetic curl and electric divergence equations, we must also do so in the connection relations in Equations (2.12). By substituting for D and H from eqns. 2.12 into eqns. 2.14, we see that the relations connecting (D', H') to (E, B) are

$$D' = \epsilon_0 E + P_b(E, B), \qquad H' = \frac{1}{\mu_0} B - M_b(E, B) \tag{2.16}$$

Instead of solving eqns. 2.10, 2.13 and 2.12 for (E, B) we can solve instead eqns. 2.10, 2.15 and 2.16 for (E, B). This constitutes a substantial simplification. It is true that eqns. 2.16 are more complicated than eqns. 2.12 but, in practice, this complication is usually only algebraic. This price is worth paying in order to simplify the differential relations in eqns. 2.13 to those appearing in eqns. 2.15. Version 2 of the electromagnetic equations in material consists, therefore, of eqns. 2.10, 2.15 and 2.16.

Maxwell's equations are what the circulation and flux laws for the electromagnetic field become if we convert from integral equations to differential equations. In particular, version 1 of Maxwell's equations constitutes the differential equivalent of the integral relations appearing in eqns. 1.37, 1.38, 1.39 and 1.40. These equations involve a surface S having a rim C that is at rest relative to the observer. In the presence of materials, the threading current in eqn. 1.39 is the total threading current – that associated with both free and bound charges. Likewise, the included charge in eqn. 1.40 is the total included charge – that associated with both free and bound charges. Let us now write down the circulation and flux laws corresponding to version 2 of Maxwell's equations. Since eqns. 2.10 have not been modified, they still correspond to eqns. 1.37 and 1.38. But eqns. 1.39 and 1.40 are replaced by

$$\left. \begin{array}{l} \displaystyle\int_C H' \cdot ds - \frac{\partial}{\partial t} \int_S D' \cdot dS \;=\; \text{threading free current} \\[4mm] \displaystyle\int_\Sigma D' \cdot dS \;=\; \text{included free charge} \end{array} \right\} \qquad (2.17)$$

Moreover, at a fixed interface between two different materials, these equations may be applied, in the manner illustrated in Figs. 1.6 and 1.8, to obtain the boundary conditions. Eqns. 1.43 are unchanged, but eqns. 1.44 are replaced by

$$n \times H'| \;=\; i_f, \qquad n \cdot D'| \;=\; q_f \qquad (2.18)$$

where i_f and q_f are the surface densities of free electric current and charge on the interface.

The facts with regard to versions 1 and 2 of the electromagnetic equations for material at rest relative to the observer are summarised in Table 2.1.

2.5 Version 3 of the electromagnetic equations for materials

Version 2 of the electromagnetic equations for material is particularly convenient when discussing the design and operation of equipment where special attention needs to be paid to free electric currents and charges on good conductors. However, when the wave character of the electromagnetic field is dominant and relative motion between the materials is not involved, it is version 3 in Table 2.1 that is particularly convenient.

In version 3 we describe free electric current and charge with the aid of free electric moment per unit volume as exhibited in eqns. 2.1, so that the total electric current and charge in the material have the volume densities given in eqns. 2.6. Moreover, in introducing a modified electric-flux-density vector and a modified magnetic vector, we now not only allow for bound electric and magnetic moments per unit volume but also for free electric and magnetic moments per unit volume.

We then introduce a doubly modified electric-flux-density vector D'' and a doubly modified magnetic vector H'' defined by (c.f. eqns. 2.14)

Table 2.1 *The electromagnetic equations for material at rest relative to the observer*

	Version 1	Version 2	Version 3
Field equations for E, B	$\operatorname{curl} E + \dfrac{\partial B}{\partial t} = 0$, $\operatorname{div} B = 0$	$\operatorname{curl} E + \dfrac{\partial B}{\partial t} = 0$, $\operatorname{div} B = 0$	$\operatorname{curl} E + \dfrac{\partial B}{\partial t} = 0$, $\operatorname{div} B = 0$
Field equations for D, H	$\operatorname{curl} H - \dfrac{\partial D}{\partial t} = J$, $\operatorname{div} D = \rho$ where $J = J_{\mathrm f} + \operatorname{curl} M_{\mathrm f} + \dfrac{\partial P_{\mathrm b}}{\partial t} + \operatorname{curl} M_{\mathrm b}$ $\rho = \rho_{\mathrm f} - \operatorname{div} P_{\mathrm b}$	$\operatorname{curl} H' - \dfrac{\partial D'}{\partial t} = J_{\mathrm f} + \operatorname{curl} M_{\mathrm f}$, $\operatorname{div} D' = \rho_{\mathrm f}$	$\operatorname{curl} H'' - \dfrac{\partial D''}{\partial t} = 0$, $\operatorname{div} D'' = 0$
Connection relations	$D = \epsilon_0 E$, $H = \dfrac{1}{\mu_0} B$	$D' = \epsilon_0 E + P_{\mathrm b}$, $H' = \dfrac{1}{\mu_0} B - M_{\mathrm b}$	$D'' = \epsilon_0 E + P_{\mathrm b} + P_{\mathrm f}$, $H'' = \dfrac{1}{\mu_0} B - M_{\mathrm b} - M_{\mathrm f}$
Boundary conditions	$n \times E \rvert = 0$, $n \cdot B \rvert = 0$ $n \times H \rvert = n \times M_{\mathrm b} \rvert + i_{\mathrm f}$, $n \cdot D \rvert = -n \cdot P_{\mathrm b} \rvert + q_{\mathrm f}$	$n \times E \rvert = 0$, $n \cdot B \rvert = 0$ $n \times H' \rvert = i_{\mathrm f}$, $n \cdot D' \rvert = q_{\mathrm f}$	$E_\parallel \rvert = 0$, $B_\perp \rvert = 0$ $H''_\parallel \rvert = 0$, $D''_\perp \rvert = 0$

$$D'' = D + P_b(E, B) + P_f(E, B), \quad H'' = H - M_b(E, B) - M_f(E, B) \quad (2.19)$$

Eqns. 2.15 are then replaced by

$$\left.\begin{aligned} \operatorname{curl} H'' - \frac{\partial D''}{\partial t} &= 0 \\[2mm] \operatorname{div} D'' &= 0 \end{aligned}\right\} \quad (2.20)$$

and the connection relations in eqns. 2.16 become

$$D'' = \epsilon_0 E + P_b(E, B) + P_f(E, B), \quad H'' = \frac{1}{\mu_0} B - M_b(E, B) - M_f(E, B)$$

$$(2.21)$$

Version 3 of the electromagnetic equations for material then consists of eqns. 2.10, 2.20 and 2.21. In version 3 of the electromagnetic equations for materials, eqns. 2.17 are replaced by

$$\left.\begin{aligned} \int_C H'' \cdot ds - \frac{\partial}{\partial t} \int_S D'' \cdot dS &= 0 \\[2mm] \int_\Sigma D'' \cdot dS &= 0 \end{aligned}\right\} \quad (2.22)$$

and the boundary conditions at a fixed interface given in eqns. 2.18 are replaced by

$$n \times H''| = 0, \qquad n \cdot D''| = 0 \quad (2.23)$$

The facts with regard to version 3 of the electromagnetic equations for material at rest relative to the observer are summarised in the last column of Table 2.1. In version 3, Maxwell's equations take the same form as in a region of space that is free of all forms of electric current and charge, but the electric-flux-density vector and the magnetic vector in use are the doubly modified ones (D'', H''). These are related to (E, B) by the connection relations appearing in eqns. 2.21.

2.6 Linear isotropic material

In simple material that is not moving relative to the observer, the free electric current density J_f and the bound electric moment per unit volume P_b are independent of the magnetic-flux-density vector B but are proportional to the local electric vector E. Also, the bound magnetic moment per unit volume M_b is independent of E, but is proportional to the local magnetic-flux-density vector B, while the free magnetic moment per unit volume is negligible. Such material is known as linear isotropic material. For it we may write

$$J_f = \sigma E, \quad P_b = \kappa \epsilon_0 E, \quad M_b = \chi \frac{1}{\mu_0} B, \quad M_f = 0 \quad (2.24)$$

where σ, κ and χ are parameters whose values describe the character of the material. The first of these equations leads to Ohm's law. Substitution for P_b and M_b from eqns. 2.24 into eqns. 2.14 shows that the second and third of the connection relations in eqns. 2.24 may be rewritten

$$D' = \epsilon E, \qquad H' = \frac{1}{\mu} B \tag{2.25}$$

where

$$\frac{\epsilon}{\epsilon_0} = 1 + \kappa, \qquad \frac{\mu}{\mu_0} = \frac{1}{1 - \chi} \tag{2.26}$$

The quantities ϵ/ϵ_0 and μ/μ_0 are known as the dielectric constant and the permeability of the material. The quantities σ and κ are known as the conductivity and electric susceptibility of the material. The quantity χ is the magnetic susceptivity of the material referred to the B vector.

With the aid of the second of eqns. 2.25, the expression for M_b in eqns. 2.24 may be written $\{\chi/(1 - \chi)\}H'$. The quantity $\chi/(1 - \chi)$ is known as the magnetic susceptibility of the material referred to the H' vector.

For linear isotropic material at rest relative to the observer, the three versions of the electromagnetic equations summarised in Table 2.1 take the form shown in Table 2.2.

2.7 The electromagnetic equations for linear isotropic non-conducting material at rest relative to the observer

In non-conducting material we have

$$\sigma = 0, \quad J_f = 0, \quad P_f = 0 \tag{2.27}$$

Both version 2 and version 3 of the electromagnetic equations in Table 2.2 are then identical. Version 2 gives

$$\left.\begin{array}{ll} \operatorname{curl} E + \dfrac{\partial B}{\partial t} = 0, & \operatorname{div} B = 0 \\[3mm] \operatorname{curl} H' - \dfrac{\partial D'}{\partial t} = 0, & \operatorname{div} D' = 0 \\[3mm] D' = \epsilon E, & H' = \dfrac{1}{\mu} B \end{array}\right\} \tag{2.28}$$

The corresponding boundary conditions at a fixed interface are:

$$\left.\begin{array}{ll} E_{\parallel} \text{ is continuous,} & B_{\perp} \text{ is continuous} \\[2mm] H'_{\parallel} \text{ is continuous,} & D'_{\perp} \text{ is continuous} \end{array}\right\} \tag{2.29}$$

Table 2.2 The electromagnetic equations for linear isotropic material at rest relative to the observer

	Version 1	Version 2	Version 3
Field equations for E, B	$\operatorname{curl} E + \dfrac{\partial B}{\partial t} = 0, \quad \operatorname{div} B = 0$	$\operatorname{curl} E + \dfrac{\partial B}{\partial t} = 0, \quad \operatorname{div} B = 0$	$\operatorname{curl} E + \dfrac{\partial B}{\partial t} = 0, \quad \operatorname{div} B = 0$
Field equations for D, H	$\operatorname{curl} H - \dfrac{\partial D}{\partial t} = J, \quad \operatorname{div} D = \rho$ $J = J_{\mathrm{f}} + \dfrac{\partial P_{\mathrm{b}}}{\partial t} + \operatorname{curl} M_{\mathrm{b}}$ $\rho = \rho_{\mathrm{f}} - \operatorname{div} P_{\mathrm{b}}$	$\operatorname{curl} H' - \dfrac{\partial D'}{\partial t} = J_{\mathrm{f}}, \quad \operatorname{div} D' = \rho_{\mathrm{f}}$	$\operatorname{curl} H'' - \dfrac{\partial D''}{\partial t} = 0, \quad \operatorname{div} D'' = 0$
Connection relations	$D = \epsilon_0 E, \quad H = \dfrac{1}{\mu_0} B, \quad P_{\mathrm{b}} = \kappa\epsilon_0 E, \quad M_{\mathrm{b}} = \chi \dfrac{1}{\mu_0} B$ $J_{\mathrm{f}} = \sigma E, \quad D = \epsilon_0 E, \quad H = \dfrac{1}{\mu_0} B$	$J_{\mathrm{f}} = \sigma E, \quad D' = \epsilon E, \quad H' = \dfrac{1}{\mu} B$	$\dfrac{\partial P_{\mathrm{f}}}{\partial t} = \sigma E, \quad D'' = \epsilon E + P_{\mathrm{f}}, \quad H'' = \dfrac{1}{\mu} B$
Boundary conditions	$n \times E| = 0, \quad n \cdot B| = 0$ $n \times H| = n \times M_{\mathrm{b}}| + i_{\mathrm{f}},$ $n \cdot D| = -n \cdot P_{\mathrm{b}}| + q_{\mathrm{f}}$	$n \times E'| = 0, \quad n \cdot B| = 0$ $n \times H'| = i_{\mathrm{f}}, \quad n \cdot D'| = q_{\mathrm{f}}$	$E_{\parallel}| = 0, \quad B_{\perp}| = 0$ $H''_{\parallel}| = 0, \quad D''_{\perp}| = 0$

These equations for linear isotropic non-conducting material at rest relative to the observer are the same as the electromagnetic equations in a region of space that is free of material and free of electric current, except that D and H are replaced by D' and H', and ϵ_0 and μ_0 are replaced by ϵ and μ (c.f. eqns. 1.41, 1.42 and 1.47). In consequence, there is a range of circumstances in which the mathematical methods for handling electromagnetic fields in free space summarised in Chapter 1 are directly adaptable for handling electromagnetic fields in linear isotropic materials at rest relative to the observer. This happens when the material is non-conducting and homogeneous. In these circumstances σ is not involved, and ϵ and μ in eqns. 2.28 are, like ϵ_0 and μ_0 in free space, independent of position. Let us illustrate this far-reaching result by considering radiation from an elementary time-varying dipole that is embedded in fixed homogeneous non-conducting material.

2.8 Radiation from a fixed time-varying dipole into fixed homogeneous linear isotropic non-conducting material

Let us suppose that, at the origin of spherical polar coordinates (r, θ, ψ), there exists a free current element that constitutes an electric dipole whose free electric moment $p(t)$ varies with time t in a prescribed manner. Let us suppose that elsewhere space is occupied by homogeneous non-conducting material of dielectric constant ϵ/ϵ_0 and permeability μ/μ_0 at rest relative to the observer. The electromagnetic field is then given by the same equations as in free space with ϵ_0 and μ_0 replaced by ϵ and μ, and with (D, H) replaced by (D', H'). This means that the velocity of propagation $(\mu_0 \epsilon_0)^{-1/2}$ in eqn. 1.48 is replaced in the homogeneous non-conducting material by a velocity of propagation $(\mu \epsilon)^{-1/2}$. It is therefore by this velocity that c in eqns. 1.65–1.70 must be replaced; in addition ϵ_0 must be replaced by ϵ.

To convert the equations for propagation in free space given in Chapter 1 into the corresponding equations for propagation in a homogeneous non-conducting material of dielectric constant ϵ/ϵ_0 and permeability μ/μ_0, it is convenient to introduce a refractive index n defined as the geometric mean of the permeability and the dielectric constant:

$$n = \left(\frac{\mu}{\mu_0} \frac{\epsilon}{\epsilon_0} \right)^{1/2}. \tag{2.30}$$

The refractive index is then the ratio of the velocity of propagation in free space to that in the homogeneous non-conducting material. For free space the velocity of propagation is

$$c = (\mu_0 \epsilon_0)^{-1/2} \tag{2.31}$$

whereas for homogeneous non-conducting material of refractive index n it is

$$c/n = (\mu \epsilon)^{-1/2} \tag{2.32}$$

Consequently, eqn. 1.69 becomes, in the presence of the homogeneous non-conducting material,

$$\Pi = \frac{1}{4\pi\epsilon} \frac{p(t - nr/c)}{r} \tag{2.33}$$

and eqns. 1.68 become

$$E = \text{grad div } \Pi - \left(\frac{n}{c}\right)^2 \frac{\partial^2 \Pi}{\partial t^2}, \qquad B = \left(\frac{n}{c}\right)^2 \text{curl} \frac{\partial \Pi}{\partial t} \tag{2.34}$$

Hence the electromagnetic field of a fixed time-varying dipole having free electric moment $p(t)$ embedded in fixed homogeneous linear isotropic non-conducting material is obtained by substituting for Π from eqn. 2.33 into eqns. 2.34. The result is obtained by replacing c by c/n and ϵ_0 by ϵ in Problem 1.7.

In the same way it follows from eqn. 1.82 that, at distance r from a small fixed loop of time-varying free electric current of magnetic moment $m(t)$, the electromagnetic field in fixed homogeneous linear isotropic non-conducting material is given by a magnetic Hertzian potential

$$\Pi' = \frac{\mu}{4\pi} \frac{m(t - nr/c)}{r} \tag{2.35}$$

in terms of which the electromagnetic field is given by (c.f. eqns. 1.81)

$$E = -\text{curl} \frac{\partial \Pi'}{\partial t}, \qquad B = \text{grad div } \Pi' - \left(\frac{n}{c}\right)^2 \frac{\partial^2 \Pi'}{\partial t^2} \tag{2.36}$$

The resulting electromagnetic field is obtained by replacing c and c/n and μ_0 by μ in Problem 1.9.

Relations corresponding to eqns. 1.70 and 1.83 may be written down to give the electromagnetic field of any prescribed distribution of time-varying free electric and magnetic moment per unit volume in homogeneous linear isotropic non-conducting material at rest relative to the observer.

We see that the effect of filling all space with homogeneous non-conducting material of refractive index n is to change the velocity of propagation from c, given by eqn. 2.31, to c/n, given by eqn. 2.32. Although this modification is trivial mathematically, it should be noted that the physical process by which it occurs needs to be thought through using version 1 of the electromagnetic equations, and is relatively complicated. Under the influence of the field radiated by a primary dipole, every element of volume of the material becomes a secondary reradiating electric and magnetic dipole for which the electric moment per unit volume P_b and the magnetic moment per unit volume M_b are given by the second and third of eqns. 2.24. The electromagnetic field at each point in space is the vector sum of the fields radiated by all the dipoles – the given primary dipole and all the secondary dipoles. The remarkably simple overall effect is to change the velocity of propagation from c to c/n.

The easy mathematical solution that we have obtained for a comparatively complicated physical problem involving induced time-varying electric and magnetic dipoles distributed throughout space has been made possible by the introduction of the modified electric-flux-density vector D' and the modified magnetic vector H' in Section 2.4, and by the use of the simple connection relations (2.24) with $\sigma = 0$, leading to eqns. 2.28. It should also be noted, however, that the assumption that the dielectric constant ϵ/ϵ_0 and the permeability μ/μ_0 are independent of position in space is also indispensible to the easy mathematical solution that has been obtained. If ϵ and μ in the last pair of eqns. 2.28 are functions of position then, on substituting for D' and H' into the second pair, the factors ϵ and $1/\mu$ are trapped inside the curl and divergence operators, and the consequent complications are major. These are the mathematical complications that connote the physical phenomena of reflection, refraction and diffraction.

2.9 Complex electromagnetic vectors

In conducting material, versions 2 and 3 of the electromagnetic equations in Table 2.1 are different. It is convenient to use version 3 when dealing with wave phenomena in conducting material at rest relative to the observer. Moreover, for linear materials, it is convenient to Fourier-analyse the time-variation of the source or sources, and then to solve the problem for the individual sinusoidal oscillations involved. At any point in space, the effect of the entire spectrum of oscillations can then be ascertained by Fourier synthesis.

The process of Fourier synthesis assumes that the principle of superposition is applicable, that is, that the equations of electromagnetic theory are linear. Maxwell's equations are linear, and so are the connection relations in eqns. 2.24. They are still linear if σ, κ and χ, and therefore the dielectric constant ϵ/ϵ_0 and the permeability μ/μ_0 in eqns. 2.26, are functions of the frequency of oscillation. A far larger range of practical materials is covered by the concepts of dielectric constant, permeability and conductivity if these quantities are permitted to be functions of the angular frequency of oscillation ω of the electromagnetic field (see Section 2.12).

When all cartesian components of all electromagnetic vectors everywhere are oscillating with the same angular frequency ω, it is convenient to describe each with the aid of a rotating vector in the complex plane. At a particular location in space, the x component of the electric vector is the real part, or reference component, of the rotating vector in the complex plane given by

$$E_x = E_x^0 \exp\left(j\omega t\right) \tag{2.37}$$

where

$$E_x^0 = |E_x^0| \exp\left(j \arg E_x^0\right). \tag{2.38}$$

A similar procedure is adopted for the y and z components of the electric vector at the location. In the same way we may introduce a complex magnetic-flux-density vector (B_x, B_y, B_z). The complex electromagnetic field at a location constitutes an ensemble of rotating vectors such as that illustrated in Fig. 2.1. Notice that (E_x, E_y, E_z) and (B_x, B_y, B_z) are not the actual electric vector and magnetic-flux-density vector at the location. It is the reference components of the rotating vectors that give the actual electromagnetic vibrations.

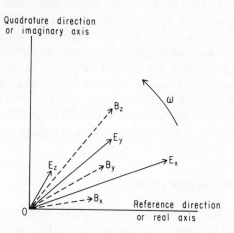

Fig. 2.1 *Illustrating, in the complex plane, rotating vectors whose reference components, or real parts, give the three-dimensional cartesian components of the electric vector and the magnetic-flux-density vector at a particular point in space*

If the positions of the rotating vectors in Fig. 2.1 at time zero correspond to (E_x^0, E_y^0, E_z^0), (B_x^0, B_y^0, B_z^0) then, in accordance with eqns. 2.37 and 2.38, these complex quantities give the amplitudes and phases of the six vibrations, and constitute what are known as their complex amplitudes. For oscillatory electromagnetic fields of angular frequency ω, it is convenient to perform most calculations in terms of the complex electromagnetic vectors. Any amplitude ratio and phase difference that may be required for a pair of vibrations are then calculated by taking the quotient of the corresponding complex electromagnetic vector components; the magnitude of the quotient gives the amplitude ratio, and the argument of the quotient gives the phase difference.

It is often convenient to suppose that the electromagnetic system under study involved no electric currents and charges at $t = -\infty$. This can be achieved by letting the angular frequency ω be complex, with a small negative imaginary part (see eqn. 2.37). Likewise, a positive imaginary part to ω corresponds to an oscillation that decays exponentially with time.

We shall frequently be concerned with situations in which all time-varying quantities may be represented as the reference components of rotating vectors in the complex plane. The rotation is described by means of a complex oscillation function exp $(j\omega t)$, with ω real or complex. This procedure involves a great simplification in the operations of differentiation and integration with respect to time. In the one case we have

$$\frac{d}{dt} \exp{(j\omega t)} = j\omega \exp{(j\omega t)} \tag{2.39}$$

In the other case we have

$$\int_{-\infty}^{t} \exp{(j\omega t)}\, dt = \frac{1}{j\omega} \exp{(j\omega t)} \tag{2.40}$$

provided that we assume that the electromagnetic system under study involved no electric currents and charges at $t = -\infty$. From eqns. 2.39 and 2.40 it follows that:

(i) differentiation with respect to time simply involves multiplication by $j\omega$, and
(ii) integration with respect to time simply involves division by $j\omega$.

2.10 The complex electromagnetic equations for linear isotropic material at rest relative to the observer

The handling of the electromagnetic equations for oscillatory behaviour, both in linear materials and in free space, is greatly simplified by the use of the complex oscillation function $\exp{(j\omega t)}$. For a given angular frequency of oscillation ω, every equation in Table 2.2 may be regarded as written in terms of complex vector fields of the type appearing in eqns. 2.37 and 2.38, and illustrated in Fig. 2.1. Although the complex electromagnetic vectors are not the actual field vectors, nevertheless they satisfy the electromagnetic equations for linear materials. This is because, for real linear equations relating complex numbers, the real parts satisfy the same equations as the complex numbers themselves. Consequently, for linear isotropic materials, the field equations, the connection relations and the boundary conditions listed in Table 2.2 may all be interpreted as equations relating the complex electromagnetic vectors for angular frequency ω. The real parts of these complex equations then take an identical form, and it is these real parts that are the actual electromagnetic equations for the vibrating electromagnetic vectors in three-dimensional space.

In accordance with eqn. 2.39, each time-derivative in Table 2.2 may now be replaced by $j\omega$. In consequence, Table 2.2 takes the form shown in Table 2.3. In Table 2.3 it will be noticed that the volume divergence equations have been omitted from the field equations, and the surface divergence equations have been omitted from the boundary conditions. This is not because these relations are untrue, but because they are redundant (c.f. eqns. 1.45 and 1.46). The divergence equations follow from the curl equations in Table 2.3 by taking the divergence and remembering that the divergence of the curl of any differentiable vector field vanishes (Identity 5, Appendix B).

In Table 2.3 it will be noticed that the boundary conditions depend on the electric vector E and on the magnetic vector H, H' or H''. On the other hand, the force per unit charge on a test charge moving with velocity v relative to the observer,

Table 2.3 *The complex electromagnetic equations for linear isotropic material at rest relative to the observer*

	Version 1	Version 2	Version 3
Field equations for **E**, **B**	curl $E + j\omega B = 0$	curl $E + j\omega B = 0$	curl $E + j\omega B = 0$
Field equations for **D**, **H**	curl $H - j\omega D = J$ where $J = J_f + j\omega P_f + \text{curl } M_b$	curl $H' - j\omega D' = J_f$	curl $H'' - j\omega D'' = 0$
Connection relations	$D = \epsilon_0 E, \quad H = \dfrac{1}{\mu_0} B$ $J_f = \sigma E, \quad P_b = \kappa \epsilon_0 E, \quad M_b = \chi \dfrac{1}{\mu_0} B$	$J_f = \sigma E, \quad D' = \epsilon E, \quad H' = \dfrac{1}{\mu} B$	$D'' = \left(\epsilon - j\dfrac{\sigma}{\omega}\right) E, \quad H'' = \dfrac{1}{\mu} B$
Boundary conditions	$n \times E| = 0$ $n \times H| = n \times M_b| + i_f$	$n \times E| = 0$ $n \times H'| = i_f$	$E_{\parallel}| = 0$ $H''_{\parallel}| = 0$

being given by expr. 1.1, depends on the electric vector E and magnetic-flux-density vector B. We see therefore that, in handling oscillatory electromagnetic fields, it is desirable to work in terms of the magnetic-flux-density vector when discussing the forces exerted on charges, but in terms of the magnetic vector when handling boundary conditions.

It is version 3 of the electromagnetic equations for materials that is particularly convenient when handling oscillatory electromagnetic fields. Special interest attaches to the form taken by the connection relations in version 3 when using a complex oscillation function. If we interpret in complex form the connection relations for version 3 in Table 2.2 and then replace the time-derivative by $j\omega$, we obtain

$$j\omega P_f = \sigma E, \quad D'' = \epsilon E + P_f, \quad H'' = \frac{1}{\mu}B \qquad (2.41)$$

If we substitute for P_f from the first of these relations into the second we obtain

$$D'' = (\epsilon - j\sigma/\omega)E, \qquad H'' = \frac{1}{\mu}B \qquad (2.42)$$

as recorded in Table 2.3. These equations may be written

$$D'' = \tilde{\epsilon}E, \qquad H'' = \frac{1}{\mu}B \qquad (2.43)$$

where
$$\tilde{\epsilon} = \epsilon - j\sigma/\omega \qquad (2.44)$$

The quantity $\tilde{\epsilon}/\epsilon_0$ is known as the complex dielectric constant of the material.

We see that, for linear isotropic conducting material at rest relative to the observer, version 3 of the electromagnetic equations, including the divergence equations, is

$$
\left.
\begin{aligned}
&\operatorname{curl} E + j\omega B = 0, \qquad &&\operatorname{div} B = 0 \\
&\operatorname{curl} H'' - j\omega D'' = 0, \qquad &&\operatorname{div} D'' = 0 \\[2mm]
&\qquad\quad D'' = \tilde{\epsilon}E, \qquad &&H'' = \frac{1}{\mu}B
\end{aligned}
\right\} \qquad (2.45)
$$

and the corresponding boundary conditions at a fixed interface are

$$
\left.
\begin{aligned}
&E_\parallel \text{ is continuous,} \qquad &&B_\perp \text{ is continuous} \\
&H''_\parallel \text{ is continuous,} \qquad &&D''_\perp \text{ is continuous}
\end{aligned}
\right\} \qquad (2.46)
$$

Comparison of eqns. 2.45 and 2.46 with eqns. 2.28 and 2.29 shows that linear isotropic material at rest relative to the observer may be handled just as conveniently if it possesses conductivity as if it is an insulator provided that

(i) we use version 3 of the electromagnetic equations,

(ii) we Fourier-analyse the time-variation and use complex electromagnetic fields [time factor $\exp{(j\omega t)}$], and

(iii) we use the complex dielectric constant $\tilde{\epsilon}/\epsilon_0$, where $\tilde{\epsilon}$ is given in eqn. 2.44.

Let us illustrate this far-reaching result by considering radiation from an elementary oscillatory dipole that is embedded in fixed homogeneous linear isotropic material that possesses conductivity.

2.11 Radiation from a fixed oscillatory dipole into fixed homogeneous linear isotropic material possessing conductivity

Let ω be the angular frequency of the oscillation, $\tilde{\epsilon}/\epsilon_0$ the complex dielectric constant of the material, and μ/μ_0 its permeability. We introduce a complex refractive index defined by (c.f. eqn. 2.30)

$$\tilde{n} = \left(\frac{\mu}{\mu_0} \frac{\tilde{\epsilon}}{\epsilon_0} \right)^{1/2} \tag{2.47}$$

Let the dipole be located at the origin and have a time-varying free electric moment that is the real part of

$$p(t) = p_0 \exp{(j\omega t)} \tag{2.48}$$

At distance r, the complex electric Hertzian potential is then (c.f. eqn. 2.33)

$$\Pi = \frac{1}{4\pi\tilde{\epsilon}} \frac{p(t - \tilde{n}r/c)}{r} \tag{2.49}$$

where, from eqns. 2.48,

$$p(t - \tilde{n}r/c) = p_0 \exp{\{j\omega(t - \tilde{n}r/c)\}} \tag{2.50}$$

It follows that

$$\Pi = \frac{1}{4\pi\tilde{\epsilon}} \frac{p_0 \exp{\{j\omega(t - \tilde{n}r/c)\}}}{r} \tag{2.51}$$

and that the corresponding complex electromagnetic field is (c.f. eqns. 2.34)

$$E = \operatorname{grad} \operatorname{div} \Pi + (\omega\tilde{n}/c)^2\, \Pi, \qquad B = j\omega(\tilde{n}/c)^2 \operatorname{curl} \Pi \tag{2.52}$$

The resulting complex electromagnetic field is that given in Problem 2.8.

Likewise, for radiation from an elementary magnetic dipole at the origin oscillating with angular frequency ω, we let the magnetic moment of the free current associated with the dipole be the real part of

$$m(t) = m_0 \exp{(j\omega t)} \tag{2.53}$$

so that

$$m(t - \tilde{n}r/c) = m_0 \exp{\{j(t - \tilde{n}r/c)\}} \tag{2.54}$$

The complex magnetic Hertzian potential at distance r is then (c.f. eqn. 2.35)

$$\Pi' = \frac{\mu}{4\pi} \frac{m \exp\{j(t - \tilde{n}r/c)\}}{r} \tag{2.55}$$

and the corresponding complex electromagnetic field is (c.f. eqns. 2.36)

$$E = -j\omega \text{ curl } \Pi', \qquad B = \text{grad div } \Pi' + (\omega\tilde{n}/c)^2 \Pi' \tag{2.56}$$

The resulting complex electromagnetic field is that given in Problem 2.9.

The principle of superposition may be used to handle any spectrum of frequencies or any prescribed spatial distribution of elementary dipoles, electric and/or magnetic.

By using eqn. 2.44 in eqn. 2.47 we see that the complex dielectric constant of a material has a positive real part n and a negative imaginary part $-a$, so that

$$\tilde{n} = n - ja \tag{2.57}$$

Hence the wave function in eqns. 2.51 and 2.55 may be written

$$\exp\{j\omega(t - \tilde{n}r/c)\} = \exp[j\omega\{t - (n - ja)r/c\}]$$
$$= \exp(-\omega ar/c)\exp\{j\omega(t - nr/c)\} \tag{2.58}$$

The first of these exponential factors indicates that there is an exponential attenuation with distance r from the elementary dipole at a rate controlled by the imaginary part of the complex refractive index $\tilde{n}$. The second exponential factor indicates that spheres of constant phase for the Hertzian potential propagate away from the dipole with the velocity c/n, that is, at a speed given by the velocity of light in free space divided by the real part of the complex refractive index $\tilde{n}$.

2.12 Models of materials possessing conduction and dielectric properties

Consider a conductor containing N free electrons per unit volume (charge e, mass m) in random thermic motion. Let us suppose that the electrons collide with obstacles such as ions and/or neutral atoms. Let the average time that an electron is in flight between collisions be $1/\nu$, so that ν is the collisional frequency. Under the influence of a time-varying electromagnetic field (E, B), let the electrons develop a drift-velocity v. This is the average velocity of a small group of electrons; it, too, is time-dependent. On the average, an electron is acted on by the following forces:

(i) a force eE due to the electric vector E,
(ii) a force $ev \times B$ due to the magnetic-flux-density vector B, and
(iii) a force $-\nu mv$ due to collisions, assuming that the drift-motion acquired by an electron prior to a collision is converted into random thermic motion by the collision so that, on the average, a drift-momentum mv is destroyed at each collision.

These forces cause an acceleration $\dot{v}$ in the drift-motion. Hence the equation of drift-motion for an electron is

$$m\dot{v} = e(E + v \times B) - mvv \tag{2.59}$$

If r is the drift-displacement of an electron, the equation of drift-motion may be written

$$m\ddot{r} + mv\dot{r} = e(E + \dot{r} \times B) \tag{2.60}$$

In practice, the drift-velocity $\dot{r}$ is very small compared with the velocity of light and, unless there is a strong imposed magnetic field, the term $e\dot{r} \times B$ on the right-hand side of eqn. 2.60 can be neglected. The equation of drift-motion then becomes

$$m\ddot{r} + mv\dot{r} = eE \tag{2.61}$$

The equation describes how electrons are accelerated by the electric field between collisions, and how the coherent motion thereby acquired is converted into random motion at collisions.

Each electron displaced through r superimposes on the system an electric moment er. If there are N free electrons per unit volume, their contribution to the time-varying free electric moment per unit volume is

$$P_f = Ner \tag{2.62}$$

Substitution for r from eqn. 2.62 into eqn. 2.61 gives

$$\ddot{P}_f + v\dot{P}_f = (Ne^2/m)E \tag{2.63}$$

With this equation one can calculate, for given time-variation of the electric vector E, what the resulting time-variation is in the free electric moment per unit volume P_f.

Consider an oscillatory time-variation of angular frequency ω, and let all oscillations be represented as the reference components of rotating vectors in the complex plane. If E and P_f are now the complex electric vector and the complex free electric moment per unit volume, they still satisfy eqn. 2.63. Moreover, each time-derivative may now be replaced by $j\omega$, so that eqn. 2.63 becomes

$$(j\omega)^2 P_f + j\omega v P_f = (Ne^2/m)E$$

or

$$P_f = -\frac{Ne^2}{m\omega^2} \frac{1}{1 - j(v/\omega)} E \tag{2.64}$$

If we assume that the free charges that are creating electric moment per unit volume are all electrons, we substitute for P_f from eqn. 2.64 into the second of eqns. 2.41; at the same time we replace ϵ by ϵ_0 if the effect of bound electrons is unimportant. We obtain

$$D'' = \left\{ 1 - \frac{Ne^2}{\epsilon_0 m\omega^2} \frac{1}{1 - j(v/\omega)} \right\} \epsilon_0 E \tag{2.65}$$

It then follows from the first of eqns. 2.43 that the complex dielectric constant

$\tilde{\epsilon}/\epsilon_0$ of linear isotropic conducting material is given, at angular frequency ω, by

$$\frac{\tilde{\epsilon}}{\epsilon_0} - 1 = -\frac{Ne^2}{\epsilon_0 m\omega^2} \frac{1}{1 - j(\nu/\omega)} \tag{2.66}$$

A conducting gas (a plasma) or a conducting liquid (an electrolyte) may possess several types of charge-carriers (electrons, positive ions, negative ions) and the ions may be of several species. The right-hand side of eqn. 2.66 then has to be summed over the various types of charge-carriers.

For a metal at radio frequencies, the angular frequency ω is small compared with the collisional frequency ν. Eqn. 2.66 may then be simplified to

$$\tilde{\epsilon} = \epsilon_0 - j\frac{Ne^2}{m\nu\omega} \tag{2.67}$$

Comparison with eqn. 2.44 shows that a linear isotropic metal at radio frequencies has a unit dielectric constant and a conductivity σ given by

$$\sigma = \frac{Ne^2}{m\nu} \tag{2.68}$$

If account were taken of the bound electrons in the material in addition to the free electrons, the dielectric constant would differ from unity. However, the precise value of the dielectric constant of a metal is unimportant at radio frequencies because the conductivity is so high that the second term on the right-hand side of eqn. 2.67 swamps the first.

For a plasma, the collisional frequency is often negligible ($\nu \ll \omega$), with the result that $\tilde{\epsilon}$ in eqn. 2.66 is real. The equation may then be written

$$\frac{\epsilon}{\epsilon_0} = 1 - \frac{\omega_N^2}{\omega^2} \tag{2.69}$$

where

$$\omega_N^2 = \frac{Ne^2}{\epsilon_0 m} \tag{2.70}$$

The quantity ω_N is known as the angular plasma frequency. From eqn. 2.69 we see that the dielectric constant of such a plasma is less than unity; this result will be discussed further in Section 14.6.

If conducting material is subjected to a strong static magnetic field of flux density B, eqn. 2.61 has to be replaced by eqn. 2.60. In consequence, eqn. 2.65 becomes a tensor relation; each cartesian component of D'' depends linearly on all three cartesian components of E. This means that isotropic conducting material, when subjected to a strong magnetic field, becomes non-isotropic; a plasma becomes what is known as a magnetoplasma. If the direction of the z axis of cartesian coordinates is taken in the direction of the imposed magnetic field, the replacement for eqn. 2.65 calculates to

$$D'' = \begin{pmatrix} 1 + \kappa_T & \kappa_H & 0 \\ -\kappa_H & 1 + \kappa_T & 0 \\ 0 & 0 & \kappa_L \end{pmatrix} \epsilon_0 E \tag{2.71}$$

where

$$\left.\begin{array}{l} \kappa_L = \dfrac{1}{j\omega} \dfrac{\omega_N^2}{\nu + j\omega} \\[2ex] \kappa_T = \dfrac{1}{j\omega} \dfrac{\omega_N^2(\nu + j\omega)}{\omega_M^2 + (\nu + j\omega)^2} \\[2ex] \kappa_H = \dfrac{1}{j\omega} \dfrac{\omega_N^2 \omega_M}{\omega_M^2 + (\nu + j\omega)^2} \end{array}\right\} \tag{2.72}$$

Here ω_N is the angular plasma frequency (eqn. 2.70), and

$$\omega_M = \frac{eB}{m} \tag{2.73}$$

The quantity $|\omega_M|$ is known as the angular gyrofrequency, or the angular cyclotron frequency. If there are several species of charge-carriers, summation signs are required in front of the terms on the right-hand sides of eqns. 2.72; in the third equation, notice that ω_M is positive for a positive charge-carrier and negative for a negative charge-carrier.

Let us now turn our attention to a linear isotropic dielectric. We are now dealing with bound electrons rather than free electrons, and we assume that they are not under the influence of a strong magnetic field. Let each atom be subjected to a time-varying electric field that causes a time-varying displacement r of a bound electron. The electron is bound by a restoring force that gives it a resonant angular frequency Ω which is observed in practice as a spectroscopic line-frequency of breadth 2ν in angular frequency. For a bound electron, the left-hand side of eqn. 2.61 is replaced by

$$m\ddot{r} + 2m\nu\dot{r} + m\Omega^2 r \tag{2.74}$$

However, consideration also has to be given to the right-hand side of eqn. 2.61.

Under the influence of an applied oscillatory electric field, free electrons in a conductor can, and often do, vibrate with an amplitude large compared with their separation. This never happens with bound electrons vibrating in a dielectric. In a dielectric each atom is in a cavity formed by the remaining atoms, and this cavity may often be treated as spherical (see Problems 2.14, 2.15 and 2.16). If an electric field E is applied to a spherical cavity in a dielectric, the electric vector in the cavity is (see Problem 2.12)

$$E' = E + P_b(3\epsilon_0) \tag{2.75}$$

where P_b is the electric moment per unit volume in the dielectric. It is this electric vector E' that has to be used on the right-hand side of eqn. 2.61 in place of E.

Consequently, the equation of motion of a bound electron is, using expr. 2.74,

$$m\ddot{r} + 2mv\dot{r} + m\Omega^2 r = eE' \tag{2.76}$$

where E' is given by eqn. 2.75.

Suppose initially that the electric field applied to the atom is static. Then the corresponding displacement of the bound electron is static. Setting $\dot{r}$ and $\ddot{r}$ in eqn. 2.76 to zero, the static displacement of the bound electron is

$$r = \frac{e}{m\Omega^2} E' \tag{2.77}$$

This displacement superimposes on the system an electric moment

$$p = er = \frac{e^2}{m\Omega^2} E' \tag{2.78}$$

For simplicity, let us assume that only one bound electron contributes significantly to the electric moment of each atom. Then p in eqn. 2.78 may be taken as the electric moment of an atom. We write this equation in the form

$$p = \alpha \epsilon_0 E' \tag{2.79}$$

where

$$\alpha = \frac{e^2}{\epsilon_0 m\Omega^2} \tag{2.80}$$

The quantity α is known as the polarisability of the atom.

Now let us suppose that the electric field applied to the dielectric varies with time. We operate on eqn. 2.76 for a bound electron in the dielectric in the same way that we operated on eqn. 2.61 for a free electron in conducting material. Eqn. 2.62 then becomes

$$P_b = Ner \tag{2.81}$$

where N is the number of atoms per unit volume. Eqn. 2.63 becomes

$$\ddot{P}_b + 2v\dot{P}_b + \Omega^2 P_b = (Ne^2/m)\{E + P_b/(3\epsilon_0)\} \tag{2.82}$$

For an oscillatory time-variation of angular frequency ω, we use complex electromagnetic vectors with a time-function $\exp(j\omega t)$, so that each time-derivative in eqn. 2.82 may be replaced by $j\omega$. The upshot is that eqn. 2.66 for conducting material is replaced in the case of dielectric material by

$$3\frac{(\tilde{\epsilon}/\epsilon_0) - 1}{(\tilde{\epsilon}/\epsilon_0) + 2} = \frac{N\alpha\Omega^2}{\Omega^2 - \omega^2 + 2j\nu\omega} \tag{2.83}$$

where α is given by eqn. 2.80.

In practice, a dielectric involves many resonators with different values of N, α, Ω and ν on the right-hand side of eqn. 2.83. This right-hand side then has to be preceded by a summation sign, and eqn. 2.83 becomes

$$3 \frac{(\tilde{\epsilon}/\epsilon_0) - 1}{(\tilde{\epsilon}/\epsilon_0) + 2} = \Sigma N \alpha g \left(\frac{\omega}{\Omega}, \frac{\nu}{\Omega} \right) \tag{2.84}$$

where

$$g(x, y) = \frac{1}{1 - x^2 + 2jxy} \tag{2.85}$$

This function is known as the dielectric dispersion function. Its real and imaginary parts are illustrated in Figure 2.2 as functions of x, and its absolute value and argument in Fig. 2.3. The lower half of Fig. 2.2 illustrates what in spectroscopy is called an absorption line. Fig. 2.4 illustrates a situation in which there are two terms on the right-hand side of eqn. 2.84. For practical dielectrics many resonances are involved, mostly at frequencies high compared with radio frequencies. At radio frequencies it is frequently possible to take the dielectric constant to be that for $\omega = 0$.

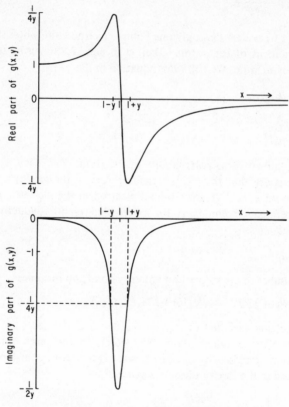

Fig. 2.2. *Real and imaginary parts of the dielectric dispersion function*

Some dielectrics have atoms or molecules with permanent electric dipole moments; they are known as polar dielectrics. Let us suppose that the dielectric has temperature T and that, as a result of thermic agitation, the dipole moments of the atoms are maintained in random directions in the absence of an applied electric field. In the presence of a static applied electric field, the directions of the dipole moments cease to be completely random and the dielectric consequently develops an electric moment per unit volume. Suppose that each atom has a permanent electric dipole moment p_0. In the presence of a static electric field E' given by eqn. 2.75, each atom has a potential energy that may be taken as (Problem 3.1)

$$W = -p_0 \cdot E' = -p_0 E' \cos \theta \qquad (2.86)$$

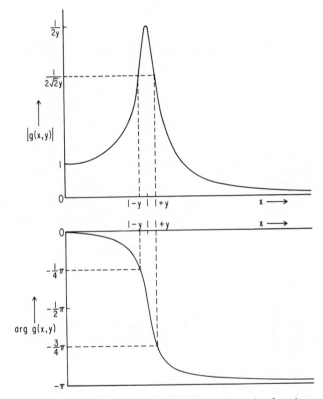

Fig. 2.3.   *Absolute value and argument of the dielectric dispersion function*

where θ is the angle between the directions of p_0 and E'. Under thermic agitation, low values of this potential energy are more probable than high values. According to statistical mechanics, the probability of finding a particular dipole moment pointing within a small solid angle $d\Omega$ at angle θ to the direction of E' is proportional to

$$\exp\left(-\frac{W}{KT}\right)d\Omega \qquad (2.87)$$

where K is Boltzmann's constant $(1\cdot380 \times 10^{-23}$ joule degree$^{-1})$ and T is the absolute temperature of the material. The component of the dipole moment in the direction of E' is $p_0 \cos \theta$. Hence the mean contribution of a dipole to the electric moment per unit volume is

$$p = \frac{\int p_0 \cos \theta \exp\left(-W/KT\right) d\Omega}{\int \exp\left(-W/KT\right) d\Omega} \tag{2.88}$$

where the integrals are taken over all directions in space.

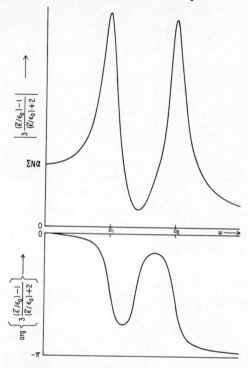

Fig. 2.4.   *Illustrating dispersion of a dielectric having more than one spectral line*

In practice, W is usually small compared with KT, and we can write

$$\exp\left(-\frac{W}{KT}\right) \doteqdot 1 - \frac{W}{KT} \tag{2.89}$$

Replacing the element of solid angle $d\Omega$ by $2\pi \sin \theta \, d\theta$ and using eqns. 2.86 and 2.89, we may evaluate the integrals in eqn. 2.88, thereby obtaining

$$p = \frac{p_0^2}{3KT} E' \tag{2.90}$$

With N atoms per unit volume, the bound electric moment per unit volume is

$$P_b = \frac{Np_0^2}{3KT} E' \tag{2.91}$$

Substitution for E' from eqn. 2.75 into eqn. 2.91 then gives

$$P_b = \frac{Np_0^2}{3KT} \left(E + \frac{P_b}{3\epsilon_0} \right) \tag{2.92}$$

This relation between P_b and E permits calculation of the electric susceptibility κ in accordance with the second of eqns. 2.24, and hence of the dielectric constant in accordance with the first of eqns. 2.26. The dielectric constant ϵ/ϵ_0 is therefore given by

$$3 \frac{(\epsilon/\epsilon_0) - 1}{(\epsilon/\epsilon_0) + 2} = \frac{Np_0^2}{3\epsilon_0 kT} \tag{2.93}$$

What we have calculated in eqn. 2.93 is the dielectric constant ϵ/ϵ_0 of a linear isotropic dielectric in which the electric dipole moment per unit volume is caused by atoms with permanent electric dipole moments that, under the influence of a static electric field, are not quite randomly oriented. If the static electric field is removed, the electric moment per unit volume disappears exponentially with a relaxation time τ. If an oscillatory electric field of angular frequency ω is applied to the material, the complex dielectric constant $\tilde{\epsilon}/\epsilon_0$ is given by

$$3 \frac{(\tilde{\epsilon}/\epsilon_0) - 1}{(\tilde{\epsilon}/\epsilon_0) + 2} = \frac{Np_0^2}{3\epsilon_0 kT} \frac{1}{1 + j\omega\tau} \tag{2.94}$$

If the dielectric contains atoms of several species having different values of N, p_0 and τ then summation signs are needed in front of the right-hand sides of eqns. 2.93 and 2.94.

2.13 Materials in motion relative to the observer

To study the properties of material in motion relative to an observer 0 use must be made of the Lorentz transformation (Section 1.15). With the aid of this transformation an observer 0 can ascertain the viewpoint of an observer $\vec{0}$ moving with any steady velocity v relative to 0. If this velocity is made identical with the velocity at a particular time of a particular volume element of the moving material, then observer $\vec{0}$ sees this volume element as stationary at this time. For observer $\vec{0}$ we may then use the methods already described in this chapter to deduce, from the electromagnet field existing at the element at the particular time, the electric and magnetic moments per unit volume existing at the element at this time. These connection relations for observer $\vec{0}$ may then be reinterpreted in the language of observer 0 using the Lorentz transformation. In this way, the connection relations for the material as seen by observer 0 may be ascertained for all volume elements of the moving material at all times. The same may be done for the boundary conditions at all moving elements of area of the interfaces between moving materials.

To deal with situations in which various elements of material are moving with various velocities at various times, a collection of moving observers $\vec{O}$ travelling steadily in various directions at various speeds is required. For any particular element at any particular time, one of these moving observers $\vec{O}$ can view the element as stationary at that time. He can report on what he sees to observer 0, who can then interpret the information with the aid of the Lorentz transformation.

This procedure is greatly simplified if all the material is moving in the same direction with the same velocity v and if this velocity does not vary with time. A single observer $\vec{O}$ moving with the steady velocity v relative to observer 0 can then view all elements as stationary for all time. But even a simple motion such as steady rotation relative to observer 0 requires a spectrum of moving observers $\vec{O}$ to report to 0 the stationary behaviour for every moving element.

Materials are not usually in motion relative to an observer with velocities v so large that the square of the ratio of v to the velocity c of light in space need be taken into account. To handle moving materials, therefore, the Lorentz transformation may be used in the form summarized in Table 1.1. For linear isotropic material in motion relative to the observer with velocity $v[(v/c)^2$ neglected] the field equations, the connection relations and the boundary conditions are summarised in Table 2.4 for the three versions of the electromagnetic equations. Version 3 is quoted only for uniform motion and for oscillatory time-variation using the complex representation of the electromagnetic field. When $v = 0$, version 3 in Table 2.4 reduces to that in Table 2.3, and versions 1 and 2 in Table 2.4 reduce to those in Table 2.2.

Versions 1 and 2 of the electromagnetic equations in Table 2.4 for motion of linear isotropic material relative to the observer are valid when the velocity v of the material is a function of position and time. When the connection relations for versions 1 and 2 are substituted into the field equations for the electric-flux-density and magnetic vectors, the velocity v appears under the space and time derivatives. This causes major complications in the solution of the electromagnetic equations unless the motion of the material is quite simple.

Problems

2.1   A sample of homogeneous material has conductivity σ, dielectric constant ϵ/ϵ_0 and permeability μ/μ_0. Its geometrical shape is that of a right cylinder of length l and normal cross-sectional areas S. Perfectly conducting electrodes, each of area S, are attached to the flat ends, and the arrangement is used as a resistor. Show that the conductance of the resistor is

$$G = \sigma S/l$$

The same arrangement is used as a capacitor for which fringing is avoided by means of guard plates. Show that the capacitance of the capacitor is

Table 2.4 The electromagnetic equations for linear isotropic materials moving with velocity v [$(v/c)^2$ neglected]

	Version 1	Version 2	Version 3 — Uniform motion, oscillatory field
Field equations for E, B	$\operatorname{curl} E + \dfrac{\partial B}{\partial t} = 0, \quad \operatorname{div} B = 0$	$\operatorname{curl} E + \dfrac{\partial B}{\partial t} = 0, \quad \operatorname{div} B = 0$	$\operatorname{curl} E + j\omega B = 0, \quad \operatorname{div} B = 0$
Field equations for D, H	$\operatorname{curl} H - \dfrac{\partial D}{\partial t} = J, \quad \operatorname{div} D = \rho$ where $J = J_f + \dfrac{\partial P_b}{\partial t} + \operatorname{curl} M_b$ $\rho = \rho_f - \operatorname{div} P_b$	$\operatorname{curl} H' - \dfrac{\partial D'}{\partial t} = J_f, \quad \operatorname{div} D' = \rho_f$	$\operatorname{curl} H'' - j\omega D'' = 0, \quad \operatorname{div} D'' = 0$
Connection relations	$D = \epsilon_0 E, \quad H = \dfrac{1}{\mu_0} B, \quad J_f = \sigma(E + v \times B) + \rho_f v$ $P_b = \kappa \epsilon_0 E + \dfrac{1}{c^2}(\kappa + \chi)\dfrac{1}{\mu_0} v \times B$ $M_b = \chi \dfrac{1}{\mu_0} B - (\kappa + \chi)\epsilon_0 v \times E$	$J_f = \sigma(E + v \times B) + \rho_f v$ $D' = \epsilon(E + v \times B) - \dfrac{1}{c^2}\dfrac{1}{\mu} v \times B$ $H' = \dfrac{1}{\mu}\left(B - \dfrac{1}{c^2} v \times E\right) + \epsilon v \times E$	$D'' = \tilde{\epsilon}(E + v \times B) - \dfrac{1}{c^2}\dfrac{1}{\mu} v \times B$ $H'' = \dfrac{1}{\mu}\left(B - \dfrac{1}{c^2} v \times E\right) + \tilde{\epsilon} v \times E$
Boundary conditions	$n \times E| - (n \cdot v)B| = 0, \quad n \cdot B| = 0$ $n \times H| + (n \cdot v)D| = i_f + n \times M_b| - (n \cdot v)P_b|$ $n \cdot D| = q_f - n \cdot P_b|$	$n \times E| - (n \cdot v)B| = 0, \quad n \cdot B| = 0$ $n \times H'| + (n \cdot v)D'| = i_f,$ $n \cdot D'| = q_f$	$n \times E| - (n \cdot v)B| = 0, \quad n \cdot B| = 0$ $n \times H''| + (n \cdot v)D''| = 0$ $n \cdot D''| = 0$

$$C = \epsilon S/l$$

The electrodes are now removed from the flat ends, and a perfectly conducting strip of width l is wrapped round the curved surface of the cylindrical sample of material. The new arrangement is used as an inductor for which non-uniformity of magnetic field within the material is avoided by using a guard inductor. Show that the inductance of the inductor is

$$L = \mu S/l$$

Discuss the statement that the conductivity σ, the capacitivity ϵ and the inductivity μ of a material are the conductance, capacitance and inductance associated with a unit cube of the material.

2.2   A rectangular column of homogeneous material has dimensions a by b by d. The material has permeability μ/μ_0, dielectric constant ϵ/ϵ_0 and conductivity σ. The column exists in air, which may be regarded as free space. Perfectly conducting strips of width b and length d are attached to a pair of opposite faces of area bd, forming a strip transmission line with a separation a between the conductors. It may be assumed that $a \ll b \ll d$, and that fringing may be neglected. Show that the inductance, capacitance and conductance per unit length of the transmission line are approximately

$$L = \frac{\mu a}{b}, \quad C = \frac{\epsilon b}{a}, \quad G = \frac{\sigma b}{a},$$

2.3   A perfectly conducting coaxial transmission line has an inner conductor of external radius a and an outer conductor of internal radius b. The space between the conductors is filled with homogeneous material of permeability μ/μ_0, dielectric constant ϵ/ϵ_0 and conductivity σ. Show that the inductance, capacitance and conductance per unit length of the transmission line are

$$L = \frac{\mu}{2\pi} \ln \frac{b}{a}, \quad C = \frac{2\pi\epsilon}{\ln (b/a)}, \quad G = \frac{2\pi\sigma}{\ln (b/a)}$$

2.4   In the previous problem, the steady voltage of the inner conductor is V and that of the outer conductor is zero, while the free electric current flowing steadily along the inner conductor and back along the outer conductor is I. At a point between the conductors distant r from the common axis, show that the electric field strength E and the magnetic field strength H' are given by

$$E = \frac{V}{\ln (b/a)} \frac{1}{r}, \quad H' = \frac{I}{2\pi r}$$

2.5   A perfectly conducting twin-wire transmission line has wires of radius a and separation d, and it may be assumed that a is small compared with d. The transmission line is embedded in infinite homogeneous material of permeability μ/μ_0, dielectric constant ϵ/ϵ_0 and conductivity σ. Show that the inductance, capacitance and conductance per unit length of the transmission line are approximately

$$L = \frac{\mu}{\pi} \ln \frac{d}{a}, \quad C = \frac{\pi \epsilon}{\ln (d/a)}, \quad G = \frac{\pi \sigma}{\ln (d/a)}$$

2.6 In the previous problem, the steady voltages of the wires are $\pm \frac{1}{2} V$. At a point outside the wires distant r_1 from the axis of the positive wire and r_2 from the axis of the negative wire, show that the electric scalar potential is approximately

$$\phi = \frac{V}{2} \frac{\ln (r_2/r_1)}{\ln (d/a)}$$

2.7 In homogeneous material of complex refractive index $\tilde{n}$ a complex electric Hertzian vector $\boldsymbol{\Pi}$ at the point (r, θ, ψ) in spherical polar co-ordinates has a time-variation proportional to $\exp (j\omega t)$. Its direction in space is everywhere parallel to the axis $\theta = 0$ and its three dimensional magnitude Π is a function of r only. Show that the complex electric scalar potential is $-\cos \theta \, \partial \Pi/\partial r$, and that the r, θ and ψ components of the complex electromagnetic field are (c.f. Problem 1.6)

$$E_r = \frac{\partial^2 \Pi}{\partial r^2} \cos \theta + \omega^2 \left(\frac{\tilde{n}}{c} \right)^2 \Pi \cos \theta$$

$$E_\theta = -\frac{1}{r} \frac{\partial \Pi}{\partial r} \sin \theta - \omega^2 \left(\frac{\tilde{n}}{c} \right)^2 \Pi \sin \theta$$

$$B_\psi = -j\omega \left(\frac{\tilde{n}}{c} \right)^2 \frac{\partial \Pi}{\partial r} \sin \theta$$

$$E_\psi = B_r = B_\theta = 0$$

where c is the velocity of light in free space.

2.8 In a system of polar co-ordinates (r, θ, ψ) there exists at the origin a small radio transmitting dipole. It is operating at angular frequency ω and consists of a short piece of straight wire of length l carrying a uniformly distributed free current whose variation with time t is given by the real part of $I = I_0 \exp (j\omega t)$, where I_0 is a complex constant. It may be assumed that, at each end of the wire, time-varying free point-charges are created by the current. Show that the complex electric moment p of the dipole is $Il/(j\omega)$. The wire is oriented along the axis $\theta = 0$, which is taken as the positive direction of the oscillatory current. The rest of space is filled with homogeneous material whose complex refractive index at angular frequency ω is $\tilde{n}$, and whose complex dielectric constant is $\tilde{\epsilon}/\epsilon_0$. Use the results of the previous problem to show that the r, θ, and ψ components of the complex electromagnetic field at the point (r, θ, ψ) at time t are (c.f. Problem 1.7)

$$E_r = \frac{1}{4\pi\tilde{\epsilon}} Il \exp \left(-j\omega \frac{\tilde{n}}{c} r \right) \left\{ \frac{\tilde{n}}{c} \frac{1}{r^2} + \frac{1}{j\omega} \frac{1}{r^3} \right\} 2 \cos \theta$$

$$E_\theta = \frac{1}{4\pi\tilde{\epsilon}} Il \exp \left(-j\omega \frac{\tilde{n}}{c} r \right) \left\{ j\omega \left(\frac{\tilde{n}}{c} \right)^2 \frac{1}{r} + \frac{\tilde{n}}{c} \frac{1}{r^2} + \frac{1}{j\omega} \frac{1}{r^3} \right\} \sin \theta$$

$$B_\psi = \frac{1}{4\pi\tilde{\epsilon}} \, Il \exp\left(-j\omega\frac{\tilde{n}}{c}r\right) \left\{ j\omega \left(\frac{\tilde{n}}{c}\right)^3 \frac{1}{r} + \left(\frac{\tilde{n}}{c}\right)^2 \frac{1}{r^2} \right\} \sin\theta$$

$$E_\psi = B_r = B_\theta = 0$$

where c is the velocity of light in free space.

2.9   In a system of polar co-ordinates (r, θ, ψ) there exists at the origin a magnetic dipole consisting of a small plane loop of area S carrying an oscillatory free current of angular frequency ω. The normal to the loop is in the direction $\theta = 0$, which is related by the right-hand-screw rule to the positive direction of the current round the wire. The current is given by the real part of $I = I_0 \exp(j\omega t)$, where I_0 is a complex constant. The rest of space is filled with homogeneous material whose complex refractive index at angular frequency ω is $\tilde{n}$ and whose permeability is μ/μ_0. Using the magnetic Hertzian potential, show that the complex electromagnetic field at the point (r, θ, ψ) has spherical polar components (c.f. Problem 1.9)

$$B_r = \frac{\mu}{4\pi} \, IS \exp\left(-j\omega\tilde{n}r/c\right) \left\{ j\omega \frac{\tilde{n}}{c} \frac{1}{r^2} + \frac{1}{r^3} \right\} 2\cos\theta$$

$$B_\theta = \frac{\mu}{4\pi} \, IS \exp\left(-j\omega\tilde{n}r/c\right) \left\{ -\omega^2 \left(\frac{\tilde{n}}{c}\right)^2 \frac{1}{r} + j\omega \frac{\tilde{n}}{c} \frac{1}{r^2} + \frac{1}{r^3} \right\} \sin\theta$$

$$E_\psi = \frac{\mu}{4\pi} \, IS \exp\left(-j\omega\tilde{n}r/c\right) \left\{ \omega^2 \frac{\tilde{n}}{c} \frac{1}{r} - j\omega \frac{1}{r^2} \right\} \sin\theta$$

$$B_\psi = E_r = E_\theta = 0$$

where c is the velocity of light in free space.

2.10   Part of space is occupied by fixed material that is composed of bound charges in atoms and molecules, and that possesses both dielectric and magnetic properties. The electric vector E and the magnetic-flux-density vector B at any point inside or outside the material are the sums of contributions due to:

(i) the distributions of free electric charge and current, and
(ii) the distributions of bound electric charge and current.

By comparing eqns. 2.14 and eqns. 1.96, show that the singly modified electric-flux-density vector D' and the singly modified magnetic vector H' at any point inside or outside the material are the unmodified electric-flux-density and magnetic vectors due to:

(i) the distributions of free electric charge and current, and
(ii) *the magnetic conjugates of* the distributions of bound electric charge and current.

2.11   A time-varying electromagnetic field (E, B, D'', H'') exists in space, part or all of which is filled with material having conduction, dielectric, and magnetic properties continuously or discontinuously distributed. A closed geometrical curve C moves through the material, and the velocity of a vector element of length ds is v relative to the observer. The moving closed curve C is spanned by a moving surface S of which a vector element of area is denoted by dS, right-hand related to *ds*. For C, show that the two circulation laws associated with the electromagnetic field may be written

$$
\begin{cases}
\int_C (E + v \times B) \cdot ds = -\dfrac{d}{dt} \int_S B \cdot dS \\[2mm]
\int_C (H'' - v \times D'') \cdot ds = \dfrac{d}{dt} \int_S D'' \cdot dS
\end{cases}
$$

where d/dt denotes the time rates of change of the fluxes across S taking into account both the rates of change of the electromagnetic field at points fixed relative to the observer and the motion of S relative to the observer.

2.12 At a point 0 in a dielectric, the electric moment per unit volume is P and the electric vector is E. A small spherical cavity is cut in the dielectric with centre at 0. Show that the electric vector in the cavity is $E + P/(3\epsilon_0)$. The shape of the small cavity is now changed to that of a disk that has small thickness. If the plane of the disk is perpendicular to P, show that the electric vector in the cavity is $E + P/\epsilon_0$. The shape of the small cavity is again changed to that of a thin cylinder. If the axis of the cylinder is parallel to P, show that the electric vector in the cavity tends to E as the cylinder becomes indefinitely thin.

2.13 At a point 0 in a magnet, the magnetic moment per unit volume is M and the magnetic-flux-density vector is B. A small spherical cavity is cut in the magnet with center at 0. Show that the magnetic-flux-density vector in the cavity is $B - (2/3)\mu_0 M$. The shape of the small cavity is now changed to that of a thin cylinder. If the axis of the cylinder is parallel to M, show that the magnetic-flux-density vector in the cavity is $B - \mu_0 M$. The shape of the small cavity is again changed to that of a thin disk. If the plane of the disk is perpendicular to M, show that the magnetic-flux-density vector in the cavity tends to B as the thickness of the disk becomes indefinitely small.

2.14 A spherical region of free space has radius R and centre at the origin 0 of cartesian co-ordinates (x, y, z). The sphere is occupied by a large number of small static electric dipoles that all have the same vector moment (p_x, p_y, p_z). The dipoles are located on a cubic lattice whose cartesian co-ordinates are $(l, m, n)a$, where (l, m, n) are integers (positive, negative or zero), and a is small compared with R. If the dipole at 0 is removed, show that the electric vector at 0 has cartesian components

$$
\frac{1}{4\pi\epsilon_0}\left[p_x \sum \frac{3x^2 - r^2}{r^5}, \; p_y \sum \frac{3y^2 - r^2}{r^5}, \; p_z \sum \frac{3z^2 - r^2}{r^5} \right]
$$

where $r^2 = x^2 + y^2 + z^2$ and the summations are over the lattice points within the sphere. Use the fact that

$$
\sum \frac{x^2}{r^5} = \sum \frac{y^2}{r^5} = \sum \frac{z^2}{r^5} = \frac{1}{3}\sum \frac{x^2 + y^2 + z^2}{r^5} = \frac{1}{3}\sum \frac{1}{r^3}
$$

to deduce that the electric field at 0 vanishes. Discuss how short the lengths of the dipoles must be for this result to apply.

2.15   In the previous problem, each dipole has charges $\pm Q$ at a distance L apart, but L is now large compared with a although still small compared with R. The positive charges constitute a uniform spherical distribution of positive charge of radius R and centre 0, and the negative charges constitute a uniform spherical distribution of negative charge of radius R and centre $0'$, the vector $\overrightarrow{0'0}$ being L. There are N dipoles per unit volume, so that the electric moment per unit volume in the sphere is $P = Np = NQL$. Show that, at the point 0, (i) the electric vector due to the positive charges vanishes, (ii) the electric vector due to the negative charges outside a sphere centre $0'$ and radius L vanishes, and (iii) the electric vector due to the negative charges inside a sphere centre $0'$ and radius L is the same as if these charges were collected at $0'$. Deduce that the electric vector at 0 due to the spherical distribution of dipoles is $-P(3\epsilon_0)$. Show that the electric vector has the same value at $0'$ or at any point on the line joining $0'$ to 0.

2.16   A dielectric consists of atoms located on a cubic lattice. Under the influence of a uniform electric field, the electronic structure of each atom is displaced through a distance small compared with the separation between adjacent atoms, thereby creating a bound electric moment per unit volume P. The associated volume and surface distributions of electric charge are calculated in accordance with eqns. 1.63, and from them is deduced the electric vector E at a point 0 in the dielectric. With centre 0, a sphere S is drawn, small enough that the electric vector E may be considered uniform in S, but large enough to contain many atoms. The point 0 is now chosen so as to coincide with an atom, and it is required to calculate the electric vector E' acting on the atom at 0. It may be assumed that the contribution of the electric dipoles outside S to the electric vector E' acting on the atom at 0 may be calculated as described in Problem 2.12. It may also be assumed that the contribution of the electric dipoles inside S to the electric vector E' acting on the atom at 0 is to be calculated as described in Problem 2.15 and not as described in Problem 2.15. Explain why

$$E' = E + P/(3\epsilon_0)$$

where P is the electric moment per unit volume of the dielectric in the vicinity of 0.

Chapter 3

Electric energy

3.1 Introduction

In electrostatics, the concept of electric potential is based on the concept of potential energy. A point charge Q_1 in free space, at rest relative to the observer, produces at a distance r an electric potential

$$\phi = \frac{1}{4\pi\epsilon_0} \frac{Q_1}{r} \tag{3.1}$$

This means that a second stationary point charge Q_2 at a distance r from Q_1 possesses potential energy $Q_2\phi$, or

$$\frac{1}{4\pi\epsilon_0} \frac{Q_1 Q_2}{r} \tag{3.2}$$

One could equally well say that the point charge Q_1 in the field of the point charge Q_2 possesses potential energy given by expr. 3.2. This quantity is the mutual potential energy of the pair of point charges at separation r in free space. The zero of potential energy in expr. 3.2 has been taken arbitrarily, but conveniently, to correspond to infinite separation. Pushing two like charges together from infinite separation against their repulsion requires work, and stores potential energy; this energy may be recovered by allowing the charges to recede to an infinite separation. For unlike charges, potential energy is stored by pulling the charges apart, and is recovered by allowing them to move towards each other again.

Potential energy is not always electrical in origin. We are familiar with potential energy of gravitational origin in connection with the raising and lowering of weights. In astronomy, there is a formula for the gravitational potential energy of two masses M_1 and M_2 at distance r apart similar to expr. 3.2 except for sign. When an electron is transported up a vertical wire carrying current, it gains potential energy of gravitational origin because its mass is raised through a certain height. Normally, however, this is completely trivial compared with the change in the potential energy of the electron arising from electrical causes. The latter is equal to the product of the magnitude of the charge on the electron and the difference of electric potential through which it is transported. Particles possess potential energy associated with their mass and also potential energy associated with their charge. The former is called gravitational energy and the latter electric energy. In electrostatics, gravitational energy frequently plays such a minor role that the expressions electric energy and potential energy become practically synonymous.

3.2 Electric energy of a charged spherical conductor in free space

Expr. 3.2 gives the potential energy of either point charge in the presence of the other, but it does not include the energy required to construct either of the point charges from charge dispersed at infinity. Moreover, this energy is infinite unless account is taken of the small but non-zero size of the point charge. Let us calculate how much energy is involved in assembling from charge Q dispersed at infinity a charge Q on a spherical conductor of radius a in free space.

When the spherical conductor is charged and a static condition has been achieved, the charge is distributed uniformly over the surface, and the electric potential at distance r ($\geqslant a$) from the centre is the same as if the charge were concentrated at the centre, namely

$$\phi = \frac{1}{4\pi\epsilon_0} \frac{Q}{r} \tag{3.3}$$

In particular, the electric potential at the surface of the sphere is

$$V = \frac{1}{4\pi\epsilon_0} \frac{Q}{a} \tag{3.4}$$

This is also the potential throughout the interior of the spherical conductor: it is the potential, or voltage, of the conductor.

Let us now consider a stage in the charging process at which the charge on the conductor is xQ, so that the potential of the conductor is xV, where V is given by eqn. 3.4 and x lies between 0 and 1. Now let us increase the charge on the conductor by a small amount dxQ. If desired, it may be supposed that the charge dxQ is distributed uniformly over the surface of a concentric sphere whose radius is allowed to decrease from infinity to a. An amount of charge dxQ is transported through a rise in electric potential of amount xV. Consequently, the work done is

$$(xV)(dxQ). \tag{3.5}$$

To charge the sphere completely, repeated transfers of charge of this type have to be made until the fraction x has increased from zero to unity. The total work done in charging the conductor is therefore

$$W_e = \int_0^1 (xV)(dxQ) \tag{3.6}$$

In this equation V and Q are constants, being the final voltage and charge on the conductor. Hence, eqn. 3.6 becomes

$$W_e = VQ \int_0^1 x\, dx$$

or

$$W_e = \tfrac{1}{2}VQ \tag{3.7}$$

This is the work that must be done in order to assemble the charge Q on the conductor from charge dispersed at infinity. It is the work recovered by allowing the charge Q to be repelled to a state of dispersal at infinity. Eqn. 3.7 therefore gives the electric potential energy, or electric energy, of the charged conductor. This energy may be expressed in terms of the charge alone by substituting for V from eqn. 3.4 into eqn. 3.7, giving

$$W_e = \frac{1}{8\pi\epsilon_0} \frac{Q^2}{a} \tag{3.8}$$

Alternatively, the electric energy of the charged sphere may be expressed in terms of its potential alone in the form

$$W_e = 2\pi\epsilon_0 a V^2 \tag{3.9}$$

3.3 The 'size' of an electron

An electron possesses a charge e $(-1{\cdot}60 \times 10^{-19}\,\mathrm{C})$ and a rest mass m $(9{\cdot}11 \times 10^{-31}\,\mathrm{kg})$. On account of the rest mass, an electron has a rest energy (Appendix C)

$$W = mc^2 \tag{3.10}$$

where c is the velocity of light in free space $(2{\cdot}998 \times 10^8\,\mathrm{m/s})$. If we picture the electron as a conducting sphere of diameter r_e, the electron has electric energy (eqn. 3.8)

$$W = \frac{1}{8\pi\epsilon_0} \frac{e^2}{\tfrac{1}{2}r_e} \tag{3.11}$$

If we identify this with the rest energy in eqn. 3.10, we obtain

$$mc^2 = \frac{1}{4\pi\epsilon_0} \frac{e^2}{r_e} \tag{3.12}$$

Since $c = (\mu_0\epsilon_0)^{-1/2}$, eqn. 3.12 gives

$$r_e = \frac{\mu_0}{4\pi} \frac{e^2}{m} \tag{3.13}$$

Substituting the numerical values appropriate for an electron, we obtain

$$r_e = 2{\cdot}82 \times 10^{-15}\,\mathrm{m} \tag{3.14}$$

This result means that, if an electron were a hollow sphere of charge having diameter r_e, its electric energy would just account for its rest energy. For this reason the quantity r_e may be described as the 'size' of the electron. The diameter r_e is somewhat paradoxically known as 'the classical radius of the electron'.

It should not be assumed that an electron in fact has a size of the order given by eqn. 3.14. The size could not be smaller because otherwise the electrostatic energy alone would exceed the rest energy. If, however, the electron is, say, a hundred times the size r_e, then eqn. 3.12 shows that the electric energy accounts for only about one per cent of the rest energy.

3.4 Total electric energy of a pair of small charged spherical conductors

Suppose that a pair of small spherical conductors are located in free space at a distance r apart as illustrated in Fig. 3.1. Let one conductor (radius $a_1 \ll r$) carry charge Q_1 and the other (radius $a_2 \ll r$) carry charge Q_2. Let us consider the process whereby this distribution of charge is constructed from a total charge $Q_1 + Q_2$ dispersed at infinity, leaving a charge $-(Q_1 + Q_2)$ on the conductor at infinity.

Fig. 3.1 *Illustrating a pair of small spherical conductors (radii a_1 and a_2) at a distance r apart ($a_1 \ll r$, $a_2 \ll r$)*
There is a surrounding conductor at infinity ('ground') carrying a total charge $-(Q_1 + Q_2)$

First, let us construct the charge Q_1 on the sphere of radius a_1. From eqn. 3.8 this requires an expenditure of energy

$$\frac{1}{8\pi\epsilon_0} \frac{Q_1^2}{a_1} \qquad (3.15)$$

In another part of space remote from this sphere, let us construct the charge Q_2 on the sphere of radius a_2. This requires an expenditure of energy

$$\frac{1}{8\pi\epsilon_0} \frac{Q_2^2}{a_2} \qquad (3.16)$$

Now let us bring the two small spheres together from a large separation to a separation r. In accordance with expr. 3.2 this requires an expenditure of energy

$$\frac{1}{4\pi\epsilon_0} \frac{Q_1 Q_2}{r} \qquad (3.17)$$

Adding the contributions from eqns. 3.15, 3.16 and 3.17, we see that the total electric energy of the charged pair of small spheres is

$$W_e = \frac{1}{8\pi\epsilon_0}\left(\frac{Q_1^2}{a_1} + 2\frac{Q_1 Q_2}{r} + \frac{Q_2^2}{a_2}\right) \tag{3.18}$$

Eqn. 3.18 shows that the electric energy of the system is a quadratic function of the charges Q_1 and Q_2. The first and last terms on the right-hand side of eqn. 3.18 are the self-energies of the two small charged spheres, and the middle term is their mutual energy at distance r apart.

3.5 Mechanical action between a pair of small charged spheres

The mechanical actions involved in a system are often conveniently calculated from the energy of the system using the principle of work. Let us illustrate this process using the pair of small charged spherical conductors for which the stored potential energy is given by eqn. 3.18.

Consider a small increment dr in the distance r between the small spheres, and let F be the force between the spheres tending to increase the separation r. Then the work done during the small displacement is Fdr. Provided that no energy is introduced into the system during the displacement, this work is equal to the decrement $-dW_e$ in the potential energy of the system, so that

$$Fdr = -dW_e \tag{3.19}$$

Taking differentials in eqn. 3.18, we obtain

$$dW_e = \frac{Q_1 Q_2}{4\pi\epsilon_0}\left(-\frac{1}{r^2}\right)dr$$

Hence the decrement in the energy of the system as a result of the increase of r to $r + dr$ is

$$-dW_e = \frac{1}{4\pi\epsilon_0}\frac{Q_1 Q_2}{r^2}dr \tag{3.20}$$

Substituting from eqn. 3.20 into eqn. 3.19, we obtain for the force between the small spheres tending to increase r

$$F = \frac{1}{4\pi\epsilon_0}\frac{Q_1 Q_2}{r^2} \tag{3.21}$$

This is, of course, the law of inverse squares and, in this simple case, could have been written down without calculation.

There are two features of the above calculation that should be noticed. First, the change in energy in increasing r to $r + dr$ in eqn. 3.18 did not involve the self-energies of the charged spheres. Only the mutual potential energy in expr. 3.2 came into play. Secondly, in taking differentials in eqn. 3.18 to obtain eqn. 3.20, the charges Q_1 and Q_2 on the conductors were kept constant. This was essential to the argument. If Q_1 or Q_2 has been allowed to change, charge would have had to be brought up form infinity. This would have required work, and would have upset eqn. 3.19.

Although it is conceptually simplest to keep the charges Q_1 and Q_2 on the conductors constant while increasing r to $r + dr$, it is not essential to do so provided that eqn. 3.19 is modified to allow for the energy introduced into the system. For example, before displacement, let V_1 and V_2 be the voltages of the conductors relative to ground (that is, the conductor at infinity). While increasing r to $r + dr$, it would be possible to keep the voltages V_1 and V_2 constant. This would require batteries having voltages V_1 and V_2 connected between the conductors and ground during the displacement. These batteries would transfer small amounts of charge to the conductors during displacement. The energy thereby introduced into the system would then have to be added to the right-hand side of eqn. 3.19. If this were done, the law of inverse squares in eqn. 3.21 would still emerge, even though the voltages of the conductors were kept constant while increasing r to $r + dr$ (see Section 3.13).

3.6 The capacitance matrix for a pair of small conducting spheres

The discussion of the electric energy of a system of conductors is facilitated by use of the concept of coefficients of self- and mutual capacitance. Let us illustrate this concept with the aid of a pair of small spherical conductors of radii a_1 and a_2 at a distance r apart $(a_1 \ll r, a_2 \ll r)$.

First let us calculate the voltages V_1 and V_2 of the spherical conductors when they carry charges Q_1 and Q_2. Because of the smallness of the conductors, the electric potential at any point on or between them can be calculated as though Q_1 and Q_2 were point charges at the centres. Hence the voltages of the conductors are approximately

$$
\left.
\begin{aligned}
V_1 &= \frac{1}{4\pi\epsilon_0}\left(\frac{Q_1}{a_1} + \frac{Q_2}{r}\right) \\[2mm]
V_2 &= \frac{1}{4\pi\epsilon_0}\left(\frac{Q_1}{r} + \frac{Q_2}{a_2}\right)
\end{aligned}
\right\}
\tag{3.22}
$$

These equations may be written

$$
\left.
\begin{aligned}
V_1 &= c_{11}Q_1 + c_{12}Q_2 \\
V_2 &= c_{21}Q_1 + c_{22}Q_2
\end{aligned}
\right\}
\tag{3.23}
$$

where

$$
\left.
\begin{aligned}
c_{11} &= \frac{1}{4\pi\epsilon_0}\frac{1}{a_1}, \quad & c_{12} &= \frac{1}{4\pi\epsilon_0}\frac{1}{r} \\[2mm]
c_{21} &= \frac{1}{4\pi\epsilon_0}\frac{1}{r}, \quad & c_{22} &= \frac{1}{4\pi\epsilon_0}\frac{1}{a_2}
\end{aligned}
\right\}
\tag{3.24}
$$

Eqns. 3.23 may be solved for Q_1 and Q_2, thereby obtaining

$$\left. \begin{aligned} Q_1 &= C_{11}V_1 + C_{12}V_2 \\ Q_2 &= C_{21}V_1 + C_{22}V_2 \end{aligned} \right\} \tag{3.25}$$

where

$$\left. \begin{aligned} C_{11} &= \frac{c_{22}}{c_{11}c_{22} - c_{12}c_{21}}, & C_{12} &= -\frac{c_{12}}{c_{11}c_{22} - c_{12}c_{21}} \\ C_{21} &= -\frac{c_{21}}{c_{11}c_{22} - c_{12}c_{21}}, & C_{22} &= \frac{c_{11}}{c_{11}c_{22} - c_{12}c_{21}} \end{aligned} \right\} \tag{3.26}$$

The values of these coefficients may be obtained by substitution from eqns. 3.24. Because the radii a_1 and a_2 of the conductors are small compared with their separation r, the term $c_{12}c_{21}$ in each denominator in eqns. 3.26 is negligible compared with $c_{11}c_{22}$. We therefore obtain approximately

$$\left. \begin{aligned} C_{11} &= 4\pi\epsilon_0 a_1, & C_{12} &= -4\pi\epsilon_0 a_1 a_2 / r \\ C_{21} &= -4\pi\epsilon_0 a_1 a_2 / r, & C_{22} &= 4\pi\epsilon_0 a_2 \end{aligned} \right\} \tag{3.27}$$

The coefficients C_{11} and C_{22} are called the coefficients of self-capacitance of the two conductors, and the common coefficients C_{12} and C_{21} the coefficient of mutual capacitance. The matrix of coefficients

$$\begin{pmatrix} C_{11} & C_{12} \\ C_{21} & C_{22} \end{pmatrix} \tag{3.28}$$

is called the capacitance matrix of the conductors, and the matrix of coefficients

$$\begin{pmatrix} c_{11} & c_{12} \\ c_{21} & c_{22} \end{pmatrix} \tag{3.29}$$

is called the inverse of the capacitance matrix. Eqns. 3.25 may be written

$$\begin{pmatrix} Q_1 \\ Q_2 \end{pmatrix} = \begin{pmatrix} C_{11} & C_{12} \\ C_{21} & C_{22} \end{pmatrix} \begin{pmatrix} V_1 \\ V_2 \end{pmatrix} \tag{3.30}$$

and the solution of these linear equations may be written (eqns. 3.23)

$$\begin{pmatrix} V_1 \\ V_2 \end{pmatrix} = \begin{pmatrix} c_{11} & c_{12} \\ c_{12} & c_{22} \end{pmatrix} \begin{pmatrix} Q_1 \\ Q_2 \end{pmatrix} \tag{3.31}$$

The capacitance matrix gives the coefficients required to express the charges on the conductors in terms of their voltages, while the inverse of the capacitance matrix gives the coefficients required to express the voltages in terms of the charges.

3.7 The capacitance matrix for any system of conductors

As illustrated in Fig. 3.2, let us now consider any system of metallic conductors numbered $1, 2, \ldots$, and let us suppose that these are surrounded by a conductor at zero potential. The latter conductor is often the conductor at infinity, that is, 'ground'. We suppose that all conductors are at rest relative to the observer.

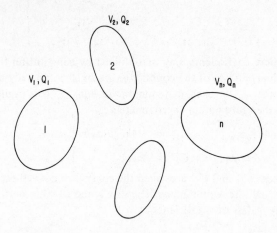

Fig. 3.2   *Illustrating a system of conductors numbered 1, 2, . . . , n possessing voltages V_1, $V_2, \ldots, V_n$ and corresponding charges $Q_1, Q_2, \ldots, Q_n$*
There is a surrounding conductor of zero voltage carrying charge $-(Q_1 + Q_2 + \ldots + Q_n)$. This surrounding conductor may be the conductor at infinity ('ground')

Consider a situation in which the voltages of the conductors are

$$1, 0, 0, \ldots \tag{3.22}$$

so that the voltage of conductor number 1 is unity and the voltages of all the other conductors are zero. Under these circumstances let the charges on the conductors be

$$C_{11}, C_{21}, C_{31}, \ldots \tag{3.33}$$

Now consider a situation in which the voltages on the conductors are

$$0, 1, 0, \ldots \tag{3.34}$$

so that there is now a unit voltage on conductor number 2 but zero voltages on all the other conductors. Under these circumstances let the charges on the conductors be

$$C_{12}, C_{22}, C_{32}, \ldots \tag{3.35}$$

Likewise when the voltages on the conductors are

$$0, 0, 1, 0, \ldots \tag{3.36}$$

let the charges on the conductors be

$$C_{13}, C_{23}, C_{33}, \ldots \tag{3.37}$$

In this way we define the coefficients of capacitance for the system of conductors.

By the principal of superposition it follows that, if we multiply the voltages (3.32) by V_1, then the charges (3.33) are also multiplied by V_1. Likewise, if the voltages (3.34) are multiplied by V_2, so also are the charges (3.35), and if the voltages (3.36) are multiplied by V_3, so also are the charges (3.37), and so on. If all these situations are superimposed in accordance with the principle of superposition, we obtain for the charges on the conductors

$$\left.\begin{aligned}
Q_1 &= C_{11}V_1 + C_{12}V_2 + C_{13}V_3 + \ldots \\
Q_2 &= C_{21}V_1 + C_{22}V_2 + C_{23}V_3 + \ldots \\
Q_3 &= C_{31}V_1 + C_{32}V_2 + C_{33}V_3 + \ldots \\
&\cdots\cdots\cdots\cdots\cdots\cdots\cdots\cdots\cdots
\end{aligned}\right\} \tag{3.38}$$

which may be written in terms of the capacitance matrix as

$$\begin{pmatrix} Q_1 \\ Q_2 \\ Q_3 \\ \cdot \end{pmatrix} = \begin{pmatrix} C_{11} & C_{12} & C_{13} & \cdots \\ C_{21} & C_{22} & C_{23} & \cdots \\ C_{31} & C_{32} & C_{33} & \cdots \\ \cdots\cdots\cdots\cdots \end{pmatrix} \begin{pmatrix} V_1 \\ V_2 \\ V_3 \\ \cdot \end{pmatrix} \tag{3.39}$$

If eqns. 3.38 are solved for $V_1, V_2, V_3, \ldots$ we obtain

$$\left.\begin{aligned}
V_1 &= c_{11}Q_1 + c_{12}Q_2 + c_{13}Q_3 + \ldots \\
V_2 &= c_{21}Q_1 + c_{22}Q_2 + c_{23}Q_3 + \ldots \\
V_3 &= c_{31}Q_1 + c_{32}Q_2 + c_{33}Q_3 + \ldots \\
&\cdots\cdots\cdots\cdots\cdots\cdots\cdots\cdots\cdots
\end{aligned}\right\} \tag{3.40}$$

or

$$\begin{pmatrix} V_1 \\ V_2 \\ V_3 \\ \cdot \end{pmatrix} = \begin{pmatrix} c_{11} & c_{12} & c_{13} & \cdots \\ c_{21} & c_{22} & c_{23} & \cdots \\ c_{31} & c_{32} & c_{33} & \cdots \\ \cdots\cdots\cdots\cdots \end{pmatrix} \begin{pmatrix} Q_1 \\ Q_2 \\ Q_3 \\ \cdot \end{pmatrix} \tag{3.41}$$

where the coefficients c_{rs} constitute the inverse of the capacitance matrix C_{rs}. The significance of the coefficients c_{rs} may be visualised as follows. When the charges on the conductors are $1, 0, 0, \ldots$, the voltages are $c_{11}, c_{21}, c_{31}, \ldots$. When the charges on the conductors are $0, 1, 0, \ldots$, the voltages on the conductors are $c_{12}, c_{22}, c_{32}, \ldots$, and so on.

In the previous section, we calculated the elements in the capacitance matrix and its inverse for a pair of small spherical conductors of radii a_1 and a_2 at a large distance r apart (eqns. 3.27 and 3.24).

3.8 Electric energy of a charged capacitor

As a preliminary to calculating the electric energy of a system of charged conductors, let us calculate the electric energy of a single charged capacitor in free space. This may be done by the method used in Section 3.2. The calculation made in Section 3.2 was, in fact, that for the energy of a charged capacitor for which one conductor is a metal sphere and the other is the conductor at infinity.

Consider an uncharged capacitor of any configuration, and let C be its capacitance. Let a charge Q be transferred from one conductor to the other, and let V be the resulting excess of the voltage of the positive conductor over that of the negative conductor. By definition of capacitance

$$C = Q/V \tag{3.42}$$

Consider a stage in the charging process at which the voltage between the conductors is a fraction x of the final voltage V. At this stage the voltage is xV, so that the charges on the conductors are $\pm xQ$. Now let a small additional amount of charge dxQ be transferred from the negative conductor to the positive conductor. The work done is given by expr. 3.5. Consequently the total work done in charging the capacitor is given by eqn. 3.6, and therefore by eqn. 3.7. Using eqn. 3.42, it follows that the electric energy of a charged capacitor can be written in any of the following equivalent forms:

$$W_e = \tfrac{1}{2}VQ = \tfrac{1}{2}CV^2 = \frac{1}{2C}Q^2 \tag{3.43}$$

The function of a capacitor is to store potential energy in electric form. For a capacitor of given capacitance, the stored potential energy is proportional to the square of its voltage and also proportional to the square of its charge. For a given voltage V, the energy stored is proportional to the capacitance C, but for a given charge Q the energy stored is inversely proportional to the capacitance.

3.9 Electric energy of a system of charged conductors

A treatment of the type used in Section 3.8 may be adapted to calculate the energy of any system of conductors in free space at rest relative to the observer. Let conductors $1, 2, \ldots$ be raised to voltages $V_1, V_2, \ldots$, and let the corresponding charges on the conductors be $Q_1, Q_2, \ldots$. At a certain stage in the charging process let the voltages and charges of the conductors be a fraction x of their final values. Now increase the charges by an additional small fraction dx of their final values. This involves bringing up from infinity charges $dxQ_1, dxQ_2, \ldots$ to potentials $xV_1, xV_2, \ldots$, respectively. The work done is

$$(xV_1)(dxQ_1) + (xV_2)(dxQ_2) + \ldots \tag{3.44}$$

To charge the system we have to allow x to go from 0 to 1, and the total work done is therefore

$$W_e = (V_1Q_1 + V_2Q_2 + \ldots)\int_0^1 x\,dx$$

or

$$W_e = \tfrac{1}{2}(V_1Q_1 + V_2Q_2 + \ldots) \tag{3.45}$$

This is the potential energy of the charged system, and is called its electric energy.

The electric energy of a system of charged conductors may be expressed entirely in terms of the voltages of the conductors by using eqns. 3.38, or entirely in terms of the charges on the conductors by using eqns. 3.40. For a system of n conductors, eqns. 3.38 and 3.40 may be written

$$\left.\begin{aligned} Q_r &= \sum_{s=1}^{n} C_{rs}V_s \\[2em] V_r &= \sum_{s=1}^{n} c_{rs}Q_s \end{aligned}\right\} \tag{3.46}$$

and eqn. 3.45 may be written

$$W_e = \frac{1}{2}\sum_{r=1}^{n} V_r Q_r \tag{3.47}$$

By substitution from eqns. 3.46 into eqn. 3.47 we derive

$$W_e = \frac{1}{2}\sum_{r=1}^{n}\sum_{s=1}^{n} C_{rs}V_r V_s = \frac{1}{2}\sum_{r=1}^{n}\sum_{s=1}^{n} c_{rs}Q_r Q_s \tag{3.48}$$

The electric energy of a system of charged conductors is therefore a quadratic function of the voltages, and for this function the coefficients are the elements in the capacitance matrix. Alternately the electric energy of the system of conductors is a quadratic function of the charges, and for this function the coefficients are the elements in the inverse of the capacitance matrix.

3.10 The reciprocity theorem of electrostatics

To a given system of conductors numbered $1, 2, \ldots$ let us apply the charges Q_1, $Q_2, \ldots$ and let the corresponding voltages of the conductors be $V_1, V_2, \ldots$. For the same system of conductors, let us alternatively apply charges $Q_1', Q_2', \ldots$ and let the corresponding voltages be $V_1', V_2', \ldots$. The reciprocity theorem of electrostatics states that

$$\left.\begin{aligned} V_1Q_1' + V_2Q_2' + \ldots &= V_1'Q_1 + V_2'Q_2 + \ldots \\ \Sigma VQ' &= \Sigma V'Q \end{aligned}\right\} \tag{3.49}$$

or

When applying the reciprocity theorem it is convenient to list the voltages and charges of the conductors according to the following scheme:

$$
\begin{array}{ll}
V_1 & V_2 \ldots \\
\underline{Q_1 \quad Q_2 \ldots} \\
V_1' & V_2' \ldots \\
Q_1' & Q_2' \ldots
\end{array}
\tag{3.50}
$$

One side of eqn. 3.49 is then formed by multiplying line 1 by line 4 and adding, while the other side is formed by multiplying line 2 by line 3 and adding. Before proving the reciprocity theorem let us examine some of its implications.

First let us consider a situation in which the voltage of conductor number 1 is V while the voltages of all the other conductors are zero; let us compare this with a situation in which the voltage of conductor number 2 is V while the voltages of all the other conductors are zero. The scheme 3.50 is then

$$
\begin{array}{llll}
V & 0 & 0 & \ldots \\
\underline{Q_1 \quad Q_2 \quad Q_3 \ldots} \\
0 & V & 0 & \ldots \\
Q_1' & Q_2' & Q_3' & \ldots
\end{array}
$$

and the reciprocity theorem gives

$$
Q_1' = Q_2
\tag{3.51}
$$

This result states that the charge on conductor number 1 due to a voltage V on conductor number 2 (the other conductors being grounded) is equal to the charge on conductor number 2 when the voltage V is applied to conductor number 1 (the other conductors being grounded).

If we now suppose that $V = 1$ then, by definition of the coefficients of capacitance, eqn. 3.51 reads

$$
C_{12} = C_{21}
$$

This result applies to any pair of conductors, and it therefore follows that

$$
C_{rs} = C_{sr}
\tag{3.52}
$$

For an example see eqns. 3.27.

Eqn. 3.52 implies that the capacitance matrix is symmetrical about its leading diagonal. The common value of C_{rs} and C_{sr} is the coefficient of mutual capacitance of conductors r and s. The elements in the leading diagonal of the capacitance matrix are the coefficients of self-capacitance of the conductors.

Now let us consider a situation in which the charge on conductor number 1 is Q, while each of the other conductors carries no net charge, and let us compare this with a situation in which the charge Q is on conductor number 2 while each of the other conductors carries no net charge. The scheme (3.50) is then

$$
\begin{array}{ccc}
V_1 & V_2 & V_3 \ldots \\
Q & 0 & 0 \ldots \\
\hline
V_1' & V_2' & V_3' \ldots \\
0 & Q & 0 \ldots
\end{array}
$$

and the reciprocity theorem gives

$$V_2 = V_1' \tag{3.53}$$

This result states that the voltage of conductor number 2 when a charge Q is applied to conductor number 1 (the other conductors being uncharged) is equal to the voltage of conductor number 1 when the charge Q is applied to conductor number 2 (the other conductors being uncharged).

If we now suppose that $Q = 1$ then, by definition of the non-diagonal elements in the inverse of the capacitance matrix, eqn. 3.53 reads

$$c_{21} = c_{12}$$

This result applies to any pair of conductors, and it therefore follows that the inverse of the capacitance matrix is symmetrical about its leading diagonal:

$$c_{rs} = c_{sr}. \tag{3.54}$$

For an example see eqns. 3.24.

Eqns. 3.52 and 3.54 also follow from each other as a matter of algebra because eqns. 3.38 and 3.40 are the solution of each other (see eqns. 3.26). Eqns 3.49, 3.52 and 3.54 are equivalent in the sense that the truth of all three may be deduced from the truth of any one.

3.11 Proof of the reciprocity theorem of electrostatics

To prove the reciprocity theorem (eqn. 3.49), consider a situation in which the voltages of the conductors are

$$(1-x)V_1 + xV_1', \quad (1-x)V_2 + xV_2'. \quad \ldots \tag{3.55}$$

where x is a positive number less than unity. By the principle of superposition the charges on the conductors are then

$$(1-x)Q_1 + xQ_1', \quad (1-x)Q_2 + xQ_2', \quad \ldots \tag{3.56}$$

These voltages and charges are such that, when $x = 0$, they are identical with the upper set in the scheme (3.50) while, when $x = 1$, they are identical with the lower set. We can therefore pass from the first set to the second set by allowing x to go from zero to unity.

Let us calculate the work done in changing from the first set of voltages and charges to the second set. When x changes to $x + dx$, the charge on conductor number 1 increases by $d\{(1-x)Q_1 + xQ_1'\}$ or

$$(Q_1' - Q_1)\,dx \tag{3.57}$$

This amount of charge therefore has to be brought from infinity, where it is at zero potential, to conductor number 1 where it is at potential

$$(1-x)V_1 + xV_1' \tag{3.58}$$

The work done is the product of expr. 3.57 and 3.58. For all conductors, therefore, the work done in increasing x to $x + dx$ is

$$\Sigma\{(1-x)V + xV'\}(Q' - Q)\,dx$$

and this can be rearranged as

$$\Sigma\{V(Q' - Q)\,dx + (V' - V)(Q' - Q)x\,dx\} \tag{3.59}$$

The total work done in changing from the first set of voltages and charges to the second set is obtained by integrating expr. 3.59 with regard to x from 0 to 1. It is therefore

$$\Sigma\{V(Q' - Q) + \tfrac{1}{2}(V' - V)(Q' - Q)\}$$

and this can be rearranged as

$$(\tfrac{1}{2}\Sigma V'Q' - \tfrac{1}{2}\Sigma VQ) + \tfrac{1}{2}(\Sigma VQ' - \Sigma V'Q) \tag{3.60}$$

This work done in changing from the first set of voltages and charges to the second is equal to the increase in the energy of the system. But, according to eqn. 3.47, the increase in the energy of the system is the first expression in parentheses in expr. 3.60. It follows that the second expression in parentheses in expr. 3.60 must vanish and this establishes the truth of eqn. 3.49.

3.12 Mechanical action on a charged conductor

Suppose that we are interested in the mechanical action on a particular conductor of a system of charged conductors at rest relative to the observer. It is frequently convenient to calculate this mechanical action from the energy of the system by means of the principle of work.

Suppose that the location and orientation of the particular conductor are specified by means of the cartesian co-ordinates (x, y, z) of a selected point of the conductor and the polar angles (θ, ψ) of a selected axis in the conductor. Displacement in the x direction then involves an increase in the linear co-ordinate x, while rotation about the polar axis involves an increase in the angular co-ordinate ψ. If F is the force on the conductor in the direction of the x axis, the work done in increasing x to $x + dx$ is

$$F\,dx$$

If all other co-ordinates are kept constant in the process, and if the charges on the conductors are also kept constant, this work done must be equal to the decrement

$$-dW_e$$

in the potential energy W_e of the system. Hence

$$F\,dx = -dW_e \qquad (3.61)$$

or

$$F = -\frac{\partial W_e}{\partial x} \qquad (3.62)$$

In evaluating this partial derivative, all co-ordinates other than x must be kept constant and, to prevent the introduction of energy into the system from outside, all charges on the conductors must also be kept constant. Likewise, if T is the torque on the conductor tending to increase the angular co-ordinate ψ, then

$$T = -\frac{\partial W_e}{\partial \psi} \qquad (3.63)$$

In evaluating this partial derivative, all co-ordinates other than ψ must be kept constant, and all charges on the conductors must also be kept constant.

Suppose that there are two conductors for which the inverse of the capacitance matrix is

$$\begin{pmatrix} c_{11} & c_{12} \\ c_{12} & c_{22} \end{pmatrix} \qquad (3.64)$$

where allowance has been made for the symmetry of the matrix required by the reciprocity theorem. Then, from the second of eqns. 3.48, the energy of the system when conductor number 1 carries a charge Q_1 and conductor number 2 carries a charge Q_2 is

$$W_e = \tfrac{1}{2}(c_{11}Q_1^2 + 2c_{12}Q_1Q_2 + c_{22}Q_2^2) \qquad (3.65)$$

We have deliberately elected to use the second of eqns. 3.48 rather than the first so that the energy W_e is expressed in terms of the charges that it is desired to keep constant when applying eqn. 3.62 or 3.63.

Let us now suppose that conductor number 1 is fixed, but conductor number 2 has a position specified by linear co-ordinates (x, y, z) and an orientation specified by angular co-ordinates (θ, ψ). Then, from eqn. 3.62, the force on conductor number 2 tending to increase the linear co-ordinate x is

$$F = -\frac{1}{2}\left(\frac{\partial c_{11}}{\partial x}Q_1^2 + 2\frac{\partial c_{12}}{\partial x}Q_1Q_2 + \frac{\partial c_{22}}{\partial x}Q_2^2\right) \qquad (3.66)$$

and from eqn. 3.63 the torque on conductor number 2 tending to increase the angular co-ordinate ψ is

$$T = \frac{1}{2}\left(\frac{\partial c_{11}}{\partial \psi}Q_1^2 + 2\frac{\partial c_{12}}{\partial \psi}Q_1Q_2 + \frac{\partial c_{22}}{\partial \psi}Q_2^2\right) \tag{3.67}$$

In evaluating the partial derivatives of c_{11}, c_{12} and c_{22}, all co-ordinates other than the designated variable co-ordinate are to be kept constant.

In eqns. 3.66 and 3.67 the terms involving the rate of change of c_{12} are always important, but the terms involving the rates of change of c_{11} and c_{22} are frequently unimportant. Hence, eqns. 3.66 and 3.67 can frequently be reduced to

$$F = -\frac{\partial c_{12}}{\partial x}Q_1Q_2 \tag{3.68}$$

$$T = -\frac{\partial c_{12}}{\partial \psi}Q_1Q_2 \tag{3.69}$$

As an example, consider the pair of small spherical conductors (radii a_1 and a_2) at a distance r apart in free space already studied in Sections 3.4–3.6. The elements of the inverse of the capacitance matrix are then given in eqns. 3.24. Because of the assumptions that $a_1 \ll r$ and $a_2 \ll r$, the coefficients c_{11} and c_{22} are independent of r. Hence the force F tending to increase r is given by eqn. 3.68 with x replaced by r. Using the value of c_{12} given in eqns. 3.24, we recover the law of inverse squares.

3.13 Calculation of mechanical action using displacements at constant voltage

In using the principle of work for calculating mechanical action from energy we have emphasised the need to keep the charges on the conductors constant in order to ensure that, during a small displacement, energy is not introduced into the system or removed from it. However, provided that proper allowance is made for the energy introduced or removed, it is quite possible to allow the charges on the conductors to change. In particular, it is possible to calculate mechanical action from energy by keeping the voltages constant instead of the charges. Moreover, it turns out to be equally simple to do so provided that the following results are understood.

Consider a system of conductors for which the voltages are denoted by V_1, V_2, ... and the corresponding charges by $Q_1, Q_2, \ldots$. In accordance with eqn. 3.47, the potential energy of the system is

$$W_e = \frac{1}{2}\Sigma VQ \tag{3.70}$$

In any change in which the voltages increase by $dV_1, dV_2, \ldots$ and the charges by $dQ_1, dQ_2, \ldots$, respectively, there is a corresponding increase dW_e in the potential energy of the system, and it is calculated by taking differentials in eqn. 3.70. We obtain

$$dW_e = \frac{1}{2}\Sigma(VdQ + QdV) \tag{3.71}$$

Now suppose that we are interested in calculating a certain force (or torque) F. Consider a displacement of the system in which the force (or torque) moves through a linear (or angular) distance dx. The work Fdx would be equal to the decrement $-dW_e$ in the potential energy of the system if the charges were kept constant. If, however, the charges on the conductors are subject to increments, then a charge dQ_1 is brought up to potential V_1, a charge dQ_2 is brought up to potential V_2, and so on. Hence there is introduced into the system an amount of energy ΣVdQ. This introduced energy must be allowed for in relating the work done Fdx and the decrement $-dW_e$ in the potential energy of the system. We therefore arrive at the equation

$$Fdx = -dW_e + \Sigma VdQ \tag{3.72}$$

which reduces to eqn. 3.61 when the charges are kept constant. By substituting for dW_e from eqn. 3.71 into eqn. 3.72 we obtain

$$Fdx = \tfrac{1}{2}\Sigma(VdQ - QdV) \tag{3.73}$$

Let us now compare eqn. 3.73 for the work done Fdx with eqn. 3.71 for the change in the potential energy of the system. By making the increments of charge vanish, we arrive at the equation

$$Fdx = -dW_e \quad \text{(constant charges)} \tag{3.74}$$

and this is identical with eqn. 3.61. But, by making the increments of voltage in eqns. 3.71 and 3.73 vanish, we arrive at the equation

$$Fdx = +dW_e \quad \text{(constant voltages)} \tag{3.75}$$

We see that, whereas the work done in a displacement in which the charges are kept constant is the decrement in the potential energy of the system, the work done in a displacement in which the voltages are kept constant is the increment in the potential energy of the system. The difference of energy in the two cases, namely

$$2dW_e \tag{3.76}$$

is that supplied by the batteries needed to maintain the constant voltages. We are therefore able to keep the voltages constant when using the method of work for calculating mechanical action from potential energy, provided that we use the positive sign in eqn. 3.75 instead of the negative sign in eqn. 3.74.

As an example, let us consider once again a pair of small spherical conductors (radii a_1 and a_2) at a distance r apart in free space. Let us calculate the force F tending to increase r by letting the distance between the conductors increase from r to $r + dr$ keeping the voltages V_1 and V_2 of the conductors constant. For this purpose we deliberately elect to use for W_e the first of eqns. 3.48, so that the energy is expressed in terms of the voltages that it is desired to keep constant. We obtain

$$W_e = \tfrac{1}{2}(C_{11}V_1^2 + 2C_{12}V_1V_2 + C_{22}V_2^2) \tag{3.77}$$

where the coefficients of self and mutual capacitance are given in eqns. 3.27.

From eqn. 3.75 we obtain

$$Fdr = +dW_e \quad \text{(constant voltages)}$$

so that

$$F = +\frac{\partial W_e}{\partial r} \tag{3.78}$$

where W_e is given by eqn. 3.77 and V_1, V_2 are to be kept constant. Hence

$$F = +\frac{1}{2}\left(\frac{\partial C_{11}}{\partial r}V_1^2 + 2\frac{\partial C_{12}}{\partial r}V_1 V_2 + \frac{\partial C_{22}}{\partial r}V_2^2\right) \tag{3.79}$$

But, because of the assumptions that $a_1 \ll r$ and $a_2 \ll r$, the coefficients of self-capacitance C_{11} and C_{22} in eqns. 3.27 are independent of r, and eqn. 3.79 therefore reduces to

$$F = +\frac{\partial C_{12}}{\partial r}V_1 V_2 \tag{3.80}$$

Substitution for the coefficient of mutual capacitance from eqns. 3.27 into eqn. 3.80 gives

$$F = 4\pi\epsilon_0 a_1 a_2 V_1 V_2/r^2 \tag{3.81}$$

This is an unusual version of the law of inverse squares. Because the radii a_1 and a_2 of the spherical conductors are small compared with their separation, we may use eqn. 3.4 to derive approximately

$$V_1 = \frac{1}{4\pi\epsilon_0}\frac{Q_1}{a_1}, \quad V_2 = \frac{1}{4\pi\epsilon_0}\frac{Q_2}{a_2} \tag{3.82}$$

Substitution for V_1 and V_2 from eqns. 3.82 into eqn. 3.81 then gives the usual version of the law of inverse squares, namely

$$F = \frac{1}{4\pi\epsilon_0}\frac{Q_1 Q_2}{r^2} \tag{3.83}$$

Had we inadvertently used a negative sign in eqn. 3.78, we would have had negative signs in eqns. 3.81 and 3.83. A law of inverse squares would have been derived, but with like charges attracting instead of repelling. Care is necessary with the signs, especially when deriving mechanical action using a displacement in which the voltages are kept constant.

3.14 Conductors in the presence of linear dielectric material

The discussion of a system of conductors developed in Sections 3.7–3.13 applies when the regions between the conductors are simply free space, but it is not restricted to this situation. The argument only depends on the linearity of eqns. 3.38 relating the charges and the voltages on the conductors. Taking the charges Q_1, Q_2, ... to be the free charges on the conductors, the results apply no matter what distribution of dielectric material may exist in the regions between the conductors so long as the dielectric behaves linearly.

For isotropic dielectric material this means that the electric-flux-density vector D' involved in version 2 of the electromagnetic equations in materials (Section 2.4) must be proportional to the electric vector E at each location between the conductors as shown in Table 2.2. There is no objection to the constant of proportionality ϵ being a function of position in the region between the conductors.

Furthermore, the dielectric material need not be isotropic so long as it is linear. Each cartesian component of D' can be linearly dependent on all three cartesian components of E, the ratio of the coefficients to ϵ_0 constituting what is known as the dielectric tensor of the material. The elements of the dielectric tensor can be functions of position. All that is ruled out are non-linear relationships between the cartesian components of the electric-flux-density vector D' and the cartesian components of the electric vector E. In particular, dielectric hysteresis is ruled out (see Section 4.11).

The presence of linear dielectric material between the conductors complicates the calculation of the coefficients of self and mutual capacitance in eqns. 3.38. But, when once these are calculated (or measured), Sections 3.7–3.13 apply without modification. In particular, the electric energy stored in a capacitor of capacitance C whose plates carry free charges $\pm Q$ and have a voltage V between them is (c.f. eqn. 3.43)

$$W_e = \tfrac{1}{2}VQ = \tfrac{1}{2}CV^2 = \frac{1}{2C}Q^2 \tag{3.84}$$

regardless of what dielectric material may be present, so long as it behaves linearly.

3.15 The conductance and resistance matrices

If the material between the conductors is not perfectly insulating, steady current can flow through it when steady voltages are maintained on the conductors. In these circumstances the conductors may be referred to as electrodes. If the electrodes themselves are perfectly conducting, each is still an equipotential surface having a specific voltage relative to a surrounding electrode, which may be the conductor at infinity. Let there be n perfectly conducting electrodes other than the surrounding electrode at zero potential, and let their steady voltages be $V_1, V_2, \ldots$. As a result, let the corresponding steady currents leaving the electrodes to pass through the intervening conducting material be $I_1, I_2, \ldots$ so that these currents have to be supplied to the electrodes by batteries maintaining the voltages of the electrodes. Let the conducting material behave linearly.

With steady voltages maintained on the electrodes, steady charges exist on them, and the relation between the voltages and the charges is described by means of the capacitance matrix for the system of electrodes. Associated with the charges on the electrodes there is an electric field between the electrodes, and this is what drives current through the conducting material between the electrodes. The relations giving the currents entering the material at the electrodes in terms of the voltages of the electrodes may be described with the aid of a matrix. This is known as the conductance matrix of the system of perfectly conducting electrodes embedded in the resistive material. The inverse of the conductance matrix is the resistance matrix. When the material between the electrodes is isotropic and the ratio of the dielectric constant ϵ/ϵ_0 to the conductivity σ is the same at all locations, the elements of the conductance matrix are obtained from those of the capacitance matrix simply by replacing ϵ by σ (c.f. Problem 2.1).

Let G_{rs} and R_{rs} $(r, s = 1, 2, \ldots, n)$ be the elements of the conductance and resistance matrices, respectively. Then (c.f. eqns. 3.46)

$$\left.\begin{aligned}
I_r &= \sum_{s=1}^{n} G_{rs} V_s \\
V_r &= \sum_{s=1}^{n} R_{rs} I_s
\end{aligned}\right\} \tag{3.85}$$

Either of these sets of linear equations is the solution of the other. The coefficient G_{rs} is the current supplied at electrode number r when electrode number s is at unit voltage and all the other electrodes are grounded. The coefficient R_{rs} is the voltage of electrode number r when unit current is supplied at electrode number s and no net current leaves or arrives at any other electrode.

There is a reciprocity theorem for a system of perfectly conducting electrodes immersed in linear conducting material that may be discussed as in Sections 3.10 and 3.11. When the voltages of the electrodes are $V_1, V_2, \ldots$ let the currents supplied at them be, respectively, $I_1, I_2, \ldots$. When the voltages are $V_1', V_2', \ldots$ let the currents be $I_1', I_2', \ldots$. Then (c.f. eqns. 3.49)

or
$$\left.\begin{aligned}
V_1 I_1' + V_2 I_2' + \ldots &= V_1' I_1 + V_2' I_2 + \ldots \\
\Sigma VI' &= \Sigma V'I
\end{aligned}\right\} \tag{3.86}$$

In particular (c.f. eqns. 3.52 and 3.54)

$$G_{rs} = G_{sr}, \quad R_{rs} = R_{sr} \tag{3.87}$$

The conductance and resistance matrices are therefore symmetrical about their leading diagonals.

However, the discussion of energy in association with a system of perfectly conducting electrodes embedded in resistive material involves a new feature. When the batteries are first connected to the electrodes to maintain the voltages, the electrodes charge up and an electric field is created between the electrodes. Potential energy is stored in the system as the electrodes charge, and this is discussed with the aid of the capacitance matrix of the system as described in Section 3.9. But when the electrodes are fully charged, steady current flows through the resistive material between the electrodes and dissipates energy at a steady rate. This power dissipation can be discussed with the aid of the conductance and resistance matrices.

The formulas for the power dissipated in the resistive material are similar to eqns. 3.45, 3.47 and 3.48, but no factor $\frac{1}{2}$ appears. The factor $\frac{1}{2}$ in eqns. 3.45, 3.47 and 3.48 for the potential energy stored in the system arose from the integral with respect to x involved in going from expr. 3.44 to eqn. 3.45. When the charging process starts, little work is done in transferring the first small fractions of charge because there then exists hardly any voltage through which they have to be raised. But when the charging process is nearly complete, the final small fractions of charge have to be raised through almost the full final voltages. On the average, the charge on each conductor has to be raised through only the average of the initial and final voltages on the conductor. This is what accounts for the factor $\frac{1}{2}$ in eqns. 3.45, 3.47 and 3.48. But for steady dissipation of energy in resistive material between perfectly conducting electrodes that are fully charged, every element of charge falls through the full voltage. If unit charge leaves electrode number 1 at voltage V_1 for ground, the work done is V_1. Consequently, if charge is leaving electrode number 1 for ground at the steady rate I_1, then work is being done at the rate $V_1 I_1$, and and this power is dissipated in the resistive material between the electrodes. Hence the total power supplied to the resistive material by the batteries maintaining the voltages of the electrodes is

or
$$\left. \begin{array}{l} P = V_1 I_1 + V_2 I_2 + \ldots \\[2mm] P = \sum_{r=1}^{n} V_r I_r \end{array} \right\} \tag{3.88}$$

These equations are similar to eqns. 3.45 and 3.47, but they refer to power rather than energy, and no factor $\frac{1}{2}$ appears.

With the aid of eqns. 3.85, the power dissipated in the resistive material may be expressed entirely in terms of the voltages of the electrodes, or entirely in terms of the currents supplied at the electrodes. We obtain (c.f. eqns. 3.48)

$$P = \sum_{r=1}^{n} \sum_{s=1}^{n} G_{rs} V_r V_s = \sum_{r=1}^{n} \sum_{s=1}^{n} R_{rs} I_r I_s \tag{3.89}$$

For a single resistor of resistance R and conductance G, eqns. 3.88 and 3.89 become (c.f. eqns. 3.43)

$$P = VI = GV^2 = RI^2 \tag{3.90}$$

Problems

3.1 An electrostatic field in free space is created by a system of fixed conductors carrying prescribed charges. At a point in the field where the electric vector is E, there exists an electric dipole consisting of prescribed charges $\pm Q$ at a small prescribed distance l apart. The vector length l points from the negative charge to the positive charge, and Ql is the vector dipole moment p. Initially p is perpendicular to E, and the total energy of the system is W_0. The dipole is now turned so that p is not perpendicular to E, and all charges are kept constant in the process. Show that the new energy of the system is

$$W_e = -p \cdot E + W_0$$

3.2   In the previous problem, the moment p of the electric dipole makes an angle θ with the electric vector E. By expressing the energy in terms of θ and differentiating with respect to θ, show that the torque T tending to increase θ is $-pE \sin \theta$ and that it may be represented vectorially as

$$T = p \times E$$

3.3   In Problem 3.1, the dipole is held so that there is a fixed angle between the vectors p and E, but the dipole is capable of a displacement in the direction of a linear co-ordinate x. In this direction the spatial rate of increase of the electric vector is $\partial E/\partial x$. Use the expression for the energy W_e derived in Problem 3.1 to deduce that the force tending to move the dipole in the direction x increasing is $p \cdot \partial E/\partial x$.

3.4   In the previous problem, the positive charge Q of the dipole has a small vector displacement l with respect to the negative charge $-Q$ as described in Problem 3.1. Show that the vector excess of the force on the positive charge over that on the negative charge is

$$Q \left(l_x \frac{\partial E}{\partial x} + l_y \frac{\partial E}{\partial y} + l_z \frac{\partial E}{\partial z} \right)$$

and hence verify that the force on the dipole is

$$F = p_x \frac{\partial E}{\partial x} + p_y \frac{\partial E}{\partial y} + p_z \frac{\partial E}{\partial z}$$

Use Maxwell's equations to show that this result may also be written

$$\begin{cases} F_x = p_x \dfrac{\partial E_x}{\partial x} + p_y \dfrac{\partial E_y}{\partial x} + p_z \dfrac{\partial E_z}{\partial x} \\[2mm] F_y = p_x \dfrac{\partial E_x}{\partial y} + p_y \dfrac{\partial E_y}{\partial y} + p_z \dfrac{\partial E_z}{\partial y} \\[2mm] F_z = p_x \dfrac{\partial E_x}{\partial z} + p_y \dfrac{\partial E_y}{\partial z} + p_z \dfrac{\partial E_z}{\partial z} \end{cases}$$

3.5   In free space there exist two electric dipoles whose vector moments are p_1 and p_2, and their lengths are small compared with their separation. The dipoles have prescribed lengths and charges, and the energy required to construct them when remotely separated from each other is W_0. The position vector of the dipole of moment p_2 with respect to the dipole of moment p_1 is now r, and the three vectors p_1, p_2 and r are co-planar. The angle from the direction of p_1 to the direction of r is θ_1, and the angle from the direction of r to the direction of p_2 is θ_2. Show that the electric energy of the pair of dipoles is

$$-\frac{p_1 p_2}{4\pi\epsilon_0 r^3} (2 \cos \theta_1 \cos \theta_2 + \sin \theta_1 \sin \theta_2) + W_0$$

3.6 For the pair of dipoles described in the previous problem, calculate (i) the force between them tending to increase r, (ii) the torque on the dipole of moment p_1 tending to increase θ_1, and (iii) the torque on the dipole of moment p_2 tending to increase θ_2. Explain why these two torques are in general unequal. By equating the torque on the complete system to zero, calculate the force acting on each dipole in a direction transverse to their join. Show that the transverse forces vanish if the dipoles are parallel or antiparallel.

Chapter 4

Energy and stress in electric fields

4.1 Introduction

An electric field in free space may be described either by means of the electric vector E or by means of the electric-flux-density vector D, and the two are related at each point by the connection relation

$$D = \epsilon_0 E \tag{4.1}$$

In electrostatics, the electric field may also be described in terms of the spatial distribution of its electric scalar potential ϕ. The electric vector is then the negative gradient of the potential:

$$E = -\operatorname{grad} \phi \tag{4.2}$$

The spatial distribution of the electric potential ϕ may be illustrated with the aid of equipotential surfaces and, if these are drawn for equal intervals of voltage, their closeness in any location gives an indication of the strength of the electric field at that location. The electric vector E may be described by means of lines of force, which intersect the equipotential surfaces at right angles. Lines of force leave from and arrive at the surface of a conductor at right angles, the surface of a conductor in electrostatics being an equipotential surface bounding an equipotential volume.

In free space the lines of force describing the electric vector E are also the lines of flux describing the electric-flux-density vector D, and the lines drawn through any small closed curve form a tube of electric flux. In electrostatics, tubes of electric flux run from a conductor at one potential to a conductor at a lower potential, and they intersect the equipotential surfaces at right angles. The tubes of electric flux and the equipotential surfaces divide the space between the conductors into elements of volume as illustrated in Fig. 4.1. Tubes of electric flux pair off equal elements of positive and negative charge on the conductors, the positive element being at the high-potential end and the negative element at the low-potential end. If the tubes of electric flux for the entire field are drawn so as to start on equal elements of positive charge and end on equal elements of negative charge, then the strength of the electric flux density in any location can be judged from the extent to which the tubes are crowded at that location.

A charged electrostatic system in free space may be discharged in many ways. If it is discharged without creating a magnetic field, the movement of charge must be along tubes of electric flux (Sections 1.5 and 1.6), and the flux of charge per unit area passing any point in the process is, in magnitude and direction, the electric-flux-density vector D at that location before discharge commences. In particular, the discharge may be effected by suitably detaching the positive charge from the positively charged conductors and allowing it to move along the lines of electric flux as illustrated in Fig. 4.2. The positive charge then occupies successive equipotential surfaces until it ultimately neutralises the negative charge on the negatively charged conductors. In this method of discharge the moving sheet of positive charge progressively sweeps the electric field away. The reverse process may be used for charging the system, and the flux density of charge passing any point during the process is then the negative of the electric-flux-density vector D existing at that location in the charged condition.

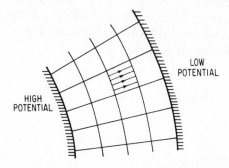

Fig. 4.1   *Illustrating analysis of an electrostatic field into elements of volume by means of tubes of electric flux intersected by equipotential surfaces*

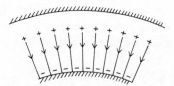

Fig. 4.2.   *Illustrating the discharging of a system along tubes of electric flux in free space*

4.2   Distribution of potential energy between statically charged conductors in free space

In the free space between a system of charged conductors that are at rest relative to the observer let us draw the equipotential surfaces and the tubes of electric flux as indicated in Fig. 4.1. Let the equipotential surfaces be drawn for small intervals of voltage, and let the tubes of electric flux begin and end on small elements of charge. The space between the conductors is then divided up into a large number of small elements of volume, each of which consists of a short portion of a thin tube of electric flux. In each element of volume the electric field is that of a parallel-plate capacitor as indicated in Fig. 4.1. Moreover, the electric field in each element of volume is the electric field for a parallel-plate capacitor with no fringing whatever, because the field in adjacent elements of volume prevents fringing. Each element of volume may therefore be thought of as an ideal parallel-plate capacitor, and the entire electric field may be regarded as an aggregate of parallel-plate capacitors fitted together like building blocks. The building blocks distributed along a particular thin tube of electric flux constitute a collection of ideal parallel-plate capacitors in series, while those distributed between a pair of close equipotential surfaces constitute a collection of ideal parallel-plate capacitors in parallel.

Any electrostatic field in the free space between a system of charged conductors consists of elementary ideal parallel-plate capacitors arranged in series along tubes of electric flux and in parallel between equipotential surfaces. Let us consider a particular one of the ideal parallel-plate capacitors such as the one for which the lines of force are illustrated in Fig. 4.1. Let it consist of a short length ds of a thin tube of electric flux of normal cross-sectional area dS. At this location in the field let E be the electric field strength and D the electric flux density. Then, for this ideal elementary capacitor, the voltage is

$$Eds \tag{4.3}$$

and the charges on the plates are

$$\pm DdS \tag{4.4}$$

It follows from the first of eqns. 3.43 that the potential energy stored in this element of volume $dsdS$ is

$$\tfrac{1}{2}(Eds)(DdS) \tag{4.5}$$

Hence the potential energy stored per unit volume at this location in the electric field is $\tfrac{1}{2}ED$. Using eqn. 4.1, it follows that potential energy is stored throughout an electric field in free space at the rate per unit volume

$$w_e = \tfrac{1}{2}ED = \tfrac{1}{2}\epsilon_0 E^2 = \frac{1}{2\epsilon_0}D^2 \tag{4.6}$$

The quantity w_e is known as the electric energy density in the electric field at the point in space where the electric field strength is E and the electric flux density is D.

The total potential energy of an electrostatic system is

$$W_e = \int w_e \, d\tau \tag{4.7}$$

where $d\tau$ is an element of volume at the location where the electric energy density is w_e, and the integral is taken throughout space. The contribution of the integral in eqn. 4.7 from the regions interior to the conductors vanishes because there is no field there, so that the integral in eqn. 4.7 may be taken throughout the space between the conductors.

As an example, consider any capacitor of voltage V and charges $\pm Q$ in free space. If it is a parallel-plate capacitor, fringing is taken into account. By means of equipotential surfaces and tubes of electric flux, we dissect the field into elements of volume. The potential energy stored in a typical element of volume is given by expr. 4.5. Now consider the space between two close equipotential surfaces for which the difference of potential is $\Delta\phi$. Then, in integrating expr. 4.5 over the space between these equipotentials, Eds is constant and equal to $\Delta\phi$. The contribution of this space to the total stored energy is therefore

$$\tfrac{1}{2}\Delta\phi \int Dds \tag{4.8}$$

Here the surface integral is taken over either of the two close equipotential surfaces and is equal to the total electric flux passing from the positive plate to the negative plate. It is therefore the total charge Q of the capacitor, so that expr. 4.8 becomes

$$\tfrac{1}{2}Q\Delta\phi \tag{4.9}$$

If we now add the contributions from all the spaces between successive equipotential surfaces, the contributions $\Delta\phi$ add up to V, the voltage of the capacitor. The total potential energy stored in the capacitor is therefore

$$W_e = \tfrac{1}{2}VQ \tag{4.10}$$

in accordance with the first of eqns. 3.43. In arriving at this result we could alternatively have integrated expr. 4.5 along a tube carrying electric flux ΔQ to obtain

$$\tfrac{1}{2}\Delta Q \int Eds = \tfrac{1}{2}V\Delta Q \tag{4.11}$$

and then summed over all tubes to obtain eqn. 4.10.

The parallelism between eqns. 3.43 and 4.6 is no accident. By dissecting the electric field into small elements of volume with the aid of equipotential surfaces and tubes of electric flux, the local electric field at each point may be visualized as that of a local parallel-plate capacitor that is fully guarded so that fringing is not involved. The voltage per unit separation for the local parallel-plate capacitor is the local electric field strength E, and the charges per unit area on the plates of the local parallel-plate capacitor are $\pm D$, where D is the local electric flux density. Hence application of eqns. 3.43 to each local parallel-plate capacitor leads directly to eqns. 4.6 for the electric energy per unit volume.

4.3   Distribution of potential energy in any electric field in free space

In the previous section, we considered the distribution of potential energy in the field of statically charged conductors that exist in free space and that are at rest relative to the observer. There was no discussion of the possibility of a static volume distribution of charge between the conductors. Likewise, the presentation in Sections 3.7–3.11 involving the capacitance matrix for a system of charged conductors excludes the possibility of a volume distribution of charge between the conductors. In particular, eqns. 3.43 for the electric energy stored in a charged capacitor assume that there is no space charge between the plates.

In spite of all this, eqns. 4.6 and 4.7 apply even if there is a static volume distribution of electric charge. In the previous section we considered a tube of electric flux as an aggregate of ideal parallel-plate capacitors in series, and it was assumed that each intermediate plate had equal and opposite charges on its two sides and therefore carried no net charge. Let us now assume that each intermediate plate does carry some net charge. This upsets the constancy of Dds as one moves along the tube, and this upsets the validity of eqns. 4.11 and 4.10. But, in each elementary capacitor, the charges on the plates are still given by expr. 4.4, the voltage between the plates is still given by expr. 4.3 and the energy stored is still given by expr. 4.5. Moreover, in the limit in which the sizes of the volume elements constituting the elementary capacitors tend to zero, the charges on the intermediate plates now become a volume distribution of charge, namely, that given by the second of eqns. 1.35. Hence eqns. 4.6 and 4.7 still apply even in the presence of space charge.

So far, we have considered only static electric fields, but eqns. 4.6 and 4.7 still apply even in the presence of a time-varying magnetic field. In these circumstances the tubes of electric flux existing at any instant of time can still be dissected into short sections each of which constitutes an ideal parallel-plate capacitor, but the capacitors located side-by-side in adjacent tubes do not now have identical voltages. This may be seen from Fig. 4.3 where, in accordance with Faraday's law of induction, the integral of the electric field strength from A_1 to B_1 differs from that from A_2 to B_2 by the time rate of change of the magnetic flux threading the loop $A_1B_1B_2A_2A_1$. This does not upset expr. 4.5 for the electric energy stored in each elementary capacitor. What it does upset is the constancy of Eds as one moves perpendicular to the electric field from one tube of electric flux to the next, and this upsets the validity of eqns. 4.9 and 4.10. But the total potential energy of the system is still given by the integral (4.7) throughout all space, and the integrand is still given by eqns. 4.6.

In short, eqns. 4.6 and 4.7 give the total electric energy of any distribution of electric charge in space, static or time-varying.

4.4   Maximum size for quasi-static behaviour

When dissecting a time-varying electric field into thin tubes of electric flux and then chopping them up into short sections each of which may be treated as an ideal parallel-plate capacitor, care is necessary to ensure that the linear dimensions of each volume element are sufficiently small that quasi-static arguments may be used in each element. This is automatically ensured in the sense that, for statements such as those appearing in eqns. 4.6 and 4.7, the linear dimensions of the volume elements are supposed to have been allowed to tend to zero in the manner customary in differential and integral calculus. However, it is often useful to be able to visualise how small the volume elements need to be in order to be able to apply quasi-static arguments. In circuit theory, each circuit element is supposed to be small enough to justify the use of quasi-static arguments.

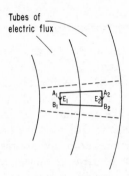

Tubes of
electric flux

Fig. 4.3 *Illustrating the fact that, in the presence of a time-varying magnetic field, the voltages of elementary capacitors located side-by-side in adjacent tubes of electric flux are not identical*

Such arguments require that the volume element, or circuit element, must be small enough for the time of transmission of electromagnetic waves across the element to be negligible. At frequency f this means that the linear dimensions of an element must be small compared with

$$\frac{c}{2\pi f} \qquad (4.12)$$

where c is the velocity of light. Taking this as the maximum size of an element for quasi-static behaviour, we obtain the numerical values shown in Table 4.1.

In dissecting an electric field into elementary parallel-plate capacitors as illustrated in Fig. 4.1, a more stringent size-limit for the volume elements is often imposed by geometry. For example, in a spherical capacitor, the linear dimensions of the volume elements must be small compared with the radii of the conductors.

4.5 Tension in tubes of electric flux in electrostatics

In the previous chapter we used the principle of work (eqns. 3.74 and 3.75) to calculate the mechanical action on a particular conductor of a statically charged system at rest relative to the observer. We calculated force and torque by considering

Table 4.1   *Maximum size for quasi-static behaviour*

Size limit	Frequency Hz	Size limit
Direct current	0	∞
Alternating current	10^2	500 km
Audio	10^4	5 km
Radio	10^6	50 m
Television	10^8	50 cm
Microwaves	10^{10}	5 mm
Light	10^{15}	5×10^{-8} m

linear and angular displacements of the charged conductor concerned, leading to results such as those given in eqns. 3.66, 3.67 and 3.79. This mechanical action in fact arises from a distribution of force over the surface of the statically charged conductor directed normally outwards. Each element of positive charge on the surface of a conductor is in the presence of an electric field directed normally outwards. Each element of negative charge on the surface of a conductor is in the presence of an electric field directed normally inwards. Consequently, each element of charge, whether positive or negative, is subject to a force normally outwards from the conductor on which it resides. It is the resultant of such forces over the surface of a conductor that constitutes the mechanical action on it. If the conductor were flexible it would expand.

The distribution of outward normal force over the surface of a statically charged conductor may be calculated by means of the principle of work employed in the previous chapter. For this purpose it is necessary to imagine that the conductor is not solid, but is hollow and flexible so that it can be distorted. A small portion of the surface can then be given a displacement in the normal direction while the rest of the surface remains fixed. In this way the normally outward force per unit area may be calculated for each location on the surface.

It is convenient to describe the outward normal forces on the various elements of area of a statically charged conductor as arising from the pull of the tubes of electric flux that terminate on the conductor. Even away from the surfaces of conductors, the tubes of electric flux may be regarded as in tension. To decide whether tubes of electric flux are in tension we must devise a means for cutting the tubes and examining what forces are then required to hold the system in equilibrium. In any electrostatic field the tubes of electric flux can be cut by introducing along any equipotential surface a double layer of flexible metal foil whose voltage coincides with that of the equipotential surface concerned, as shown in Fig. 4.4. To keep the two layers of foil in position, force is required between them, suitably distributed. The force per unit area needed between the two flexible metallic sheets at any location is the tension per unit cross-sectional area of the tube of electric flux corresponding to that location. Thus all tubes of electric flux are in tension, and it is the pull that they exert on the conductors to which they are attached that constitutes the mechanical action of the electric field on the conductors.

To calculate the tension in a tube of electric flux at a particular location in free space we can proceed as follows. Consider a short length x of the tube at the location, and let S be the normal cross-sectional area of the tube at this location. Let E be the electric field strength at this location, and D the electric flux density. This section of the tube by itself is an ideal parallel-plate capacitor of capacitance

$$C = \epsilon_0 S/x \tag{4.13}$$

having voltage

$$V = Ex \tag{4.14}$$

and charges $\pm Q$ where

$$Q = DS \tag{4.15}$$

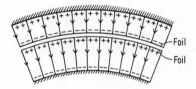

Fig. 4.4. *Illustrating the cutting of tubes of electric flux by means of a double layer of flexible metallic foil along an equipotential surface*

From eqns. 3.43, the energy of this elementary capacitor, if expressed in terms of the charge, is

$$W_e = \frac{1}{2C}Q^2 = \frac{1}{2}\frac{x}{\epsilon_0 S}Q^2 \tag{4.16}$$

and if expressed in terms of the voltage is

$$W_e = \frac{1}{2}CV^2 = \frac{1}{2}\frac{\epsilon_0 S}{x}V^2 \tag{4.17}$$

The tension in the tube is the force of attraction between the plates of this capacitor, and this is $-F$ where F is the force tending to increase x. This force may be calculated from eqn. 3.74 as

$$F = -\frac{\partial W_e}{\partial x} \quad \text{(constant } Q\text{)} \tag{4.18}$$

or from eqn. 3.75 as

$$F = +\frac{\partial W_e}{\partial x} \quad \text{(constant } V\text{)} \tag{4.19}$$

To apply eqn. 4.18, we express the energy W_e in terms of the charge Q as in eqns. 4.16 and obtain

$$F = -\frac{1}{2}\frac{1}{\epsilon_0 S}Q^2 \tag{4.20}$$

To apply eqn. 4.19 we express the energy W_e in terms of the voltage V as in eqns. 4.17 and obtain

$$F = -\frac{1}{2}\frac{\epsilon_0 S}{x^2}V^2 \tag{4.21}$$

The results (4.20) and (4.21) are identical by virtue of eqns. 4.14 and 4.15. The tension per unit cross-sectional area T_e of the tube is $-F/S$; it may be written in any of the forms

$$T_e = \frac{1}{2}ED = \frac{1}{2}\epsilon_0 E^2 = \frac{1}{2\epsilon_0}D^2 \tag{4.22}$$

Force per unit area is dimensionally the same as energy per unit volume. Eqns. 4.6 and 4.22 show that the electric energy per unit volume and the tension per unit cross-sectional area of tubes of electric flux are not only dimensionally the same but they are numerically identical.

Where a tube of electric flux terminates on a statically charged conductor, the tension per unit cross-sectional area of the tube is also the outward normal force per unit area on the conductor. At any point of a conductor this is therefore given by exprs. 4.22 where E and D are the electric field strength and the electric flux density just outside the conductor at this point. Notice that the outward normal force per unit area is not simply the product of the charge per unit area D on the conductor and the outward normal electric field strength E just outside the conductor. This is because the electric field strength just inside a statically charged conductor is zero. It is the average of the electric field strength E just outside and the zero field strength just inside that has to be multiplied by the charge per unit area D on the conductor.

The argument for deriving the tension per unit cross-sectional area in eqns. 4.22 from the energy per unit volume may, if desired, be reversed. We picture the system as being discharged as shown in Fig. 4.2. At a point of the surface of displaced positive charge where the undisturbed electric field strength is E and the undisturbed electric flux density is D, there is a charge DdS on an element of area dS and it is subject to a mean force per unit charge $\frac{1}{2}E$ in the normal direction. The work recovered in allowing the positive charge to displace a further normal distance ds is therefore

$$\frac{1}{2}E(DdS)\,ds \tag{4.23}$$

As the surface of displaced positive charge sweeps away the electric field, therefore, energy is recovered at the rate $\frac{1}{2}ED$ per unit volume of electric field destroyed. In the corresponding charging process, normal displacement of the surface of positive charge in the opposite direction would require that work be done at the rate $\frac{1}{2}ED$ per unit volume of electric field created. This is achieved by applying normally directed force $\frac{1}{2}ED$ per unit area to the surface of the positive charge being displaced.

4.6   Sideways pressure of tubes of electric flux in electrostatics

Tubes of electric flux, besides being in tension, exert a sideways pressure on each other. This sideways pressure may be regarded as the cause of fringing in a parallel-plate capacitor. The existence of sideways pressure by tubes of electric flux can be demonstrated by placing two rectangular parallel-plate capacitors edge-to-edge as indicated in Fig. 4.5. Forces are required to hold the two capacitors together against the sideways pressure of the tubes of electric flux in one capacitor on those in the other. The forces between the capacitors may be calculated by applying the law of inverse squares to the charge distributions on the plates, but it is simpler to perform the calculation using the concept of sideways pressure between tubes of electric flux.

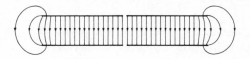

Fig. 4.5.   *Two rectangular parallel-plate capacitors placed edge-to-edge to demonstrate the sideways pressure of tubes of electric flux*

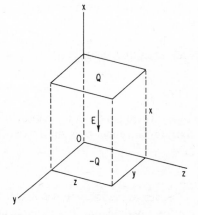

Fig. 4.6.   *Illustrating calculation of the sideways pressure of a short length x of a thin rectangular tube of electric flux (cross-sectional dimensions y by z)*

At a point in an electrostatic field in free space let us again consider a short length x of a thin tube of electric flux. It will now be convenient to regard the tube as having a rectangular cross-section of dimensions y by z as shown in Fig. 4.6 so that, in eqns. 4.13–4.17, the cross-sectional area of the tube is

$$S = yz \tag{4.24}$$

Instead of allowing x to increase to $x + dx$, let us now allow y to increase to $y + dy$, both x and z remaining constant. If p_e is the sideways pressure tending to increase y, the work done by this pressure in allowing y to increase to $y + dy$ is

$$(zx)p_e \, dy \tag{4.25}$$

Hence (c.f. eqn. 4.18)

$$p_e zx = -\frac{\partial W_e}{\partial y} \quad (\text{constant } Q, x, z) \tag{4.26}$$

or (c.f. eqn. 4.19)

$$p_e zx = +\frac{\partial W_e}{\partial y} \quad (\text{constant } V, x, z) \tag{4.27}$$

Use in eqns. 4.26 or 4.27 of eqns. 4.13–4.17 together with eqn. 4.24 then gives

$$p_e = \tfrac{1}{2}ED = \tfrac{1}{2}\epsilon_0 E^2 = \frac{1}{2\epsilon_0}D^2 \tag{4.28}$$

The same result is obtained if the transverse dimension z of the tube in Fig. 4.6 is allowed to increase, rather than the transverse dimension y.

At a point in an electric field in free space the sideways pressure of tubes of electric flux is the same in all directions perpendicular to the field, and is given by eqns. 4.28. The sideways pressure of the tubes is equal in magnitude to the tension per unit cross-sectional area of the tubes given by eqns. 4.22.

4.7   The electric stress tensor in free space

Using cartesian co-ordinates (x, y, z), let us consider the forces acting on the short length x of the thin rectangular tube of electric flux shown in Fig. 4.6 (cross-sectional dimensions y by z). There is a tension per unit area acting across the faces perpendicular to the x axis; for the upper face, this force per unit area has cartesian components (eqns. 4.22).

$$(+\tfrac{1}{2}ED, 0, 0) \tag{4.29}$$

There is also a pressure acting across the faces perpendicular to the y axis in Fig. 4.6; for the left-hand face the force per unit area has cartesian components (eqns. 4.28)

$$(0, -\tfrac{1}{2}ED, 0) \tag{4.30}$$

There is likewise a pressure acting across the faces perpendicular to the z axis; for the right-hand face the force per unit area has cartesian components

$$(0, 0, -\tfrac{1}{2}ED) \tag{4.31}$$

The three vectors in eqns. 4.29, 4.30 and 4.31 may be written as the columns of the electric stress tensor

$$
\begin{pmatrix}
\tfrac{1}{2}ED & 0 & 0 \\
0 & -\tfrac{1}{2}ED & 0 \\
0 & 0 & -\tfrac{1}{2}ED
\end{pmatrix}
\tag{4.32}
$$

The elements in the leading diagonal of a stress tensor describe the force per unit area acting across a surface in the direction perpendicular to that surface; a positive value indicates tension and a negative value indicates pressure. The non-diagonal elements describe the forces per unit area acting across a surface in the direction parallel to that surface; they are shearing stresses. The zeros in the tensor (4.32) connote absence of shearing stress in an electric field; no forces act across any face of the rectangular parallelopiped in Fig. 4.6 in directions parallel to the face, so that there is no torque tending to rotate the volume element of electric field. Forces acting on the element tend to stretch it parallel to electric field and to compress it in the perpendicular direction.

The tensor (4.32) may be written as the sum of two tensors:

$$
\begin{pmatrix}
ED & 0 & 0 \\
0 & 0 & 0 \\
0 & 0 & 0
\end{pmatrix}
\tag{4.33}
$$

plus

$$
\begin{pmatrix}
-\tfrac{1}{2}ED & 0 & 0 \\
0 & -\tfrac{1}{2}ED & 0 \\
0 & 0 & -\tfrac{1}{2}ED
\end{pmatrix}
\tag{4.34}
$$

The second of these describes a pressure $-\tfrac{1}{2}ED$ in all directions, both parallel to the tubes of electric flux as well as perpendicular to them. This isotropic pressure then as to be corrected by a tension per unit cross-sectional area along the tubes described by the tensor (4.33).

The tensors (4.33) and (4.34) are referred to a coordinate system for which the x axis is along the tube of electric flux as shown in Fig. 4.6. For a co-ordinate system with a different orientation the cosine rule for resolving vectors has to be applied. This does not affect the pressure tensor (4.34) but the tensor (4.33) becomes

$$
\begin{pmatrix}
E_x D_x & E_y D_x & E_z D_x \\
E_x D_y & E_y D_y & E_z D_y \\
E_x D_z & E_y D_z & E_z D_z
\end{pmatrix}
\tag{4.35}
$$

The first column of this tensor, for example, implies that the force per unit area across a plane $x = $ constant has cartesian components

$$E_x(D_x, D_y, D_z) \tag{4.36}$$

This is the force per unit area exerted by the region $x < 0$ on the region $x > 0$. It is in the direction of the D vector but its magnitude is less than ED by the cosine of the angle between the E vector and a normal to the plane $x =$ constant. This simply allows for the fact that the plane $x =$ constant is not now a normal cross-section of the tube of electric flux. It follows that the tensor (4.35) implies that there is a tension along the tubes of electric flux of amount ED per unit cross-sectional area. To this has to be added an isotropic pressure $\frac{1}{2}ED$ in accordance with the tensor (4.34).

The sum of the stress tensors (4.34) and (4.35) can be interpreted as implying that there exists a stress across any surface drawn in an electrostatic field in free space. However, the existence of this stress can only be demonstrated experimentally for surfaces that are equipotential surfaces or for surfaces that are composed of lines of electric flux. The former demonstration may be made as illustrated in Fig. 4.4 and the latter as indicated in Fig. 4.5. Dissection of the field would in general require the introduction not only of surface distributions of electric charge but also of surface distributions of magnetic current (Section 1.14).

4.8   Distribution of potential energy in linear dielectric material

In dielectric material at rest relative to the observer, application of an electric field distorts the equilibrium distribution of electric charge in atoms and molecules so as to create a bound electric moment per unit volume P_b. The cartesian components of P_b at any location in the dielectric material are determined by the cartesian components of the electric vector E at that location. For a linear isotropic dielectric (Section 2.6) this relation takes the form

$$P_b = \kappa \epsilon_0 E \tag{4.37}$$

where κ is the electric susceptibility of the material at the location under consideration; κ may be a function of position. For a linear non-isotropic dielectric each cartesian component of P_b depends linearly on all three cartesian components of E, so that

$$\left. \begin{aligned} P_{bx} &= \epsilon_0 (\kappa_{xx} E_x + \kappa_{xy} E_y + \kappa_{xz} E_z) \\ P_{by} &= \epsilon_0 (\kappa_{yx} E_x + \kappa_{yy} E_y + \kappa_{yz} E_z) \\ P_{bz} &= \epsilon_0 (\kappa_{zx} E_x + \kappa_{zy} E_y + \kappa_{zz} E_z) \end{aligned} \right\} \tag{4.38}$$

The κ coefficients constitute the electric susceptibility tensor of the material at the location under consideration. The direction of P_b is in general different from that of E.

Consider a simple model of a dielectric consisting of distributions of positive and negative charge $\pm \rho$ per unit volume. In the absence of an applied electric field, the positive and negative distributions of charge overlap precisely and neutralise each other. Under the influence of an applied electric field (not necessarily uniform), let the positive distribution have a vector displacement r (not necessarily uniform) relative to the negative distribution. Corresponding elements of positive and negative charge then constitute an electric dipole, and the bound electric moment per unit volume is

$$P_b = \rho r \tag{4.39}$$

Work is required to polarise the dielectric material in this way, and the work done constitutes the potential energy of the polarised dielectric. This potential energy is distributed throughout the volume of the dielectric and is additional to the potential energy per unit volume of the applied electric field.

To calculate the additional potential energy per unit volume due to polarisation of the dielectric let us suppose that the material behaves in a linear (but not necessarily isotropic) manner. Let the electric field be gradually applied in such a way that, at each location, the electric vector at a certain stage is a fraction x of its final value. Then the same is true for the bound electric moment per unit volume because the dielectric behaves linearly. At a particular location in the material let E be the final value of the electric vector and let P_b, given by eqn. 4.39, be the final value of the bound electric moment per unit volume. At a stage in the polarisation process when the electric vector is

$$xE \tag{4.40}$$

the bound electric moment per unit volume is

$$xP_b = x\rho r \tag{4.41}$$

Let us now increase x by a small amount dx so that there is brought into existence an additional electric moment per unit volume

$$dxP_b = dx\rho r \tag{4.42}$$

This involves giving a charge per unit volume ρ a vector displacement dxr in the presence of an electric vector xE. The work done per unit volume is

$$\rho(xE) \cdot (dxr) \tag{4.43}$$

or, using eqn. 4.39,

$$E \cdot P_b x dx \tag{4.44}$$

By integrating with regard to x from zero to unity we see that the work done per unit volume in polarising the dielectric material is

$$\tfrac{1}{2}E \cdot P_b \tag{4.45}$$

This is additional to the work done per unit volume in creating the polarising electric field itself. The latter is given by the first of eqns. 4.6, and may be written

$$\tfrac{1}{2}E \cdot D \tag{4.46}$$

where

$$D = \epsilon_0 E \tag{4.47}$$

By addition of exprs. 4.45 and 4.46 it follows that the total electric potential energy per unit volume, allowing for both that of the polarised dielectric and that of the polarising electric field, is

$$w_e = \tfrac{1}{2}E \cdot (D + P) \tag{4.48}$$

or

$$w_e = \tfrac{1}{2}E \cdot D' \tag{4.49}$$

where

$$D' = D + P_b \tag{4.50}$$

The vector D' appearing in eqns. 4.49 and 4.50 is the modified electric-flux-density vector used in version 2 of the electromagnetic equations for materials (Section 2.4). This is the version that is most convenient for handling non-conducting dielectric material. In it the bound charges and currents are suppressed in Maxwell's magnetic curl and electric divergence equations (Table 2.1). Moreover, the connection formula (4.47) becomes, in accordance with eqn. 4.50,

$$D' = \epsilon_0 E + P_b \tag{4.51}$$

where P_b is given by eqn. 4.37 for linear isotropic dielectric material and by eqns. 4.38 for linear non-isotropic dielectric material.

For linear isotropic dielectric material, substitution from eqn. 4.37 into eqn. 4.51 gives

$$D' = \epsilon E \tag{4.52}$$

where

$$\epsilon = \epsilon_0 (1 + \kappa) \tag{4.53}$$

and ϵ/ϵ_0 is the dielectric constant of the material (eqns. 2.26). Eqn. 4.49 then gives

$$w_e = \frac{1}{2}ED' = \frac{1}{2}\epsilon E^2 = \frac{1}{2\epsilon}D'^2 \tag{4.54}$$

Comparison of eqns. 4.54 with eqns. 4.6 shows that, for linear isotropic dielectric material, the potential energy of polarisation is taken into account merely by using ϵ in place of ϵ_0 and using the modified electric flux density D' in eqn. 4.52 in place of the unmodified electric flux density D in eqn. 4.47.

For linear non-isotropic dielectric material, substitution from eqns. 4.38 into eqn. 4.51 gives

$$\left. \begin{aligned} D'_x &= \epsilon_{xx}E_x + \epsilon_{xy}E_y + \epsilon_{xz}E_z \\ D'_y &= \epsilon_{yx}E_x + \epsilon_{yy}E_y + \epsilon_{yz}E_z \\ D'_z &= \epsilon_{zx}E_x + \epsilon_{zy}E_y + \epsilon_{zz}E_z \end{aligned} \right\} \tag{4.55}$$

where the ratios of the ϵ coefficients to ϵ_0 constitute the dielectric tensor of the material. If eqn. 4.53 is interpreted as a tensor equation with 1 denoting the unit tensor, it describes the relation between the dielectric tensor and the susceptibility tensor of the material. Eqn. 4.49 may be written

$$w_e = \tfrac{1}{2}(E_x D_x' + E_y D_y' + E_z D_z')$$ (4.56)

and we may then substitute for (D_x', D_y', D_z') from eqns. 4.55 to obtain

$$w_e = \tfrac{1}{2}[\epsilon_{xx} E_x^2 + \epsilon_{yy} E_y^2 + \epsilon_{zz} E_z^2 + (\epsilon_{yz} + \epsilon_{zy})E_y E_z$$

$$+ (\epsilon_{zx} + \epsilon_{xz})E_z E_x + (\epsilon_{xy} + \epsilon_{yx})E_x E_y]$$ (4.57)

An argument in terms of field quantities similar to that used in connection with voltages and charges in Section 3.11, shows (see Problems 4.1 and 4.2) that the dielectric tensor of a material is symmetrical about its leading diagonal, so that

$$\epsilon_{xy} = \epsilon_{yx}, \quad \epsilon_{yz} = \epsilon_{zy}, \quad \epsilon_{zx} = \epsilon_{xz}$$ (4.58)

Eqn. 4.57 may therefore be written

$$w_e = \tfrac{1}{2}[\epsilon_{xx} E_x^2 + \epsilon_{yy} E_y^2 + \epsilon_{zz} E_z^2 + 2\epsilon_{yz} E_y E_z + 2\epsilon_{zx} E_z E_x + 2\epsilon_{xy} E_x E_y]$$
(4.59)

In terms of (i) the cartesian components of the electric vector at a given location in linear non-isotropic dielectric material, and (ii) the components of the dielectric tensor at that location, eqn. 4.59 gives the electric energy per unit volume allowing for both the polarising electric field and the polarised dielectric.

The model of a dielectric that we have used is greatly oversimplified. We have assumed that the dielectric consists of two volume distributions of positive and negative charge that neutralise each other in the unpolarised state. In fact a dielectric consists of an aggregate of atoms. The dielectric can be polarised by distorting the distribution of electronic charge in each atom in relation to the positive charge on the nucleus, but these two distributions of positive and negative charge are not in fact identical in the unpolarised state.

Energy is stored in the assemblage of atomic particles that constitutes the unpolarised dielectric. This includes mechanical, thermodynamical and quantum-mechanical energy, for discussion of which reference may be made to books on these subjects. The physics of materials involves many complications, the full understanding of which requires study of classical and quantum mechanics in addition to electromagnetism. In calculating the energy involved in polarising a dielectric we have made the assumption that none of the energy of the unpolarised dielectric is made available to the electric field during the process of polarisation.

Some materials possess atoms or molecules with permanent electric dipole moments that are randomly oriented in the absence of an imposed electric field (polar dielectrics). Under the influence of an imposed electric field the random orientation of the dipoles is slightly modified, thereby creating an electric moment per unit volume as described in Section 2.12 (eqns. 2.86–2.94). The equations that we have developed in this section for the energy stored in a non-polar dielectric apply equally well for a polar dielectric, provided that no non-linear behaviour is involved.

4.9   Electric stress in linear dielectric material

The discussion of electric stress in an electric field in free space that was developed in Sections 4.5, 4.6 and 4.7 on the basis of eqns. 4.6 may be readily adapted to give the electric stress in linear isotropic dielectric material. It is a matter of replacing ϵ_0 by ϵ and D by D'. The tubes of modified electric flux density D' possess a tension per unit cross-sectional area (c.f. eqns. 4.22)

$$T_e = \frac{1}{2}ED' = \frac{1}{2}\epsilon E^2 = \frac{1}{2\epsilon}D'^2 \tag{4.60}$$

and exert a sideways pressure (c.f. eqns. 4.28)

$$p_e = \frac{1}{2}ED' = \frac{1}{2}\epsilon E^2 = \frac{1}{2\epsilon}D'^2 \tag{4.61}$$

If the x axis is directed along a tube at a particular location in the material, the electric stress tensor is (c.f. exprs. 4.33 and 4.34)

$$\begin{pmatrix} ED' & 0 & 0 \\ 0 & 0 & 0 \\ 0 & 0 & 0 \end{pmatrix} \tag{4.62}$$

plus

$$\begin{pmatrix} -\frac{1}{2}ED' & 0 & 0 \\ 0 & -\frac{1}{2}ED' & 0 \\ 0 & 0 & -\frac{1}{2}ED' \end{pmatrix} \tag{4.63}$$

If the x axis does not have this special orientation, the tensor (4.62) becomes (c.f. the tensor (4.35))

$$\begin{pmatrix} E_x D'_x & E_y D'_x & E_z D'_x \\ E_x D'_y & E_y D'_y & E_z D'_y \\ E_x D'_z & E_y D'_z & E_z D'_z \end{pmatrix} \tag{4.64}$$

For a linear non-isotropic dielectric material the tensor (4.64) still applies but the pressure tensor (4.63) must be written (c.f. eqn. 4.49)

$$\begin{pmatrix} -\frac{1}{2}E \cdot D' & 0 & 0 \\ 0 & -\frac{1}{2}E \cdot D' & 0 \\ 0 & 0 & -\frac{1}{2}E \cdot D' \end{pmatrix} \tag{4.65}$$

so that the total electric stress tensor is then the sum of the tensors (4.64) and (4.65). It must be remembered, however, that D'_x, D'_y and D'_z in exprs. 4.64 and 4.65 are now to be interpreted with the aid of eqns. 4.55.

For non-isotropic dielectric material, the stress tensor (4.64) is not symmetrical about its leading diagonal. This means that the electric shearing stresses on a small rectangular piece of the material constitute a torque. For the piece of material to be in equilibrium this torque due to electric shearing stresses must be balanced by a torque due to other shearing stresses. In non-isotropic material there are electric, magnetic and mechanical stress tensors that are separately unsymmetrical, although their sum is symmetrical.

4.10   Non-electrical stress in material

There are important forces between atoms that come into play when material is deformed. These constitute what is usually described as the mechanical stress in the strained material. We have disregarded the possibility that the material may be strained and stressed in this sense, and have calculated only the stress associated with (i) the electric polarisation of the material, and (ii) the electric field that causes this polarisation.

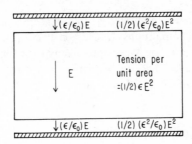

Fig. 4.7.   *Illustrating the change in tension of a tube of electric flux where it crosses an interface between free space and a dielectric of uniform dielectric constant ϵ/ϵ_0*

As an illustration, let us consider a parallel-plate capacitor between whose plates there exists linear homogeneous isotropic non-conducting material of dielectric constant ϵ/ϵ_0 as shown in Fig. 4.7. We neglect fringing and suppose that there are small gaps between the dielectric material and the plates. Let E be the electric field strength in the dielectric, so that the modified electric flux density there is

$$D' = \epsilon E \tag{4.66}$$

This is also the electric flux density in the gaps because there is no free charge on the surface of the dielectric (second of eqns. 2.18). Taking the gaps as equivalent to free space, the electric field strength there must be

$$(\epsilon/\epsilon_0)E \tag{4.67}$$

in order that ϵ_0 times this expression may give the quantity on the right-hand side of eqn. 4.66. From expr. 4.67 it follows that the tension per unit cross-sectional area of the tubes of electric flux in the gaps is (eqns. 4.22)

$$\tfrac{1}{2}\epsilon_0 \{(\epsilon/\epsilon_0)E\}^2 \;=\; \tfrac{1}{2}\frac{\epsilon^2}{\epsilon_0}E^2 \tag{4.68}$$

whereas the tension per unit cross-sectional area of the tubes of electric flux in the dielectric material is (eqns. 4.54)

$$\tfrac{1}{2}\epsilon E^2 \tag{4.69}$$

Exprs. 4.68 and 4.69 show that there is a difference between the tension per unit cross-sectional area in the gaps and that in the dielectric material. This difference in tension is taken up by mechanical stress in the material. This stress would cause distortion of the material, but this we have neglected. Elastic deformation of dielectric material by an electric field is what is known as electrostriction.

> What we have called mechanical stress in a material in fact arises from the forces between the atoms and molecules of which the material is constructed. In reality these forces are largely electrical in origin. Nevertheless, discussion of them belongs to classical and quantum mechanics rather than to electromagnetism.

Energy storage and stress in an electric field and in a polarised dielectric is not restricted to electrostatic conditions. Under non-electrostatic conditions, however, motional energy also has to be taken into account. In Chapters 5 and 6 we shall see that tubes of magnetic flux are in tension and exert sideways pressure in a manner similar to tubes of electric flux.

4.11   Non-linear dielectric material

Expr. 4.43 illustrates the fact that the work done in causing an increment dP_b in the bound electric moment per unit volume at a particular location in a dielectric at a stage when the electric vector at that location is E is

$$dP_b \cdot E \tag{4.70}$$

The total work done in changing the electric moment per unit volume from P_{b1} to P_{b2} is therefore

$$\int_{P_{b1}}^{P_{b2}} E \cdot dP_b \tag{4.71}$$

Expr. 4.45 is the value of this integral when (i) the initial electric vector and the initial electric moment per unit volume are zero, (ii) the final electric vector and the final electric moment per unit volume are (E, P_b), and (iii) there is a linear relationship between the electric vector at each location in the material and the corresponding electric moment per unit volume.

For non-linear dielectric material, expr. 4.71 still gives the work done in changing the electric moment per unit volume from P_{b1} to P_{b2}, but evaluation of the integral is now more complicated. It is necessary to know what the non-linear relation is between P_b and E. Moreover, this relation may be different in going from P_{b1} to P_{b2} from what it is in going from P_{b2} back to P_{b1}, a phenomenon known as dielectric hysterisis. Energy can then be absorbed by the dielectric material in such a cycle, causing what is known as dielectric hysteresis loss.

If desired, the work done per unit volume in changing the electric moment per unit volume from P_{b1} to P_{b2} in non-linear dielectric material may be combined with the electric energy per unit volume of the polarising electric field. The latter is simply $\frac{1}{2}ED$ where

$$D = \epsilon_0 E \tag{4.72}$$

The change in the energy density of the polarising electric field as E changes from E_1 to E_2 and D changes from D_1 to D_2 may, however, be written

$$\int_{D_1}^{D_2} E \cdot dD \tag{4.73}$$

The sum of exprs. 4.71 and 4.73 is then

$$\int_{D_1'}^{D_2'} E \cdot dD' \tag{4.74}$$

where

$$D' = D + P_b = \epsilon_0 E + P_b \tag{4.75}$$

Here D' is the modified electric-flux-density vector used in the version 2 of the electromagnetic equations for materials (Section 2.4). To evaluate the integral (4.74) the non-linear relationship between D' and E must be ascertained, usually experimentally.

Care is necessary in using the concept of energy in connection with a non-linear system. If a non-linear dielectric is to be polarised, work must be done and energy expended to perform the operation. But the material may be taken through numerous alternative sequences during the process of polarisation, corresponding to different functional relations between D' and E. Different amounts of work are done in achieving the final state of polarisation by these different routes, and different amounts of energy are expended. Consequently, associated with a given state of polarisation of a non-linear dielectric, there is no unique potential energy that is independent of the previous history of the material.

4.12   Density of power dissipation in conducting materials

In Sections 4.1–4.3, we studied the storage of potential energy per unit volume in an electric field in free space, and in Section 4.8 we examined the additional potential energy per unit volume stored in the presence of polarisable dielectric material. Let us now suppose that the material, besides having dielectric properties, also has conduction properties. Then the electric vector E at a particular location in the material accelerates free electrons in the material. The consequent ordered motion of the free electrons is then converted into random motion as they make collisions with atoms and molecules. By this process electric energy is first converted into ordered kinetic energy of the electrons, and then into disordered kinetic energy, that is, into heat. Let us enquire how much energy per unit volume is dissipated as heat per unit time.

We suppose that, at a particular location in the conducting material, there are N free electrons per unit volume each of charge e. Let the electric vector at this location be E and the magnetic-flux-density vector be B. Under the influence of the electromagnetic field, a free electron is accelerated until it is knocked out of the coherent motion by a collision. In this way the free electrons develop a drift-velocity v (Section 2.12), so that the density of free electric current at this location in the conducting material is

$$J = Nev \qquad (4.76)$$

The average force exerted by the electromagnetic field on a free electron is

$$F = e(E + v \times B) \qquad (4.77)$$

so that the average force on the free electrons per unit volume is

$$Ne(E + v \times B) \qquad (4.78)$$

The average rate at which work is being done by the electromagnetic field on the free electrons per unit volume is therefore

$$Ne(E + v \times B) \cdot v \qquad (4.79)$$

and the magnetic contribution vanishes because $v \times B$ is perpendicular to v. Using eqn. 4.76, it follows that the average rate at which work is being done by the electromagnetic field on the free electrons is

$$p = E \cdot J \qquad (4.80)$$

Let us now suppose that the drift-velocity of the free electrons is steady, so that no motional energy is being created or dissipated on the average. In these circumstances the work done by the electromagnetic field on a free electron between one collision and the next is used to increase its thermic motion after the second collision. Consequently, eqn. 4.80 also gives the average rate at which heat is generated in the conducting material per unit volume.

In the above discussion the conduction properties have been regarded as arising only from free electrons in the material. In some materials, however, free positively charged ions are involved as well — for example, in an ionised gas or plasma. In a fluid such as the electrolyte of a battery, the negative electrons become attached to neutral atoms, so that the free electric current is composed of equal numbers of positively and negatively charged heavy ions moving in opposite directions. If several types of charge carriers are involved in the conduction process, summation signs have to be used on the right-hand side of eqn. 4.76 and in exprs. 4.78 and 4.79, but this does not modify eqn. 4.80.

For linear isotropic material of conductivity σ at rest relative to the observer the relation between the free current density vector and the electric vector is

$$J = \sigma E \tag{4.81}$$

Eqn. 4.80 then gives for the power dissipation per unit volume of the material

$$p = EJ = \sigma E^2 = \frac{1}{\sigma} J^2 \tag{4.82}$$

These are simply eqns. 3.90 applied to a cylindrical element of the conducting material having unit length in the direction of current flow and unit cross-sectional area normal to the direction of current flow.

For linear non-isotropic conducting material at rest relative to the observer use must be made of a conductivity tensor such that (c.f. eqns. 4.55)

$$\left.\begin{aligned}
J_x &= \sigma_{xx}E_x + \sigma_{xy}E_y + \sigma_{xz}E_z \\
J_y &= \sigma_{yx}E_x + \sigma_{yy}E_y + \sigma_{yz}E_z \\
J_z &= \sigma_{zx}E_x + \sigma_{zy}E_y + \sigma_{zz}E_z
\end{aligned}\right\} \tag{4.83}$$

where (c.f. eqns. 4.58)

$$\sigma_{yz} = \sigma_{zy}, \quad \sigma_{zx} = \sigma_{xz}, \quad \sigma_{xy} = \sigma_{yx} \tag{4.84}$$

Eqn. 4.80 may be written

$$p = E_x J_x + E_y J_y + E_z J_z \tag{4.85}$$

and substitution from eqns. 4.83 then gives for the power dissipation per unit volume in linear non-isotropic conducting material

$$p = \sigma_{xx}E_x^2 + \sigma_{yy}E_y^2 + \sigma_{zz}E_z^2 + 2\sigma_{yz}E_yE_z + 2\sigma_{zx}E_zE_x + 2\sigma_{xy}E_xE_y \tag{4.86}$$

Eqns. 4.80, 4.82, and 4.86 for the energy dissipated per unit time per unit volume in assocation with the conduction properties of the material may be compared with eqns. 4.49, 4.54, and 4.59 for the electric energy stored per unit volume in association with the dielectric polarisation of the material and with the polarising electric field. The absence of the factor $\frac{1}{2}$ in the dissipation formulas should be noted (c.f. eqns. 3.88, 3.89 and 3.90).

4.13 Density of power generation in an electromagnetic field

In the previous section, let us now suppose that the free electrons are driven by a force F per unit charge, possibly not of electromagnetic origin. Then the average rate at which work is being done on them per unit volume by the driving mechanism is (c.f. eqn. 4.80)

$$F \cdot J \tag{4.87}$$

Let us suppose that the drift-velocity of the free charges is steady, so that no motional energy is being created or dissipated on the average. Then the driving force F per unit charge is simply being used to offset the effect of whatever electromagnetic field (E, B) is present. This means that

$$F = -(E + v \times B) \tag{4.88}$$

Substituting from eqn. 4.88 into expr. 4.87, and using the fact that v and J are in the same direction, we obtain for the average rate at which work is being done by the driving mechanism per unit volume the value

$$-E \cdot J \tag{4.89}$$

Comparison of this expression with that on the right-hand side of eqn. 4.80 shows that, whereas a positive value of $E \cdot J$ connotes dissipation of power per unit volume, a negative value of $E \cdot J$ connotes generation of power per unit volume.

For example, in the electrolyte of a battery delivering steady electric current, $E \cdot J$ is positive almost everywhere, corresponding to dissipation of power in the internal resistance of the battery. But immediately adjacent to the surfaces of the electrodes $E \cdot J$ is negative. Here power is being generated by a chemical process involving the separation of electrons from their atoms. Adjacent to the electrodes, electromagnetic energy is being created at the expense of the quantum-mechanical reservoir of energy associated with the assemblage of atoms and molecules.

Problems

4.1   There exists in space a distribution of dielectric material that is linear but is not necessarily either isotropic or homogeneous. Using the second version of the electromagnetic equations in materials (Table 2.1), a comparison is made between two situations in the first of which the material is subject to an electrostatic field $[E_1, D_1']$ and in the second to an electrostatic field $[E_2, D_2']$. By considering a situation in which the material is subject to an electrostatic field $[(1 - x)E_1 + xE_2, (1 - x)D_1' + xD_2']$ and allowing x to go from 0 to 1, show that the work done per unit volume in converting field 1 into field 2 is (c.f. Section 3.11)

$$E_1 \cdot (D_2' - D_1') + \tfrac{1}{2}(E_2 - E_1) \cdot (D_2' - D_1')$$

By equating this to the increase in the energy per unit volume, deduce that

$$E_1 \cdot D_2' = E_2 \cdot D_1'$$

Verify the truth of this relation for isotropic material of dielectric constant ϵ/ϵ_0.

4.2   In the previous problem the linear material is non-isotropic and has a dielectric tensor ϵ. Express the result $E_1 \cdot D'_2 = E_2 \cdot D'_1$ in terms of the elements in the dielectric tensor and the cartesian components of E_1 and E_2. By taking $E_1 = (1, 0, 0)$ and $E_2 = (0, 1, 0)$ deduce that $\epsilon_{xy} = \epsilon_{yx}$. Hence establish the truth of all of eqns. (4.58).

Chapter 5

Motional energy in an electromagnetic field in free space

5.1 Introduction

Besides possessing potential energy associated with its position or geometry, a system also possesses energy associated with its motion. This is the energy required to create the motion from rest relative to the observer; it is the energy that the observer would recover in stopping the motion. The motional energy that a system possesses by virtue of its mass is familiar in mechanics and is called its kinetic energy. However, a system also possesses motional energy by virtue of its electric charge. This is known as magnetic energy. Magnetic energy plays a major role in electromagnetism. But kinetic energy as understood in mechanics also plays a role in electromagnetism. In a particle accelerator, such as a Van de Graff machine or a cyclotron, the motional energy of the particles associated with their mass plays a major role. On the other hand, in an inductor, the motional energy of the conduction electrons associated with their mass is nearly always trivial compared with that associated with their charge. There are many situations in electromagnetism in which motional energy is completely dominated by magnetic energy. But there are also many situations in which it is impossible to overlook the contribution of kinetic energy to the total motional energy.

5.2 The concept of motional energy associated with moving charge

Let us acquire some insight into the concept of motional energy associated with electric charge by first studying a single tube of magnetic flux in free space. Let the tube have the form of a long circular cylinder of radius a. Let the length l of the cylinder be so great that its external magnetic field is negligible compared with the internal field, while the internal field may be regarded as uniform and parallel to the axis.

The tube of magnetic flux is maintained by means of electric current flowing solenoidally over its curve surface. Specifically, we imagine the moving charge to consist of a rotating hollow circular non-conducting cylinder of radius a possessing a uniformly distributed total charge Q. Let the angular velocity of the cylinder be $\omega = v/a$, so that the perimeter is moving relative to the observer with a peripheral velocity v [$(v/c)^2$ negligible]. Electrostatically, this moving charge is supposed to be neutralised by a coincident hollow circular cylinder of radius a carrying a uniformly distributed total charge $-Q$ that is not rotating relative to the observer. We assume at present that the mass of the moving cylinder is negligible.

The moving cylinder carries a charge Q round through one revolution in time $2\pi/\omega = 2\pi a/v$. Hence the solenoidal electric current round the curved surface of the tube of magnetic flux is

$$I = \frac{Q}{2\pi a}v \tag{5.1}$$

Measured per unit length, the electric current round the cylinder is

$$H = \frac{I}{l} = \frac{Q}{2\pi al}v \tag{5.2}$$

so that the magnetic flux density within the cylinder is

$$B = \mu_0 H = \frac{\mu_0 Q}{2\pi al}v \tag{5.3}$$

and the magnetic flux threading the cylinder is

$$\Phi = \pi a^2 B = \frac{\mu_0 aQ}{2l}v \tag{5.4}$$

This magnetic flux is illustrated in the left half of Fig. 5.1, together with the variation of the magnetic flux density with perpendicular distance r from the axis. Also shown are the lines of magnetic vector potential A (circular by symmetry) and the variation of A with r. The latter is obtained by applying the circulation law for magnetic vector potential (eqn. 1.56) to a circle of radius r, so that $2\pi rA$ is equated to the magnetic flux threading the circle. In particular, at the location of the moving charged cylinder, we have

$$A = \frac{1}{2\pi a}\Phi = \frac{\mu_0 Q}{4\pi l}v \tag{5.5}$$

Note that the magnetic vector potential is proportional to the speed of rotation.

Let us now enquire what force F must be applied round the periphery of the moving cylinder, at right angles to the axis, in order to produce a steady increase in the speed of rotation. The quantities I, H, B, Φ and A are now all increasing proportionally to dv/dt in accordance with eqns. 5.1–5.5. Moreover, time-variation of the magnetic field brings into play an electric field E in accordance with Faraday's law of induction. By symmetry, this electric field is in circles round the axis of the cylinder as shown in the right half of Fig. 5.1. The direction of E is related by the left-hand-screw rule to the direction of the time-rate of increase of the threading magnetic flux because it is the time-rate of decrease of the latter that gives the right-hand related circulation of the electric vector (eqn. 1.37). If E is the electric field strength, it varies with r as shown at the bottom of Fig. 5.1; this radial variation is obtained by applying Faraday's law to a circle of radius r, with E in the direction shown at the top of Fig. 5.1, so that $2\pi r E$ is equated to the time rate of increase of the magnetic flux threading the circle. In particular, at the location of the moving charged cylinder, the tangential electric field strength acting in the direction opposite to that of acceleration of the charge is

$$E = \frac{1}{2\pi a}\frac{d\Phi}{dt} = \frac{\mu_0 Q}{4\pi l}\frac{dv}{dt} \tag{5.6}$$

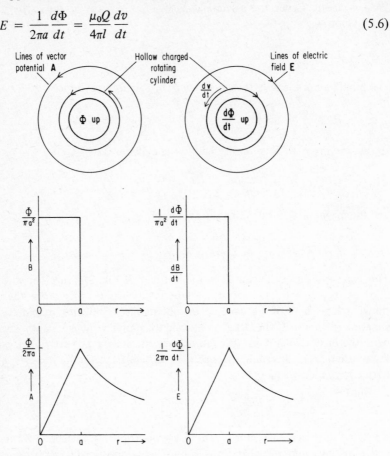

Fig. 5.1   *Illustrating a straight tube of magnetic flux of circular cross-section in which the flux is increasing steadily with time*

To create the desired steady acceleration, the effect of this force per unit charge must be overcome.

Comparison of eqns. 5.5 and 5.6 shows that, at $r = a$,

$$E = \frac{dA}{dt} \tag{5.7}$$

Moreover, the two diagrams at the bottom of Fig. 5.1 show that eqn. 5.7, with d/dt replaced by $\partial/\partial t$, is true for all values of r. This is a consequence of the circulation law for the vector $E + \partial A/\partial t$ (eqn. 1.57), combined with the symmetry of the field and the fact that E and A are oppositely directed as shown at the top of Fig. 5.1.

To create steady acceleration of the charge, we must apply, in the direction of acceleration, a peripheral force F sufficient to overcome the force per unit charge that was calculated in eqn. 5.6. The magnitude of the necessary force is

$$F = QE = \frac{\mu_0 Q^2}{4\pi l}\frac{dv}{dt} \tag{5.8}$$

Using eqn. 5.7, this force may also be written

$$F = Q\frac{dA}{dt} \tag{5.9}$$

or

$$F = \frac{dp_m}{dt} \tag{5.10}$$

where

$$p_m = QA \tag{5.11}$$

The magnetic vector potential A at any point in an electromagnetic field is the electromagnetic momentum per unit charge of a test charge at the point (eqn. 1.52). Eqn. 5.11 therefore states that the peripheral electromagnetic momentum of the cylinder is QA, and eqn. 5.10 states that the time-rate of increase of this electromagnetic momentum gives the force needed to accelerate the rotation. From eqn. 5.8, we see that a similar hollow cylinder having no charge but having a uniformly distributed mass

$$\frac{\mu_0 Q^2}{4\pi l} \tag{5.12}$$

would required the same peripheral force to create the same acceleration.

Let us now calculate the rate at which work is being done on the moving charged cylinder in accelerating it, and consequently the rate of increase of its motional energy. From eqn. 5.8, this power is

$$Fv = \frac{\mu_0 Q^2}{4\pi l}v\frac{dv}{dt} = \frac{d}{dt}\left\{\frac{\mu_0 Q^2}{8\pi l}v^2\right\} \tag{5.13}$$

The motional energy of the charged cylinder is therefore

$$W_m = \frac{\mu_0 Q^2}{8\pi l}v^2 \tag{5.14}$$

Since this motional energy is entirely associated with charge, it is magnetic energy. We may express this magnetic energy in terms of the electric current I round the cylinder by substituting for v from eqn. 5.1 into eqn. 5.14. We obtain

$$W_m = \tfrac{1}{2}LI^2 \tag{5.15}$$

where

$$L = \mu_0 \frac{\pi a^2}{l} \tag{5.16}$$

We recognise L as the inductance of the inductor formed by the cylinder in circumstances when the internal magnetic field dominates the external magnetic field (Problem 2.1).

In arriving at the expressions for magnetic energy given in eqns. 5.14 and 5.15 we have used a straight cylindrical tube, long compared with its radius, but nevertheless one of finite length l. The magnetic flux Φ must in fact emerge from one end of the cylinder and return to the other end to form closed tubes of magnetic flux. Viewed on a scale large compared with l, the cylinder is a magnetic dipole that produces an external field as illustrated in Fig. 1.9. Time-variation in the rate of rotation of the cylinder therefore causes radiation (Section 1.13). If the charged cylindrical tube is started impulsively with angular velocity v_0/a then, even with perfectly smooth bearings, its rotation would die away exponentially on account of radiation. We can drastically reduce this radiation by connecting the two open ends of the cylindrical tube together to form an endless tube; in so doing, we must suppose that the tube is flexible, so that the charge Q can retain its spinning motion. The inductance is then given quite accurately by eqn. 5.16, and eqn. 5.15 still applies.

5.3 Motional energy involving both mass and charge

We have supposed so far that the spinning tube possesses charge but no mass. Let us now suppose that, in addition to a charge Q uniformly distributed over the surface, there is also a mass M uniformly distributed over the surface. To accelerate the tube, an additional peripheral force $M dv/dt$ is now required, so that eqn. 5.9 is replaced by

$$F = M \frac{dv}{dt} + Q \frac{dA}{dt} \tag{5.17}$$

Eqn. 5.10 is consequently replaced by

$$F = \frac{dp}{dt} \tag{5.18}$$

where

$$p = p_k + p_m \tag{5.19}$$

and

$$p_k = mv, \quad p_m = QA \tag{5.20}$$

The quantities p_k and p_m are, respectively, the mechanical and electromagnetic contributions to the total peripheral momentum p of the tube, and eqn. 5.18 states that the peripheral force required to accelerate the spinning motion is equal to the time-rate of increase of the total peripheral momentum, mechanical plus electrical.

The rate at which the system is acquiring motional energy is the rate at which work is being done on it, and this is (eqn. 5.17)

$$Fv = Mv\frac{dv}{dt} + Qv\frac{dA}{dt} \tag{5.21}$$

If, in the last term of this equation, we substitute for A in terms of v from eqn. 5.5 and then for v in terms of I from eqn. 5.1, we obtain

$$Fv = Mv\frac{dv}{dt} + LI\frac{dI}{dt} \tag{5.22}$$

where L is the inductance of the tube given by eqn. 5.16. Eqn. 5.22 may now be written

$$Fv = \frac{d}{dt}(\tfrac{1}{2}Mv^2 + \tfrac{1}{2}LI^2) \tag{5.23}$$

and it follows that the total motional energy of the system is

$$W_M = W_k + W_m \tag{5.24}$$

where

$$W_k = \tfrac{1}{2}Mv^2, \quad W_m = \tfrac{1}{2}LI^2 \tag{5.25}$$

The quantity W_k is the kinetic energy of the system and W_m is the magnetic energy.

Eqn. 5.24 illustrates the necessity of adding the kinetic and magnetic contributions to obtain the total motional energy of a system, and eqn. 5.19 illustrates the corresponding necessity of adding the mechanical and electromagnetic contributions to obtain the total momentum; the total driving force is the sum of contributions associated with mass and charge.

There are many electromechanical systems where both kinetic energy and magnetic energy play important roles (electric motors, electric generators, loudspeakers, microphones, etc.). However, there are also many situations where, although both forms of motional energy are technically involved, nevertheless one swamps the other for practical purposes. Suppose, for example, that we attempt to construct the spinning tube just analysed by bending a length of hollow plastic tubing into a ring and connecting the two ends together through the hollow spindle of a massless wheel that can be driven by a motor. After making suitable allowances for elastic forces, frictional forces, etc., the effect of the kinetic energy $\tfrac{1}{2}Mv^2$ in eqn. 5.23 and of the force Mdv/dt in eqn. 5.17 could be demonstrated. On the other hand, however much we charged the plastic tube it would be virtually impossible to measure any effect from the magnetic energy $\tfrac{1}{2}LI^2$ in eqn. 5.23 and the force QdA/dt in eqn. 5.17. The effect would still be too small to measure easily when the electrostatic field close to the system reached the value at which the insulation of air at normal temperature and pressure breaks down (about 3×10^6 v/m).

Now suppose that we replace the closed hollow plastic tube by a closed hollow copper tube that is held fixed, and that we then drive a current of electrons solenoidally round the copper tube so as to use it as a toroidal inductor. Only the conduction electrons in the copper are now taking part in the motion. The effect of the kinetic energy $\tfrac{1}{2}Mv^2$ in eqn. 5.23 and of the force Mdv/dt in eqn. 5.17 has therefore been drastically reduced. On the other hand, the corresponding magnetic effect can be fully maintained. Indeed it can be greatly increased, because the negative charge of the moving conduction electrons is now neutralised by fixed positively charged particles in the copper, thereby removing the electrostatic field that was causing breakdown in the insulation of the air. Consequently, with the electronic current arrangement, it would be easy to demonstrate the effect of the magnetic energy $\frac{1}{2}LI^2$ in eqn. 5.23 and of the force QdA/dt in eqn. 5.17. But it would be very difficult to detect the corresponding effects associated with the mass of the conduction electrons.

For a metallic circuit carrying an electronic conduction current, the kinetic energy is proportional to the number of electrons involved. But the factor Q^2 in eqn. 5.14 shows that the magnetic energy is proportional to the square of the number of electrons involved. In a metal, the density of conduction electrons is so high that, for all practical electric circuits, the magnetic energy of the electrons swamps the kinetic energy. By contrast, for orbital motion of electrons in an atom, motional energy is almost entirely associated with their mass rather than their charge. Between these extremes we have an ionised gas, or plasma, for which account must be taken both of magnetic energy and of kinetic energy in formulating the total motional energy.

The above discussion refers to steady acceleration of the moving charge constituting the electric current. If the current is flowing solenoidally round the curved surface of a tube of magnetic flux, there is a steady increase of the threading magnetic flux and of the magnetic energy stored in the inductor. However, there is an associated circular electric field as shown in Fig. 5.1, and potential energy is stored in this as described in Chapter 4. But this electric field and its stored electric energy remain constant if the acceleration remains constant. All the growth of energy stored by the inductor is then associated with growth in the threading magnetic flux.

With non-uniform acceleration, however, change does take place in the electric field, and therefore also in the stored electric energy. Suppose, for example, that the charged cylinder introduced in the previous section is stationary prior to time zero, and that a steady peripherial force is then applied to produce acceleration. Steady acceleration is not acquired instantaneously because the circular electric field illustrated in Fig. 5.1 first has to be created. As the cylinder gets into motion, electromagnetic waves are launched from the moving charged surface both inwards into the tube and outwards from it. Behind the electromagnetic wavefront that propagates away from the inductor with the velocity of light, the circular electric field is established in a time of the order of l/c where l is a length characteristic of the linear dimensions of the inductor. When once the steady circular electric field has been established then, as described in the previous paragraphs, nearly all the work done on the system is used to increase the motional energy of the inductor. Even at this stage, however, a little more work has to be done on the moving charge by the forces driving it than is needed to increase the motional energy. This is because accelerated charge radiates (Problem 1.8), so that steady radiation from the steadily accelerating charge continues. However, the rate at which motional energy is being stored in the inductor is itself increasing steadily as the velocity increases (eqn. 5.13). Consequently, at times large compared with l/c, it becomes progressively more true that all the work done on the spinning charged tube is used for acceleration.

A similar situation occurs in mechanics but is not always mentioned in elementary treatments. Suppose that a heavy solid wheel is mounted on smooth bearings and is accelerated from rest by a steady torque in the form of peripheral force applied uniformly at the rim. The wheel does not develop uniform angular acceleration instantaneously because of the elastic properties of the material of which it is constructed. The rim starts a little before the hub, the intervening time being that taken by an elastic wave to travel from the rim to the hub. While the wheel is taking the strain, potential energy is being stored in it. It is only when this initial process is complete that the work done on the wheel is used primarily to generate kinetic energy. Moreover, if the flywheel is surrounded by air that is set in motion by the wheel, a loss-mechanism is involved that is analogous to radiation in the electromagnetic system.

5.4 Magnetic energy of any stationary tube of magnetic flux in free space

The above discussion is not restricted to a tube of magnetic flux having uniform circular cross-sectional area. Consider any thin tube conveying a magnetic flux Φ in free space as shown in Fig. 1.7, and let the tube be stationary relative to the observer. Such a tube is endless, and its normal cross-sectional area in general varies from place to place. At a location where the normal cross-sectional area of the tube is S, the magnetic flux density is

$$B = \frac{\Phi}{S} \tag{5.26}$$

If the tube exists by itself, the electric current per unit length flowing solenoidally round it at this location is

$$H = \frac{1}{\mu_0} \frac{\Phi}{S} \tag{5.27}$$

The common direction of B and H is along the tube in the direction of the threading magnetic flux, and if n is a unit inward normal to the surface of the tube, the electric current per unit length over the surface is (eqns. 1.36)

$$i = n \times H \tag{5.28}$$

where the magnitude of H is given by eqn. 5.27. The current round a short length dl of the tube at the place where the normal cross-sectional area is S is

$$H\,dl = \frac{\Phi}{\mu_0} \frac{dl}{S} \tag{5.29}$$

so that the solenoidal electric current round the entire tube is

$$I = \oint H \, dl = \frac{\Phi}{\mu_0} \oint \frac{dl}{S} \tag{5.30}$$

The inductance of the tube is therefore

$$L = \frac{\Phi}{I} = \mu_0 \bigg/ \oint \frac{dl}{S} \tag{5.31}$$

If the magnetic flux Φ threading the tube is increasing steadily with time, so is the electric current round each element of length dl (eqn. 5.29), and so is the total solenoidal current round the entire tube (eqn. 5.30). But the time-variation of the magnetic flux brings into play an electric field E in accordance with Faraday's law of induction. Consider a normal cross-section of area S, and let ds be a vector element of length of its perimeter in the direction related by the right-hand-screw rule to the direction of the threading magnetic flux. Then Faraday's law states that (eqn. 1.37)

$$\oint E \cdot ds = -\frac{d\Phi}{dt} \tag{5.32}$$

This voltage acts round the perimeter of each normal cross-section of the tube, and the negative sign implies that the voltage is in the direction opposite to the solenoidal current. To create a steady increase in the current, the voltage in eqn. 5.32 must be overcome by a voltage V applied in the direction of the current. This driving voltage is

$$V = -\oint E \cdot ds \tag{5.33}$$

From eqns. 5.32 and 5.33 it follows that

$$V = \frac{d\Phi}{dt} \tag{5.34}$$

so that the driving voltage is the same for every cross-section of the tube. Eqn. 5.34 states that application of a driving voltage V solenoidally round a tube causes a steady time-rate of increase in the magnetic flux Φ threading the tube in the direction related by the right-hand-screw rule.

This result may also be expressed in terms of the solenoidal current I round the tube by using the inductance L of the tube derived in eqn. 5.31. Eqn. 5.34 becomes

$$V = \frac{d}{dt}(LI) \tag{5.35}$$

Because the tube is not distorting, the inductance L is independent of time, and eqn. 5.35 becomes

$$V = L\frac{dI}{dt} \tag{5.36}$$

The driving voltage V is brought to bear on the electric current flowing solenoidally round the tube by means of the force necessary to overcome the electric field E appearing in eqn. 5.32. Let us suppose that the solenoidal current per unit length round the tube consists of a charge per unit area q on the surface of the tube moving with velocity v in the direction of the element of length ds of the perimeter of a normal cross-section of the tube. Then the solenoidal electric current H per unit length appearing in eqn. 5.27 is

$$H = qv \qquad (5.37)$$

and, expressed as a vector, this surface density of electric current is (eqn. 5.28)

$$i = n \times H = qv \qquad (5.38)$$

The force per unit area that must be applied to the electric charge on the surface of the tube to overcome the electric field E in eqn. 5.32 is

$$F = -qE \qquad (5.39)$$

and the rate at which this force is doing work on the tube per unit area of the surface is

$$F \cdot v = -qv \cdot E \qquad (5.40)$$

Integrating round the perimeter of a normal cross-section, we see that the rate at which the work is being done on the tube per unit length is

$$F \cdot v \, ds = -q \oint v \cdot E \, ds$$

$$= -q \oint vE \cdot ds$$

$$= -qv \oint E \cdot ds$$

$$= HV \qquad (5.41)$$

from eqns. 5.33 and 5.37. If we now integrate round the length of the closed tube of magnetic flux, we see that the total rate at which work is being done on the tube is

$$V \oint H dl = VI \qquad (5.42)$$

from the first of eqns. 5.30.

We have shown that, to cause a steady rate of increase V of the magnetic flux Φ threading the tube as described in eqn. 5.34, work has to be done on the tube at a total rate

$$VI = I \frac{d\Phi}{dt} \qquad (5.43)$$

In accordance with the first of eqns. 5.31, this may also be written

$$VI = I\frac{d}{dt}(LI) \tag{5.44}$$

Because the inductance L is independent of time, we may rewrite eqn. 5.44 as

$$VI = \frac{d}{dt}(\tfrac{1}{2}LI^2) \tag{5.45}$$

The rate at which work is being done on the tube is therefore equal to the time-rate of increase of its magnetic energy

$$W_m = \tfrac{1}{2}LI^2 \tag{5.46}$$

By means of the first of eqns. 5.31, this magnetic energy may also be expressed in terms of the magnetic flux Φ threading the tube or as a product of the mutually threading electric current I and magnetic flux Φ:

$$W_m = \tfrac{1}{2}I\Phi = \tfrac{1}{2}LI^2 = \frac{1}{2L}\Phi^2 \tag{5.47}$$

If the tube were created by means of electric current flowing through a metallic sheet covering the surface of the tube, the metal would in practice possess some resistance. If R is the resistance of the circuit, a voltage RI would be required merely to keep the current constant, and this would consume energy at the rate RI^2 (eqns. 3.90). It would be the excess of the applied voltage V over RI that would be available to increase the threading magnetic flux, so that eqn. 5.35 would be replaced by

$$V - RI = \frac{d}{dt}(LI) \tag{5.48}$$

and eqn. 5.36 by

$$V = L\frac{dI}{dt} + RI \tag{5.49}$$

Eqn. 5.45 would then become

$$VI = \frac{d}{dt}(\tfrac{1}{2}LI^2) + RI^2 \tag{5.50}$$

showing that the rate at which work is being done on the tube exceeds that dissipated in the resistance by the time rate of increase of the magnetic energy $\tfrac{1}{2}LI^2$ stored in the tube.

The above equations may be compared with the corresponding equations in mechanics for a body of mass M being accelerated by a force F. If at time t the velocity of the body is v and its momentum is p then (c.f. the first of eqns. 5.31)

$$M = \frac{p}{v} \tag{5.31m}$$

Newton's first law of motion states that (c.f. eqn. 5.34)

$$F = \frac{dp}{dt} \tag{5.34m}$$

Substitution for p from eqn. 5.34m into eqn. 5.31m gives (c.f. eqn. 5.35)

$$F = \frac{d}{dt}(Mv) \tag{5.35m}$$

and, if M is independent of time, this becomes (c.f. eqn. 5.36)

$$F = M\frac{dv}{dt} \tag{5.36m}$$

The rate at which the force F is doing work on the body is Fv and this may be written either as (c.f. eqn. 5.43)

$$Fv = v\frac{dp}{dt} \tag{5.43m}$$

or as (c.f. eqn. 5.44)

$$Fv = v\frac{d}{dt}(Mv) . \tag{5.44m}$$

If M is independent of time, this becomes (c.f. eqn. 5.45)

$$Fv = \frac{d}{dt}(\tfrac{1}{2}Mv^2) \tag{5.45m}$$

which states that the rate at which work is being done on the body is the time rate of increase of its kinetic energy (c.f. eqn. 5.46)

$$W_k = \tfrac{1}{2}Mv^2 \tag{5.46m}$$

This kinetic energy may also be expressed in terms of the momentum p of the body or in terms of the product of v and p (c.f. eqns. 5.47):

$$W_k = \tfrac{1}{2}vp = \tfrac{1}{2}Mv^2 = \frac{1}{2M}p^2 \tag{5.47m}$$

In the presence of a frictional resistance equal to μ times the velocity, it would only be the excess of F over μv that would be available to increase momentum. Eqn. 5.35m would be replaced by (c.f. eqn. 5.48)

$$F - \mu v = \frac{d}{dt}(Mv) \tag{5.48m}$$

Likewise, eqn. 5.36m would be replaced by (c.f. eqn. 5.49)

$$F = M\frac{dv}{dt} + \mu v \tag{5.49m}$$

and eqn. 5.45m by (c.f. eqn. 5.50)

$$Fv = \frac{d}{dt}(\tfrac{1}{2}Mv^2) + \mu v^2 \ . \tag{5.50m}$$

5.5 Distribution of magnetic energy in a magnetic field in free space

In the previous section we thought of a stationary tube of magnetic flux in free space bounded by solenoidally flowing charge to which forces were applied to increase steadily the threading magnetic flux and consequently the magnetic energy stored in the tube. In addition to considering this phenomenon for the complete tube we can study the way in which cause and effect are distributed along the length of the tube.

At a location along the tube where the normal cross-sectional area is S, the solenoidal electric current per unit length is H, given by eqn. 5.27, and the rate at which work is being done on the tube per unit length by the driving forces is HV, appearing in eqn. 5.41. Replacing V by $d\Phi/dt$ from eqn. 5.34, we see that the rate at which work is being done on the tube per unit length is

$$HV = H\frac{d\Phi}{dt} \tag{5.51}$$

Substitution for Φ in terms of H from eqn. 5.27 then gives

$$HV = \mu_0 SH\frac{dH}{dt} = \frac{d}{dt}(\tfrac{1}{2}\mu_0 H^2 S) \ . \tag{5.52}$$

Consequently, for a short length l of the tube, the rate at which work is being done on it is

$$HVl = \frac{d}{dt}(\tfrac{1}{2}\mu_0 H^2 Sl) \tag{5.53}$$

It follows that the magnetic energy stored in a short length l of the tube at the location where the normal cross-sectional area is S is

$$\tfrac{1}{2}\mu_0 H^2 Sl \tag{5.54}$$

Since Sl is the volume of this section of the tube, we may say that, at a location of the tube where the current per unit length is H, the magnetic energy stored per unit volume is

$$w_m = \tfrac{1}{2}\mu_0 H^2$$

This result may also be expressed in terms of the magnetic flux density B in the tube at this location, or as a product of H and B:

$$w_m = \tfrac{1}{2}HB = \tfrac{1}{2}\mu_0 H^2 = \frac{1}{2\mu_0}B^2 \tag{5.55}$$

The quantity w_m is known as the magnetic energy density in the magnetic field at the point where the magnetic field strength is H and the magnetic flux density is B.

The statement that expr. 5.54 gives the magnetic energy stored in a short length l of the tube at a location where the normal cross-sectional area is S becomes an approximation unless the rest of the tube is present to act as a guard inductor. Nevertheless, a separate accounting can be made of the energy supplied by the driving forces over the length l and over the rest of the tube, and expr. 5.54 is the energy supplied to the length l. We may also note that expr. 5.54 may be written

$$\frac{1}{2}\left(\frac{\mu_0 S}{l}\right)\left(Hl\right)^2 \tag{5.56}$$

Since Hl is the current round the length l of the tube and $\mu_0 S/l$ is the inductance of this length if properly guarded, expr. 5.56 shows that eqn. 5.46 applies to any portion of a tube if properly guarded.

In terms of the magnetic energy density w_m given by eqns. 5.55 we may calculate total magnetic energy W_m by volume integration. The total stored magnetic energy is

$$W_m = \int w_m \, d\tau \tag{5.57}$$

where $d\tau$ is an element of volume and the integral is taken throughout the magnetic field.

5.6 Magnetic energy of an inductor

The magnetic energy of an inductor consisting of a single tube of magnetic flux in free space is given by eqns. 5.47. We shall show that these equations apply to any electric circuit in free space whose configuration does not vary with time, provided that the current may be regarded as uniform round the circuit.

Fig. 5.2   *Illustrating flow of current through an inductor as a result of voltage applied by a generator*

Let us suppose that the magnetic flux $\Phi(t)$ threading an electric circuit is varying with time t as a result of a generator of voltage $V(t)$ applied in the circuit as shown in Fig. 5.2. Let the resulting uniform electric current round the circuit be $I(t)$. Let the positive direction of this current point round the circuit in the direction in which the voltage $V(t)$ acts, and let this direction be related by the right-hand-screw rule to the positive direction of the threading magnetic flux $\Phi(t)$. Neglecting the resistance of the circuit, Faraday's law of induction gives

$$V = \frac{d\Phi}{dt} \tag{5.58}$$

The power supplied by the generator is

$$P = VI \tag{5.59}$$

and, on substituting for V from eqn. 5.58, eqn. 5.59 can be written

$$P = I\frac{d\Phi}{dt} \tag{5.60}$$

If L is the inductance of the circuit, we have

$$\Phi = LI \tag{5.61}$$

Provided that L is independent of t, eqn. 5.60 gives

$$P = LI\frac{dI}{dt} \tag{5.62}$$

or

$$P = \frac{dW_m}{dt} \tag{5.63}$$

where

$$W_m = \tfrac{1}{2}LI^2$$

Using eqn. 5.61, we may write the stored magnetic energy as

$$W_m = \tfrac{1}{2}I\Phi = \tfrac{1}{2}LI^2 = \frac{1}{2L}\Phi^2 \tag{5.64}$$

The function of an inductor is to store motional energy electrically. For an inductor of given inductance, the stored motional energy is proportional to the square of the electric current and also proportional to the square of the threading magnetic flux. For a given electric current I the stored motional energy is proportional to the inductance L, but for a given threading magnetic flux Φ the stored motional energy is inversely proportional to the inductance.

Any magnetic field in free space may be regarded as an aggregate of thin tubes of magnetic flux each of which may be dissected into building blocks, or elements of volume, as indicated in Fig. 5.3. At a point in the field where the magnetic field strength is H and the magnetic flux density is B, the local building block possesses the field of a properly guarded solenoidal inductor. The current per unit length for this inductor is H and the threading magnetic flux per unit area is B. Application of eqns. 5.64 to this inductor gives eqns. 5.55.

If one knows the inductance L of an inductor, the magnetic energy that it stores when carrying an electric current I is $\tfrac{1}{2}LI^2$ in accordance with eqns. 5.64. But the magnetic energy that the inductor stores is also given by eqn. 5.57. Equating these expressions for the magnetic energy, we obtain

$$L = \frac{2}{I^2} \int w_m \, d\tau \tag{5.65}$$

where the integral is to be taken throughout the magnetic field of the inductor. Eqn. 5.65 sometimes provides a convenient way of calculating the inductance of and inductor.

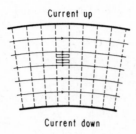

Current up

Current down

Fig. 5.3 *Illustrating dissection of a magnetic field into building blocks each of which is an ideal solenoidal inductor*

5.7 Quasi-static treatment of magnetic energy

If we multiply eqn. 5.60 by dt, the right-hand side becomes $Id\Phi$, and the left-hand side is the amount of energy provided during the interval dt by the generator connected into the circuit. It follows that the work that must be done to increase by a small amount $d\Phi$ the magnetic flux threading a circuit carrying a right-hand related electric current I is

$$Id\Phi \tag{5.66}$$

This work has to be done by a generator connected into the circuit. Although it in fact takes time to do the work, time does not appear explicitly in expr. 5.66.

Expr. 5.66 may be compared with the expression

$$VdQ \tag{5.67}$$

encountered in electrostatics. This is the work that must be done to increase by a small amount dQ the electric charge on a conductor whose potential relative to ground is V. To perform this operation, an amount of charge dQ has to be transferred from ground to the conductor through a potential rise of V, and the work done is given by expr. 5.67. Again, time does not appear explicitly in expr. 5.67 although, in practice, time would be required to do the work.

In Chapters 3 and 4, many results concerning electric energy were deduced from expr. 5.67. In the same way, many results concerning magnetic energy can be deduced from expr. 5.66. As an example, let us rederive eqns. 5.64 for the magnetic energy stored in an inductor of prescribed inductance.

Consider an inductor of any configuration and let its inductance have the fixed value L. Let I be the electric current round the inductor, and let Φ be the resulting magnetic flux threading the inductor in the direction related by the right-hand-screw rule to the direction of the current. Let us imagine that the electric current was initially zero, and consider how much work had to be done to increase it to I. At a stage when the electric current is a fraction x of its final value I, the magnetic flux threading the inductor is also a fraction x of its final value Φ. At this stage, the electric current is xI, and the right-hand related threading magnetic flux is $x\Phi$. Now let us increase the threading magnetic flux by a small amount $dx\Phi$. Then, according to expr. 5.66, the work done is

$$(xI)(dx\Phi)$$

or

$$I\Phi x dx$$

Integrating with respect to x from zero to unity we obtain for the total work done in establishing the system

$$W_m = \tfrac{1}{2}I\Phi \tag{5.68}$$

In agreement with eqns. 5.64, this is the magnetic energy of the system when fully established.

5.8 Magnetic energy of a system of inductors carrying steady electric currents

The magnetic energy of any system of electric circuits that carry steady electric currents and that are at rest relative to the observer may be derived by a process analogous to that used in Section 3.9 for deriving the electric energy of a system of statically charged conductors. In free space let there be n loops of wire numbered $1, 2, \ldots$, and let the electric currents round them be $I_1, I_2, \ldots$. Let the magnetic fluxes threading the loops in the directions related by the right-hand-screw rule to the electric currents be $\Phi_1, \Phi_2, \ldots$, respectively. Imagine that the system of mutually threading electric currents and magnetic fluxes is being created in such a way that, at a certain stage, the currents and fluxes are a fraction x of their final values. Using expr. 5.66, the energy necessary to increase x to $x + dx$ is (c.f. expr. 3.44)

$$(I_1\Phi_1 + I_2\Phi_2 + \ldots)x\, dx$$

Integrating with respect to x from zero to unity, we obtain for the total work done in establishing the system

$$W_m = \tfrac{1}{2} \sum_{r=1}^{n} I_r\Phi_r \tag{5.69}$$

This expression for the magnetic energy of a system of loops of electric current may be compared with eqn. 3.47 for the electric energy of a system of charged conductors.

Eqn. 5.69 may be written as a quadratic function of the electric currents by using the inductance matrix L_{rs} of the system. This is such that (c.f. the first of eqns. 3.46)

$$\Phi_r = \sum_{r=1}^{n} L_{rs}I_s \tag{5.70}$$

Alternatively, eqn. 5.69 may be written as a quadratic function of the threading magnetic fluxes by using the inverse of the inductance matrix. This is such that (c.f. the second of eqn. 3.46)

$$I_r = \sum_{r=1}^{n} l_{rs}\Phi_s \tag{5.71}$$

These equations are simply the algebraic solution of eqns. 5.70. Substitution into eqn. 5.69 from eqns. 5.70 and 5.71 then gives for the magnetic energy of the system (c.f. eqns. 3.48)

$$W_m = \tfrac{1}{2} \sum_{r=1}^{n} \sum_{r=1}^{n} L_{rs}I_sI_s = \tfrac{1}{2} \sum_{r=1}^{n} \sum_{r=1}^{n} l_{rs}\Phi_r\Phi_s \tag{5.72}$$

5.9 The reciprocity theorem of magnetostatics

A reciprocity theorem for magnetostatics may be derived by a process analogous to that used in Sections 3.10 and 3.11 for the reciprocity theorem of electrostatics. For a system of loops in free space, let us first apply steady currents $I_1, I_2, \ldots$ producing steady right-hand related magnetic fluxes $\Phi_1, \Phi_2, \ldots$, respectively. For the same system of loops let us now apply currents $I'_1, I'_2, \ldots$ producing threading magnetic fluxes $\Phi'_1, \Phi'_2, \ldots$. Then (c.f. eqn. 3.49)

$$\Sigma I\Phi' = \Sigma I'\Phi \tag{5.73}$$

where Σ denotes summation over the loops. In particular (c.f. eqns. 3.52 and 3.54)

$$L_{rs} = L_{sr}, \quad l_{rs} = l_{sr} \tag{5.74}$$

The first of eqn. 5.74 implies that the magnetic flux threading loop number r due to an electric current I round loop number s (the currents round the remaining loops being zero) is equal to the magnetic flux threading loop number s when the electric current I flows round loop number r (the currents round the remaining loops being zero). The second of eqn. 5.74 implies that the electric current round loop number r when a magnetic flux Φ is threaded through loop number s (there being no magnetic flux threading the other loops) is equal to the electric current round loop number s when the magnetic flux Φ is threaded through loop number r (there being no magnetic flux threading the other loops).

For a simple transformer there are only two magnetically coupled electric circuits, loops 1 and 2. The coefficient of self-inductance of the primary circuit is L_{11} and that of the secondary circuit is L_{22}, while the coefficient of mutual inductance is L_{12}, which is the same as L_{21}. With electric currents I_1 in the primary circuit and

I_2 in the secondary circuit, the magnetic energy stored in the transformer is, from the first of eqns. 5.72,

$$W_m = \tfrac{1}{2}(L_{11}I_1^2 + 2L_{12}I_1I_2 + L_{22}I_2^2) \tag{5.75}$$

The same value can be obtained by integrating throughout the field the magnetic energy density given in eqns. 5.55 in accordance with eqn. 5.57.

5.10 Calculation of forces of magnetic origin from changes in magnetic energy

The calculation of forces of magnetic origin from changes in magnetic energy follows the same lines as the calculation in Sections 3.12 and 3.13 of forces of electric origin from changes in electric energy.

In free space consider a system of electric circuits round which there are flowing steady electric currents denoted by $I_1, I_2, \ldots$ and through which there thread, in the directions related to the currents by the right-hand-screw rule, magnetic fluxes denoted by $\Phi_1, \Phi_2, \ldots$, respectively. In accordance with eqn. 5.69 the magnetic energy of the system is

$$W_m = \tfrac{1}{2}\Sigma I\Phi \tag{5.76}$$

Consider a situation in where there are increments $d\Phi_1, d\Phi_2, \ldots$ in the magnetic fluxes threading the circuits, and increments $dI_1, dI_2, \ldots$ in the electric currents round the circuits. By taking differentials in eqn. 5.76 we deduce that the corresponding increase in the magnetic energy of the system is

$$dW_m = \tfrac{1}{2}\Sigma(Id\Phi + \Phi dI) \tag{5.77}$$

Now suppose that the locations of the loops, including their orientations, are specified by a number of co-ordinates, linear and/or angular, and that we are interested in calculating the force (or torque) F tending to increase a particular linear (or angular) co-ordinate x. Consider a displacement of the system in which the co-ordinate x increases to $x + dx$ and all other co-ordinates remain fixed. Then the work done is Fdx and this is equal to the decrement $-dW_m$ in the magnetic energy if no energy enters or leaves the system, for example, if the magnetic fluxes threading the circuits are kept constant (expr. 5.66). But when there is an increment $d\Phi_1$ in the magnetic flux threading the current I_1, a generator in this circuit has to supply energy $I_1 d\Phi_1$ in accordance with expr. 5.66. The same applies to the other circuits; consequently, there is introduced into the system an amount of magnetic energy $\Sigma Id\Phi$. This introduced energy must be allowed for in relating the work done Fdx to the decrement $-dW_m$ in the magnetic energy of the system. We therefore arrive at the equation

$$Fdx = -dW_m + \Sigma Id\Phi \tag{5.78}$$

By substituting for dW_m from eqn. 5.77 into eqn. 5.78, we obtain

$$Fdx = \tfrac{1}{2}\Sigma(Id\Phi - \Phi dI) \tag{5.79}$$

Now let us compare expr. 5.79 for the work done with expr. 5.77 for the increment in the magnetic energy of the system. By making the increments of magnetic flux vanish, we arrive at the result (c.f. eqn. 3.74)

$$Fdx = -dW_m \quad \text{(constant magnetic fluxes)} \tag{5.80}$$

On the other hand, by making the increments of electric current vanish in eqns. 5.77 and 5.79, we obtain (c.f. eqn. 3.75)

$$Fdx = +dW_m \quad \text{(constant electric currents)} \tag{5.81}$$

We thus see that, whereas the work done in a displacement in which the magnetic fluxes are kept constant is the decrement in the magnetic energy of the system, the work done in a displacement in which the electric currents are kept constant is the increment in the magnetic energy of the system. The difference of energy in the two cases, namely

$$2dW_m \tag{5.82}$$

is that supplied by the electric generators needed in the circuits to maintain the electric currents constant in the situation corresponding to eqn. 5.81.

Eqn. 5.80 may be written

$$F = -\frac{\partial W_m}{\partial x} \tag{5.83}$$

where the partial derivative means not only that all co-ordinates other than x are to be kept constant but also that all the magnetic fluxes threading the circuits are to be kept constant. Likewise, eqn. 5.81 may be written

$$F = +\frac{\partial W_m}{\partial x} \tag{5.84}$$

where the partial derivative means not only that all co-ordinates other than x are to be kept constant but also that all electric currents round the circuits are to be kept constant.

If the circuits are superconducting so that no resistance has to be overcome in maintaining the electric currents, no generators whatever are required in the circuits during displacements in which the threading magnetic fluxes are kept constant (expr. 5.66). The system is then isolated from any source or sink of energy, and eqns. 5.80 and 5.83 state that the work done when a co-ordinate x is allowed to increase to $x + dx$ is equal to the decrement in the energy of the system. However, for circuits that are not superconducting, electric generators are needed merely to overcome the resistance. A displacement of the system in which the threading magnetic fluxes remain constant requires changes in the electric currents round the circuits. If the circuits possess resistance, there is a corresponding change in the power delivered by the generators that are driving the electric currents against the resistance. However, this complication does not upset the application of eqns. 5.80 and 5.83.

In a displacement of the system in which the electric currents remain constant, the threading magnetic fluxes change, and generators in the circuits have to supply or withdraw energy in accordance with expr. 5.66. These can be the same generators used to maintain the constant electric currents against resistance. This makes application of eqns. 5.81 and 5.84 seem easier to understand than application of eqns. 5.80 and 5.83. Application of eqns. 5.80 and 5.83 is easy to understand if one thinks in terms of superconducting circuits, but it is a correct procedure even in the presence of resistance.

Let us consider the mechanical action between the primary and secondary circuits of a transformer in free space. For simplicity, let us suppose that the two circuits are coaxial circles whose planes are at a perpendicular distance x apart, and that we wish to calculate the force F tending to increase x. Let us elect to use eqn. 5.84 for which the electric currents round the primary and secondary circuits must be kept constant. With this in mind, we deliberately express the magnetic energy of the system in terms of the electric currents by using the first of the two expressions for W_m in eqns. 5.72, thereby arriving at eqn. 5.75. We then apply eqn. 5.84, obtaining

$$F = \frac{1}{2}\left(\frac{\partial L_{11}}{\partial x}I_1^2 + 2\frac{\partial L_{12}}{\partial x}I_1 I_2 + \frac{\partial L_{22}}{\partial x}I_2^2\right) \tag{5.85}$$

The self-inductances L_{11} and L_{22} of the primary and secondary circuits frequently depend only slightly on the separation x, and consequently eqn. 5.85 can often be reduced to

$$F = \frac{\partial L_{12}}{\partial x}I_1 I_2 \tag{5.86}$$

If, instead of a linear displacement, one of the circuits of the transformer were given a rotary displacement described by variation of an angle ψ, eqn. 5.86 would be replaced by

$$T = \frac{\partial L_{12}}{\partial \psi}I_1 I_2 \tag{5.87}$$

where T is the torque tending to increase the angle ψ.

5.11 The magnetic stress tensor in free space

Tubes of magnetic flux in free space are in tension and also exert a sideways pressure. This may be deduced from the previous section by arguments analogous to those used in Sections 4.5 and 4.6 to derive, respectively, the tension per unit cross-sectional area and the sideways pressure in tubes of electric flux in free space. Only the results will be stated.

The tension of a tube of magnetic flux per unit cross-sectional area at a location where the magnetic field strength is H and the magnetic flux density is B calculates to (c.f. eqns. 4.22)

$$T_m = \tfrac{1}{2}HB = \tfrac{1}{2}\mu_0 H^2 = \frac{1}{2\mu_0}B^2 \tag{5.88}$$

and the sideways pressure calculates to (c.f. eqns. 4.28)

$$p_m = \tfrac{1}{2}HB = \tfrac{1}{2}\mu_0 H^2 = \frac{1}{2\mu_0}B^2 \tag{5.89}$$

The action of these forces may be visualised by considering a toroidal inductor in free space carrying a tube of magnetic flux maintained by solenoidal electric current. If the material of which the inductor is made were completely flexible, the inductor would explode under the pressure described by eqns. 5.89. It is only the mechanical stress in an inductor that prevents it from blowing up. To visualise the tension in the tubes of magnetic flux per unit cross-sectional area described by eqns. 5.88, let us suppose that the toroid is made of material that will not strain, but that it is cut in half along an azimuthal plane as shown in Fig. 5.4. The tension in the tubes of magnetic flux then tends to hold the two halves together.

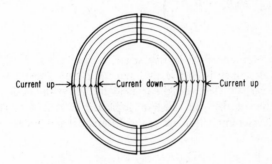

Current up→➤▲▲▲▲◄—Current down→➤˅˅˅˅◄—Current up

Fig. 5.4   *Illustrating how the two halves of a toroidal inductor that is bisected along an azimuthal plane tend to be held together by tension in the tubes of magnetic flux (c.f. the pull of a magnet)*

The tension per unit cross-sectional area in eqns. 5.88, and the sideways pressure in eqns. 5.89, may be described by means of the magnetic stress tensor (c.f. expr. 4.32)

$$\begin{pmatrix} \tfrac{1}{2}HB & 0 & 0 \\ 0 & -\tfrac{1}{2}HB & 0 \\ 0 & 0 & -\tfrac{1}{2}HB \end{pmatrix} \tag{5.90}$$

provided that the cartesian co-ordinate system is oriented so that the x axis is directed along the tube of magnetic flux at the location under consideration. Moreover, the tensor (5.90) may be expressed as the sum of two tensors (c.f. expr. 4.33 and 4.34):

$$\begin{pmatrix} HB & 0 & 0 \\ 0 & 0 & 0 \\ 0 & 0 & 0 \end{pmatrix} \tag{5.91}$$

plus

$$\begin{pmatrix} -\frac{1}{2}HB & 0 & 0 \\ 0 & -\frac{1}{2}HB & 0 \\ 0 & 0 & -\frac{1}{2}HB \end{pmatrix} \tag{5.92}$$

of which the second describes a uniform pressure $-\frac{1}{2}HB$ in all directions, both parallel to the tubes of magnetic flux as well as perpendicular to them. Referred to a cartesian co-ordinate system for which the x axis is not necessarily along a tube of magnetic flux, the pressure tensor (5.92) is unchanged, but the tensor (5.91) becomes (c.f. expr. 4.35)

$$\begin{pmatrix} H_xB_x & H_yB_x & H_zB_x \\ H_xB_y & H_yB_y & H_zB_y \\ H_xB_z & H_yB_z & H_zB_z \end{pmatrix} \tag{5.93}$$

In an electromagnetic field in free space, account must be taken of both the magnetic stress described in this section and the electric stress described in Section 4.7. In a time-varying electromagnetic field the linear dimensions of all volume elements must be less than the size-limit indicated in Table 4.1. This is ensured by the limiting processes implicit in use of the differential and integral calculus.

Problems

5.1   From the fact that $ev \times B$ gives the force exerted on an electron of charge e moving in free space with velocity v in a static magnetic field for which the flux density vector is B, deduce that the force exerted on a steady electric current I flowing round a closed circuit C in free space is

$$F = I \oint ds \times B$$

where B is the external magnetic-flux-density vector at a point of the circuit where a vector element of length in the direction of the current is ds.

5.2   In the previous problem, the circuit C is now subjected to a displacement in which a vector element of length ds suffers a small vector displacement ds' and the current I is kept constant. Show that the work done by the forces on the circuit during the displacement is

$$I \oint B \cdot (ds' \times ds)$$

and verify that this work may be written

$$Id\Phi$$

where $d\Phi$ is the increment in the external magnetic flux Φ threading the circuit in the direction related by the right-hand-screw rule to the direction of I.

5.3   In the previous problem, the circuit C is now a small loop carrying a steady current I, so that it has a magnetic moment $m = IS$, where S is the vector area of the loop. The external magnetic field has a flux density vector B at the location of the loop. This field, although static, is not necessarily uniform, so that B is a function of the cartesian co-ordinates (x, y, z) in space. Show that the work done by the forces on the loop in any linear or angular displacement in which the current is kept constant is $d(m \cdot B)$. Deduce that the torque T tending to increase the angle θ measured from the direction of B to the direction of m is $-mB \sin\theta$ and that this torque may be expressed vectorially as

$$T = m \times B$$

5.4   By considering in the previous problem a linear rather than an angular displacement, show that the force F exerted on the loop by the field has cartesian components

$$\begin{cases} F_x = m_x\dfrac{\partial B_x}{\partial x} + m_y\dfrac{\partial B_y}{\partial x} + m_z\dfrac{\partial B_z}{\partial x} \\[2mm] F_y = m_x\dfrac{\partial B_x}{\partial y} + m_y\dfrac{\partial B_y}{\partial y} + m_z\dfrac{\partial B_z}{\partial y} \\[2mm] F_z = m_x\dfrac{\partial B_x}{\partial z} + m_y\dfrac{\partial B_y}{\partial z} + m_z\dfrac{\partial B_z}{\partial z} \end{cases}$$

Use Maxwell's equations to show that this result may be written

$$F = m_x\frac{\partial B}{\partial x} + m_y\frac{\partial B}{\partial y} + m_z\frac{\partial B}{\partial z}$$

5.5   A magnetostatic field is created by a system of coils carrying prescribed electric currents. At a point in the field where the magentic-flux-density vector is B, there exists a small loop of wire carrying a prescribed electric current I. The vector area of the loop is S, pointing in the direction related by the right-hand-screw rule to the direction of the current I. Initially S is perpendicular to B, and the total energy of the system is then W_0. The loop is now turned so that S is not perpendicular to B, and all electric currents are kept constant in the process. Use expr. 5.66 to show that the new energy of the system is

$$W_m = IS \cdot B + W_0$$

5.6   In the previous problem, a generator is needed in the circuit to maintain the current round the loop constant while the loop is being turned. Use the results of Problem 5.2 to show that the energy supplied by the generator is

$$2IS \cdot B$$

5.7   By introducing an angular co-ordinate θ and linear co-ordinates (x, y, z), and by differentiating the magnetic energy W_m in Problem 5.5 with respect to these co-ordinates, keeping the electric currents constant, verify the expressions for torque T and force F derived in Problems 5.3 and 5.4.

5.8   In free space there exist two small loops of wire of vector areas S_1 and S_2 carrying fixed currents I_1 and I_2, respectively. The directions of the vector areas are related by the right-hand-screw rule to the directions of the corresponding currents. The position vector of the loop of area S_2 with respect to the loop of area S_1 is r, and r is large compared with the linear dimensions of S_1 and S_2. The three vectors S_1, S_2, and r are coplanar. The angle from the direction of S_1 to the direction of r is θ_1, and the angle from the direction of r to the direction of S_2 is θ_2. Prove that the magnetic energy of the system is

$$\frac{\mu_0 S_1 S_2}{4\pi r^3} I_1 I_2 (2 \cos \theta_1 \cos \theta_2 + \sin \theta_1 \sin \theta_2) + W_0$$

where W_0 is a constant as long as I_1 and I_2 are constant.

5.9   Use the expression for energy derived in the previous problem to calculate (i) the force between the two loops tending to increase r, (ii) the torque on the loop of area S_1 tending to increase θ_1, and (iii) the torque on the loop of area S_2 tending to increase θ_2. Explain why these two torques are in general unequal. By equating the torque on the complete system to zero, calculate the force acting on either loop in a direction transverse to their join. Show that the transverse forces vanish if the axes of the loops are parallel or antiparallel.

Chapter 6

Magnetic energy and magnetic stress in magnetic materials

6.1 Introduction

In Sections 4.2–4.7 we discussed energy and stress in an electric field in free space. In Sections 4.8–4.11 we considered what modifications are required in the presence of dielectric material. In the previous chapter we discussed in Sections 5.5–5.11 how energy and stress are distributed throughout a magnetic field in free space, and we now need to enquire what modifications are required in the presence of magnetic material.

6.2 Faraday's law of induction in the presence of material

Faraday's law of induction for a closed curve at rest relative to the observer is stated in eqn. 1.37. Its differential form is the electric curl equation appearing at the top of Table 2.1. Validity of the law is totally independent of what material may be present. Regardless of whether the material is homogeneous or non-homogeneous, isotropic or non-isotropic, linear or non-linear, Faraday's law of induction is applicable.

In Section 5.6 we considered an electric circuit small enough to have the same time-varying current $I(t)$ at all points (see Table 4.1), and we supposed that a generator having time-varying voltage $V(t)$ acts around the circuit as shown in Fig. 5.2. Neglecting losses due to resistance, we applied Faraday's law of induction in eqn. 5.58 to deduce the rate at which work is being done by the generator and found it to be (eqn. 5.60)

$$I \frac{d\Phi}{dt} \tag{6.1}$$

From this we deduced in Section 5.7 that the work that must be done to increase by a small amount $d\Phi$ the magnetic flux threading a circuit carrying a right-hand related electric current I is

$$Id\Phi \qquad\qquad\qquad\qquad\qquad\qquad\qquad (6.2)$$

Every aspect of this argument applies in the presence of material of any kind. The corresponding statement in electrostatics is that the work that must be done to increase by a small amount dQ the electric charge on a conductor at potential V above ground is

$$VdQ \qquad\qquad\qquad\qquad\qquad\qquad\qquad (6.3)$$

and this statement, too, is true regardless of what material may be present.

In spite of the wide validity of the statements associated with exprs. 6.2 and 6.3, care is necessary in their application. For the time being, let us rule out the presence of non-linear materials (see Sections 4.11 and 6.11). In Chapters 3 and 4 we used expr. 6.3 to calculate the increase in the electric energy of a system. We did this even in the presence of linear dielectric material. We discussed the energy involved in polarising the dielectric but, as mentioned in Section 4.8, we took no account of energy stored in the atoms of the dielectric in the unpolarised state. We assumed that none of the quantum mechanical energy associated with the assemblage of atomic particles that constitutes the unpolarised dielectric is made available to the electric field during the process of polarisation. For dielectric material this assumption is satisfactory, and it makes it possible to assume that the increase in the energy of a system in charging it is simply the work done by the generators needed to transfer the charge. Work done calculated from expr. 6.3 is identical with increase in the electric energy of the system.

However, we must not jump to the conclusion that the same is true for expr. 6.2 if magnetic material is present. The magnetisation of magnetic material is quantum-mechanically more complicated than the polarisation of dielectric material.

In so-called diamagnetic materials, the electronic structures of atoms acquire a rotation (the Larmor precession) in the presence of an applied magnetic field. In this phenomenon, which is quite weak, none of the quantum-mechanical energy associated with the assemblage of atomic particles in the unmagnetised state is made available to the magnetic field during magnetisation. In these circumstances the work done by generators connected in the electric circuits creating the magnetisation is identical with increased motional energy of the system, provided that the effect of resistance is disregarded. For diamagnetic materials, therefore, we may calculate work done from expr. 6.2 and treat it as increased motional energy of the system.

But the situation is quite different for paramagnetic materials (also a weak phenomenon) and for ferromagnetic materials (magnets). In these materials the magnetic moment is caused by the quantum-mechanical phenomenon of electron spin. During the process of magnetisation of a magnet, energy is made available to the magnetic field from the quantum-mechanical reservoir of energy associated with the assemblage of atoms in the unmagnetised state. Consequently, the magnetic energy acquired by a magnet during magnetisation exceeds (indeed greatly exceeds) the energy supplied by the generators connected in the electric circuits creating the magnetisation. What expr. 6.3 gives is the energy supplied by the generators. It does not include the energy simultaneously made available quantum-mechanically. For a magnet, it is necessary to distinguish between the energy supplied electro-magnetically and the energy supplied quantum-mechanically.

6.3   An electromagnet

The effect of magnetic material may conveniently be illustrated, both theoretically and experimentally, with the aid of a toroidal inductor at rest relative to the observer. Such an inductor consists of a thin isolated circular tube of magnetic flux which is maintained in existence by means of electric current flowing solenoidally over its surface, uniformly distributed round the perimeter. If the interior of the inductor is free space, a solenoidal current H per unit length round the inductor gives a magnetic flux density B within the inductor, where

$$B = \mu_0 H \qquad\qquad (6.4)$$

If H is the magnetic vector, then the magnetic-flux-density vector is

$$B = \mu_0 H \qquad\qquad (6.5)$$

and the surface density of current flowing solenoidally round the tube is $n \times H$ where n is a unit normal to the surface directed into the tube (eqns. 1.36).

If, however, the inductor is filled with homogeneous magnetic material as indicated in Fig. 6.1, the material develops a magnetic moment per unit volume M. Electric currents in the atoms of which the material is constructed result in a small cylindrical piece with axis parallel to the direction of M being equivalent to an inductor carrying in free space an electric current per unit length M, right-hand related to the direction of M. This electric current bound in the material contributes to the magnetic flux round the toroidal inductor as effectively as does the free electric current in the winding. This can be demonstrated experimentally by providing a narrow gap in the magnetic material as indicated in Fig. 6.1. The magnetic flux Φ round the inductor, and the corresponding magnetic flux density B, may then be measured in the gap and compared with the associated free electric current per unit length applied by the winding. The relation between the two is dependent on the magnetic properties of the material within the inductor.

If the magnetic moment per unit volume M in the material is in the same direction as H, the additional bound solenoidal current in the material is equivalent to a current per unit length M flowing solenoidally over its curved surface. This current does not have to be provided by winding, which therefore only has to provide a current per unit length

$$H' = H - M \qquad\qquad (6.6)$$

If M is not in the same direction as H, the additional bound surface electric current per unit length due to the magnetic moment per unit volume is $n \times M$ where n is a unit normal to the surface of the material drawn into the field (eqns. 1.77). The surface density of free electric current that has to be provided by the winding is then $n \times H'$ where

$$H' = H - M \tag{6.7}$$

This is the modified magnetic vector used in version 2 of the electromagnetic equations in materials in order to transfer the complication associated with bound magnetic moment per unit volume from the Maxwell magnetic curl equation to the magnetic connection relation (eqns. 2.14).

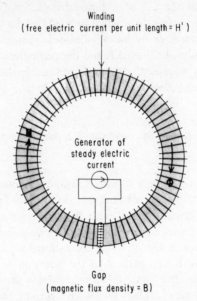

Winding
(free electric current per unit length = H')

Generator of
steady electric
current

Gap
(magnetic flux density = B)

Fig. 6.1   *Illustrating production of a magnetic flux density B in a gap in a magnetic circuit energised by a toroidal winding possessing a free electric current per unit length equal to H'*

For diamagnetic materials, the bound magnetic moment per unit volume M points closely in the direction opposite to H so that, to produce the same magnetic flux density in the toroidal inductor, more electric current per unit length has to be provided by the winding than if the material were not present. The amount of this additional electric current is, however, very small, and diamagnetic materials are often treated as non-magnetic.

For paramagnetic materials, the bound magnetic moment per unit volume M points in closely the same direction as H so that, to produce the same magnetic flux density in the toroidal inductor, less electric current per unit length has to be provided by the winding than if the magnetic material were not present. However, the reduction in the electric current to be provided by the winding is often small, and many paramagnetic materials are frequently treated as though they were non-magnetic.

The principal exceptions are the ferromagnetic solids such as iron, cobalt and nickel. With a ferromagnetic material interior to the inductor in Fig. 6.1, application of a free solenoidal current per unit length via the winding can produce a bound solenoidal current per unit length on the material a thousand times bigger. Most of the magnetic flux threading the inductor is then created by the magnetic moment per unit volume in the magnetic material. Nevertheless, the large bound solenoidal current per unit length on the material can be controlled by the small free current per unit length in the winding. With ferromagnetic material in the toroidal inductor, an enormous magnetic flux is obtained round the inductor for a comparatively small free electric current per unit length applied by the winding. This behaviour is essential to the operation of many electrical machines.

6.4   Work done on a linear isotropic electromagnet by a generator in the winding

Consider an electromagnet of the type illustrated in Fig. 6.1. Ohmic losses may be calculated as described in Sections 3.15 and 4.12, and in this chapter we neglect them. For simplicity we also assume for the time being that the magnetic material behaves in a linear isotropic manner. The magnetic moment per unit volume M developed under the influence of a magnetic-flux-density vector B is then (eqns. 2.24)

$$M = \chi \frac{1}{\mu_0} B \tag{6.8}$$

where χ is the magnetic susceptibility of the material referred to the B vector. The modified magnetic vector H' used in version 2 of the electromagnetic equations is then (eqns. 2.25)

$$H' = \frac{1}{\mu} B \tag{6.9}$$

where (eqns. 2.26)

$$\frac{\mu}{\mu_0} = \frac{1}{1 - \chi} \tag{6.10}$$

This is the permeability of the material.

Let us suppose that the inductor in Fig. 6.1 is filled with linear isotropic magnetic material of permeability μ/μ_0. Let the electric current in the winding be zero initially, and let us calculate how much work is done by the generator as the current is slowly increased to a steady value I. At the stage when the electric current is a fraction x of its final value I, the magnetic flux threading the inductor is also a fraction x of its final value Φ. At this stage the electric current is xI, and the right-hand-related magnetic flux is $x\Phi$. Now let us increase the threading magnetic flux by a small amount $dx\Phi$. According to expr. 6.2, the work done is

$$(xI)(dx\Phi)$$

or

$$I\Phi x dx$$

Integrating with respect to x from zero to unity, we obtain for the total work done by the generator in Fig. 6.1 during the process of magnetisation of the magnetic material

$$W'_m = \frac{1}{2}I\Phi \tag{6.11}$$

If the inductance of the inductor is L, we may write

$$W'_m = \frac{1}{2}I\Phi = \frac{1}{2}LI^2 = \frac{1}{2L}\Phi^2 \tag{6.12}$$

Eqns. 6.12 are to be compared with eqns. 5.64 calculated in the absence of magnetic material. The argument leading to eqn. 5.68 is the same as the argument leading to eqn. 6.11. But in free space we were able to argue that the work done by the generator is the increase in the magnetic energy of the system. In the presence of magnetic material, however, we must not jump to this conclusion. All that is certain at this stage is that eqns. 6.12 give the work done by the generator in Fig. 6.1 during the process of establishing the magnetic field in the inductor.

If desired, eqns. 6.12 may be expressed per unit volume of the inductor. Each elementary tube of magnetic flux of normal cross-sectional area dS may be dissected into elements of length ds to form elements of volume $dsdS$. If each thin tube existed by itself, the free electric current per unit length round it would be H', given by eqn. 6.6. For the element of volume, we have a free electric current equal to $H'ds$ and a threading magnetic flux equal to BdS. Using these values in eqns. 6.12, we deduce that the work done by the generator connected in the winding, if expressed per unit volume of the inductor, is

$$w'_m = \frac{1}{2}H'B = \frac{1}{2}\mu H'^2 = \frac{1}{2\mu}B^2 \tag{6.13}$$

Eqns. 6.13 may be compared with eqns. 5.55 applicable in free space. But, in the presence of the magnetic material, we must not jump to the conclusion that eqns. 6.13 give the magnetic energy stored per unit volume, because additional energy may be supplied from the quantum-mechanical reservoir of energy associated with the assemblage of atoms in the unmagnetised state. What eqns. 6.13 give, per unit volume, is the work done via the winding during the process of establishing the magnetic field in the inductor.

6.5   A model of diamagnetic material

Diamagnetic behaviour of materials is a weak phenomenon arising from the fact that the electronic charge structure of an atom possessing no permanent magnetic dipole moment nevertheless develops rotation in the presence of an applied magnetic field. The speed of rotation, or precession, is proportional to the applied field, and the resulting small magnetic moment is in the direction reverse to that of the applied field.

A proper discussion of diamagnetism depends on the quantum mechanics of the atom. However, a crude model of the electronic structure of an atom for this purpose is a pair of coincident rings, each of radius a and mass m, each carrying a charge e, and each rotating on its axis in opposite directions with angular velocity ω_0. To this structure let us apply a magnetic field of flux density B parallel to the axis. If we allow the magnetic field to increase from zero then, as time progresses, the angular velocity of a ring is changed from ω_0 to ω. Since each ring has area πa^2, the threading magnetic flux increases at the rate

$$\frac{\partial B}{\partial t}\,\pi a^2 \tag{6.14}$$

Consequently, an electric field strength E acts round the perimeter of a ring, and Faraday's law of induction gives (eqn. 1.37)

$$E\,2\pi a \;=\; -\frac{\partial B}{\partial t}\,\pi a^2 \tag{6.15}$$

where the directions of E and B are related by the right-hand-screw rule. The ring rotating right-handed about the magnetic field is therefore subjected to a tangential retarding force

$$eE \;=\; -\tfrac{1}{2}ea\,\frac{\partial B}{\partial t} \tag{6.16}$$

In accordance with Newton's law of motion, this is equal to the time rate of change of the angular momentum $ma\omega$ of the ring. Consequently,

$$\frac{\partial}{\partial t}\,(ma\omega) \;=\; -\tfrac{1}{2}ea\,\frac{\partial B}{\partial t} \tag{6.17}$$

By integrating this equation with respect to time we see that, when the magnetic field has increased to B, the angular velocity has decreased to

$$\omega \;=\; \omega_0 - \frac{1}{2}\frac{eB}{m} \tag{6.18}$$

For the ring rotating left-handed about the magnetic field, the angular velocity has changed from $-\omega_0$ to

$$\omega \;=\; -\omega_0 - \frac{1}{2}\frac{eB}{m} \tag{6.19}$$

Hence the complete structure is now rotating left-handed about the magnetic field with angular velocity

$$\frac{1}{2}\frac{eB}{m} \tag{6.20}$$

This is the Larmor precession. With e numerically negative, the precession is actually right-handed about the magnetic field.

A ring of radius a and charge e rotating on its axis with angular velocity ω constitutes a circular electric current $e\omega/(2\pi)$ with loop area πa^2. Hence the pair of rings with angular velocities given in eqns. 6.18 and 6.19 constitute an electric current

$$\frac{e}{2\pi}\left\{\left(\omega_0 - \frac{1}{2}\frac{eB}{m}\right) + \left(-\omega_0 - \frac{1}{2}\frac{eB}{m}\right)\right\}$$

or

$$-\frac{e^2}{2\pi m}B \tag{6.21}$$

Since the loop area is πa^2, the resulting magnetic moment is

$$-\pi a^2 \frac{e^2}{2\pi m}B \tag{6.22}$$

Because this magnetic moment is in the direction of the magnetic field we may write it vectorially as

$$-\frac{e^2 a^2}{2m}B \tag{6.23}$$

With N atoms per unit volume, the resulting magnetic moment per unit volume is

$$M = -\frac{Ne^2 a^2}{2m}B \tag{6.24}$$

It follows from eqn. 6.8 that the magnetic susceptibility of diamagnetic material is

$$\chi = -\frac{\mu_0 Ne^2 a^2}{2m} = -\frac{Ne^2}{\epsilon_0 m}\frac{a^2}{2c^2} \tag{6.25}$$

where $c = (\mu_0 \epsilon_0)^{-1/2}$ is the velocity of light. The permeability of diamagnetic material is then given by eqn. 6.10. Since χ is numerically small for diamagnetic materials, we obtain

$$\frac{\mu}{\mu_0} \doteq 1 + \chi = 1 - \frac{Ne^2}{\epsilon_0 m}\frac{a^2}{2c^2} \tag{6.26}$$

From eqns. 6.18 and 6.19, the kinetic energy of the pair of rings is

$$\frac{1}{2}ma^2\left\{\left(\omega_0 - \frac{1}{2}\frac{eB}{m}\right)^2 + \left(-\omega_0 - \frac{1}{2}\frac{eB}{m}\right)^2\right\}$$

or

$$ma^2\left\{\omega_0^2 + \left(\frac{1}{2}\frac{eB}{m}\right)^2\right\} \tag{6.27}$$

The extra kinetic energy per atom as a result of applying the magnetic field is therefore

$$ma^2 \left(\frac{1}{2}\frac{eB}{m}\right)^2 \tag{6.28}$$

Consequently, the extra kinetic energy per unit volume of the material is

$$Nma^2 \left(\frac{1}{2}\frac{eB}{m}\right)^2 \tag{6.29}$$

or (eqn. 6.24)

$$-\tfrac{1}{2}MB \tag{6.30}$$

or (eqn. 6.25)

$$-\frac{\chi}{2\mu_0}B^2 \tag{6.31}$$

Because, for diamagnetic material, χ is negative (eqn. 6.25) and M is in the direction opposite to B (eqn. 6.24), exprs. 6.30 and 6.31 are positive, being equal to expr. 6.29. Each of exprs. 6.29, 6.30 and 6.31 gives the extra kinetic energy per unit volume stored in diamagnetic material as a result of the Larmor precession created by the applied magnetic field. Less crude atomic models also lead to eqn. 6.24 and to expr. 6.29 but with a more precise meaning for the dimension a when e and m are taken as the charge and mass of an electron.

Because each atomic loop of area πa^2 carries an electric current given by expr. 6.21, it stores magnetic energy. To obtain the total change in motional energy stored in an atom as a result of Larmor precession, both the magnetic contribution and the kinetic contribution must be taken into account. Yet only the kinetic contribution is included in expr. 6.27 and, therefore, in expr. 6.29. Why?

In eqn. 6.15, let us write Faraday's law of induction in terms of the magnetic vector potential A instead of the magnetic-flux-density vector B (eqns. 1.56). Both the electric field of a ring and its magnetic vector potential are symmetrical round the axis of rotation, and we take the directions of E and A to be the same. It then follows from eqn. 1.57 that $(E + \partial A/\partial t)2\pi a$ vanishes, so that

$$E = -\frac{\partial A}{\partial t} \tag{6.32}$$

Hence

$$eE = -\frac{\partial p_m}{\partial t} \tag{6.33}$$

where

$$p_m = eA \tag{6.34}$$

The quantity p_m is the peripheral electromagnetic momentum of the ring (eqns. 5.20). Moreover, if p_k is the mechanical momentum of the ring, Newton's law of motion gives

$$\frac{\partial p_k}{\partial t} = eE \qquad (6.35)$$

It then follows from eqns. 6.33 and 6.35 that

$$\frac{\partial p_k}{\partial t} = -\frac{\partial p_m}{\partial t} \qquad (6.36)$$

This equation conveys the same information as eqn. 6.17. Integration of eqn. 6.36 gives

$$p_k + p_m = \text{constant} \qquad (6.37)$$

showing that the ring acquires peripheral momentum associated with its mass at the expense of peripheral momentum associated with its charge. Likewise, the kinetic energy W_k of Larmor precession is only acquired at the expense of its magnetic energy W_m:

$$W_k + W_m = \text{constant} \qquad (6.38)$$

Larmor precession does not change the total motional energy associated with the atoms.

We may therefore describe the effect of Larmor precession in diamagnetic material as follows. Let us suppose that the material occupies the interior of a toroidal inductor as shown in Fig. 6.1. When a specified electric current flows in the winding, the rotating electronic structures create magnetic flux. Moreover, this magnetic flux is directed so as to reduce the magnetic flux density in the inductor somewhat (eqn. 6.24). Consequently, to produce a specified magnetic flux density B in the inductor, a slightly higher winding-current is required in the presence of the diamagnetic material than would be needed in its absence. The associated increase in the motional energy stored in the inductor is not, however, used for increased magnetic energy; it is converted into the kinetic energy of Larmor precession in accordance with eqn. 6.38. There is no magnetic energy per unit volume stored in the inductor other than

$$\frac{1}{2\mu_0} B^2 \qquad (6.39)$$

where B is the magnetic flux density due to all the electric currents flowing in free space – the current in the winding and the currents associated with Larmor precession (see Problem 6.11). The upshot is that the total motional energy per unit volume stored in the inductor is sum of the magnetic energy per unit volume given by expr. 6.39 and the kinetic energy per unit volume given by expr. 6.29.

The magnetic energy per unit volume given by expr. 6.39 may be written (eqn. 6.4)

$$\tfrac{1}{2} HB \qquad (6.40)$$

The kinetic energy per unit volume created by Larmor precession is (expr. 6.30)

$$-\tfrac{1}{2}MB \tag{6.41}$$

which is in fact positive (eqns. 6.24 and 6.29). Addition of exprs. 6.40 and 6.41 gives

$$\tfrac{1}{2}(H - M)B \tag{6.42}$$

or (eqn. 6.6)

$$\tfrac{1}{2}H'B \tag{6.43}$$

Hence the total motional energy per unit volume created by application of a magnetic field to diamagnetic material is

$$w_M = \frac{1}{2}H'B = \frac{1}{2}\mu H'^2 = \frac{1}{2\mu}B^2 \tag{6.44}$$

where μ/μ_0 is the permeability of the material given by eqn. 6.26.

If the diamagnetic material occupies the interior of the toroidal inductor illustrated in Fig. 6.1, then H' is the free electric current per unit length applied by the winding (see expr. 6.6), and B is the magnetic flux density created in the inductor. In terms of these quantities, eqns. 6.44 give the motional energy stored in the inductor per unit volume. Comparison with eqns. 5.55 shows that the effect of the Larmor precession in diamagnetic materials is taken into account by employing μ in place of μ_0 and by employing the modified magnetic vector H' belonging to version 2 of the electromagnetic equations for materials in place of the magnetic vector H belonging to version 1 (see Table 2.2).

Comparison of eqns. 6.44 with eqns. 6.13 shows that, for diamagnetic material,

$$w_M = w'_m \tag{6.45}$$

This relation states that the motional energy stored per unit volume in the diamagnetic material in Fig. 6.1 is equal to the work done per unit volume by the generator in the winding during the process of creating the magnetic field. Eqn. 6.45 does not state that the work done by the generator is equal to the magnetic energy stored in the inductor. This is because kinetic energy is involved in the Larmor precession. Expr. 6.39 shows that there is, in fact, no additional magnetic energy stored in the inductor as a result of presence of the diamagnetic material. The additional motional energy is kinetic. Per unit volume, it is given by expr. 6.29.

We thus see that, with diamagnetic material filling the inductor in Fig. 6.1, the work done by the generator connected in the winding during magnetisation is identical with increase in the motional energy of the system, but the part of motional energy associated with the material is kinetic rather than magnetic.

Numerically, the increase in the energy of diamagnetic material associated with the kinetic energy of Larmor precession is only about one part in a million. The susceptibility in eqn. 6.25 evaluates to

$$\chi \sim -10^{-6} \tag{6.46}$$

Consequently, for many practical purposes, diamagnetic materials may be treated as non-magnetic.

It may be mentioned that, in a magnetoplasma (an ionised gas in the presence of a steady imposed magnetic field), the free electrons in thermic motion spiral about the imposed magnetic field and give the gas a diamagnetic behaviour (see Problem 6.12). However, this free magnetic moment per unit volume, like the bound magnetic moment per unit volume in other diamagnetic materials, is numerically unimportant in most situations encountered in practice.

6.6   Electron spin

Let us now turn to magnetic materials in which the magnetic moment per unit volume is associated with electron spin. An electron (charge e, mass m) has an angular momentum per unit mass that is identical with its magnetic moment per unit charge. Consequently, if the angular momentum is represented by a vector l pointing along the spin axis in the direction related to the spin by the right-hand-screw rule, then the magnetic moment of the electron is

$$(e/m)l \tag{6.47}$$

The numerical magnitude of this magnetic moment is

$$9{\cdot}28 \times 10^{-24} \, \mathrm{A\,m^2} \tag{6.48}$$

Because of their magnetic moments, a pair of adjacent electrons interact magnetically. But they have a far stronger interaction between their spins. Spin-interaction is a quantum mechanical phenomenon that has no natural counterpart in classical mechanics. The spin of an electron is controlled quantum-mechanically. The magnetic moment is controlled by the spin in accordance with expr. 6.47.

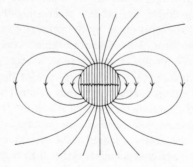

Fig. 6.2   *Field of magnetic flux of a uniformly rotating hollow sphere having electric charge distributed uniformly over its surface*

Suppose that an electron, with its spin, were pictured as a hollow rotating sphere of charge, so that its field of magnetic flux density is as shown in Fig. 6.2. This is a motion qualitatively similar to the Larmor precession discussion in the previous section. If the spinning charged sphere were subjected to a magnetic field in the direction of the spin vector and if this magnetic field were increasing with time, Faraday's law of induction would bring forces into play that would modify the rate of spin (c.f. eqns. 6.17 and 6.18). No such modification takes place with electron spin because the angular momentum is controlled quantum-mechanically. One could say that the quantum-mechanical forces influencing the electron spin constitute a control-motor that counteracts the effect of Faraday's law of induction so far as electron spin is concerned.

However, it is quantum mechanically inappropriate to discuss a single electron in isolation. A piece of material involves a large number of electrons most of which are grouped in atoms or molecules. In many situations each atom or molecule can almost be treated as an isolated quantum-mechanical system, but for ferromagnetic materials even this assumption is unsatisfactory. According to quantum mechanics, the electrons in most atoms and molecules are grouped in pairs, with the spins in each pair constrained to point in opposite directions so that their magnetic moments practically cancel. However, in ferromagnetic material, the electron spins do not completely cancel in pairs; there is roughly one uncancelled electron spin per atom. Moreover, because the interactions between adjacent spins are predominantly quantum-mechanical, there is strong alignment of the spins in a whole domain of the ferromagnetic material, and therefore strong alignment of the associated magnetic moments in accordance with expr. 6.47. Such domains have linear dimensions of the order of 10^{-4} m–10^{-3} m. Even an unmagnetised piece of iron possesses such domains, but the directions of magnetisation in the numerous small domains are more-or-less randomly oriented. The piece of iron is said to be magnetised if the directions of magnetisation of the domains are not more-or-less randomly oriented. It becomes a strong magnet if the directions of magnetisation of the domains are almost the same everywhere.

Magnetisation of materials associated with electron spin therefore involves a situation in which the material develops a coherent angular momentum per unit volume L as a result a non-random orientation of the net electron spins of the various atoms. In accordance with expr. 6.47, the material then possesses a magnetic moment per unit volume given by

$$M = (e/m)L \tag{6.49}$$

6.7   A model of magnetic material whose properties depend on electron spin

For simplicity, let us assume initially that the material behaves in a linear isotropic manner, so that eqns. 6.8, 6.9 and 6.10 apply. For materials whose magnetic properties depend on electron spin, the magnetic susceptibility χ is positive. For paramagnetic materials

$$\chi \sim 10^{-4} \tag{6.50}$$

but for ferromagnetic materials χ is not much less than unity insofar as these materials may be considered to behave linearly:

$$1 - \chi \sim 10^{-3} - 10^{-4} \tag{6.51}$$

It follows from eqn. 6.10 that the permeability of paramagnetic material is only slightly greater than 1, but that of ferromagnetic material is of the order of $10^3 - 10^4$.

Consider the arrangement shown in Fig. 6.1, and let the toroidal inductor be filled with homogeneous linear isotropic material whose magnetic properties depend on electron spin. Consider a thin tube of magnetic flux having normal cross-sectional area dS, and let us select a short section of length ds. Let the electrons in this section have a net angular momentum per unit volume L due to spin, and consequently an associated magnetic moment per unit volume M given by eqn. 6.49. Then the small cylindrical piece of material may be regarded as a well guarded solenoidal inductor carrying, in free space, an electric current

$$Mds$$

round its curved surface, and having a threading magnetic flux

$$BdS$$

The stored motional energy in the cylindrical piece of material is therefore given by (eqns. 5.55)

$$\tfrac{1}{2}(Mds)(BdS)$$

or, measured per unit volume of the material,

$$\tfrac{1}{2}MB \tag{6.52}$$

From eqns. 5.55, the magnetic energy per unit volume of the magnetic field applied to the material, excluding any energy associated with the material itself, is

$$\tfrac{1}{2}HB \tag{6.53}$$

Adding exprs. 6.52 and 6.53 we see that, not counting any motional energy that may be associated with the mass of an electron, the total motional energy per unit volume stored in the toroidal inductor of Fig. 6.1 is

$$w_m = \tfrac{1}{2}(H + M)B \tag{6.54}$$

The additional term $\tfrac{1}{2}MB$ in eqn. 6.54 is positive for magnetic moment per unit volume arising from electron spin. It connotes the fact that, in the toroidal inductor depicted in Fig. 6.1, there is more organised motional energy per unit volume with the magnetic material present than there would be if it were absent. Nevertheless, to create this situation, less work has to be done by the generator connected in the winding. From the first of eqns. 6.13, the work done per unit volume by the generator connected in the winding is

$$w'_m = \tfrac{1}{2}H'B \tag{6.55}$$

or, using eqn. 6.6,

$$w'_m = \tfrac{1}{2}(H - M)B \tag{6.56}$$

As a result of presence of the magnetic material, the energy per unit volume provided by the free electric current in the winding is reduced in accordance with eqn. 6.56 whereas the organised motional energy developed in the toroidal inductor is increased in accordance with eqn. 6.54. More organised motional energy is produced than is supplied electromagnetically.

The extra energy per unit volume is

$$w_m - w_m' = MB \tag{6.57}$$

In terms of the net angular momentum L per unit volume appearing in eqn. 6.49, the additional energy per unit volume may be written

$$w_m - w_m' = L\omega_M \tag{6.58}$$

where ω_M is the angular gyrofrequency associated with the magnetic field, namely

$$\omega_M = eB/m \tag{6.59}$$

The expression $w_m - w_m'$ in eqn. 6.58 represents organised spin energy per unit volume acquired by the magnetic material but not provided electromagnetically via the winding. Moreover, the extra energy is non-trivial. For typical ferromagnetic material only about a millijoule of energy has to be supplied via the winding in order to have a joule of organised spin energy in the ferromagnetic material. Some 99·9% of the organised motional energy in ferromagnetic material is supplied from another source. This source is the quantum mechanical reservoir of energy associated with the assemblage of atomic particles that constitutes the ferromagnetic material.

An electromagnet is a power-assisted device. Energy supplied electromagnetically via the winding and not dissipated as heat is used primarily to control how much organised spin energy is supplied quantum-mechanically. Moreover, if energy is withdrawn electromagnetically via the winding, the organised spin energy that was supplied quantum-mechanically is returned to the quantum-mechanical reservoir of energy associated with the assemblage of atomic particles. All that is returned electromagnetically via the winding is what was originally supplied via the winding, less thermic losses.

In ferromagnetic material, electron spin makes a major contribution to the magnetic flux created. Moreover, the energy density $w_m - w_m'$ in eqn. 6.58 is of great importance for the quantum-mechanical energy budget of the material. Nevertheless, the only energy density that enters into the electromagnetic energy budget of the system is w_m' given by eqn. 6.55. This may be written, with the aid of eqn. 6.9, in any of the equivalent forms

$$w_m' = \frac{1}{2}H'B = \frac{1}{2}\mu H'^2 = \frac{1}{2\mu}B^2. \tag{6.60}$$

6.8 Definition of magnetic energy

Eqns. 6.60, relevant to linear isotropic material whose magnetic properties arise from electron spin, are to be compared with eqns. 6.44, relevant to linear isotropic material whose magnetic properties arise from Larmor precession. We elect to write both sets of equations as

$$w_m = \frac{1}{2}H'B = \frac{1}{2}\mu H'^2 = \frac{1}{2\mu}B^2 \qquad (6.61)$$

and to define w_m as the magnetic energy of the material per unit volume. In so doing, the following points must be borne in mind:

(i) For diamagnetic material, w_m is in reality the sum of the magnetic energy per unit volume that would exist in the absence of the material and the kinetic energy per unit volume associated with Larmor precession.

(ii) For paramagnetic and ferromagnetic materials, w_m excludes the magnetic energy that is made available from the quantum-mechanical reservoir of energy associated with the assemblage of atomic particles and that has to be returned to that reservoir when the material is demagnetised.

In both cases, w_m is energy per unit volume that is made available electromagnetically to the material from generators connected in electric circuits. In both cases, w_m is energy per unit volume that, on demagnetisation, will be returned electromagnetically via the electric circuits, less thermic losses.

The magnetic energy of material, as we have defined it, includes the kinetic energy of bound electrons arising from the Larmor precession created by any applied magnetic field, including a time-varying applied magnetic field. But it does not include the kinetic energy of vibration of bound electrons associated with a time-varying applied electric field. Motion of the latter type is connected with behaviour of the material as a dielectric. At radio frequencies and below, the kinetic energy of vibration of bound electrons in a dielectric is usually unimportant. But if it is to be included in the total motional energy, this must be done additionally.

The magnetic energy of a system such as that illustrated in Fig. 6.1 includes the motional energy that is associated with movement of electric charge round the electric circuit or circuits. But it does not include the kinetic energy associated with movement of mass round the circuit or circuits. If the motional energy of an electric current associated with the mass of the free charges is to be taken into account, this must be done additionally (see Section 5.3). For practical electric circuits, kinetic energy of the electrons is in fact important. However, kinetic energy is quite important for the dynamics of a conducting fluid in motion in the presence of an electromagnetic field, including the dynamics of an ionised gas, or plasma.

If the magnetic material is non-isotropic, the magnetic moment per unit volume is not in the same direction as the applied magnetic field, and eqn. 6.6 has to be replaced by eqn. 6.7. The consequence is that, for linear non-isotropic material, the first of eqns. 6.13 and the first of eqns. 6.60 become

$$w'_m = \tfrac{1}{2}H' \cdot B \qquad (6.62)$$

with the result that the definition of magnetic energy per unit volume in the first of eqns. 6.61 becomes (c.f. eqn. 4.49)

$$w_m = \tfrac{1}{2}H' \cdot B \tag{6.63}$$

However, if we wish to express the magnetic energy density w_m entirely in terms of the components of H' or entirely in terms of the components of B, we must remember that the relation between these vector components now takes the form (c.f. eqns. 4.55)

$$\left.\begin{aligned}
B_x &= \mu_{xx}H'_x + \mu_{xy}H'_y + \mu_{xz}H'_z \\
B_y &= \mu_{yx}H'_x + \mu_{yy}H'_y + \mu_{yz}H'_z \\
B_z &= \mu_{zx}H'_x + \mu_{zy}H'_y + \mu_{zz}H'_z
\end{aligned}\right\} \tag{6.64}$$

where (c.f. eqns. 4.58; see Problems 6.9 and 6.10)

$$\mu_{xy} = \mu_{yx}, \quad \mu_{yz} = \mu_{zy}, \quad \mu_{zx} = \mu_{xz} \tag{6.65}$$

If we write eqn. 6.63 in the form

$$w_m = \tfrac{1}{2}(H'_x B_x + H'_y B_y + H'_z B_z) \tag{6.66}$$

we may substitute for (B_x, B_y, B_z) from eqns. 6.64 and obtain (c.f. eqns. 4.59)

$$\begin{aligned}
w_m = \tfrac{1}{2}(&\mu_{xx}H'^2_x + \mu_{yy}H'^2_y + \mu_{zz}H'^2_z \\
&+ 2\mu_{yz}H'_y H'_z + 2\mu_{zx}H'_z H'_x + 2\mu_{xy}H'_x H'_y) \tag{6.67}
\end{aligned}$$

Alternatively, the energy density w_m in eqn. 6.66 may be written entirely in terms of (B_x, B_y, B_z) by using as coefficients the elements in the inverse of the tensor.

With the definition of magnetic energy that we have adopted, the magnetic energy per unit volume stored in magnetic material is given by eqn. 6.63 regardless of whether the magnetic properties of the material arise from electron spin or from precession of bound electrons, and regardless of whether the material is isotropic or non-isotropic, homogeneous or non-homogeneous, so long as it behaves linearly. The total magnetic energy in the system, exclusive of any energy loaned from the quantum-mechanical reservoir of the system and repayable to that reservoir, is

$$W_m = \int w_m \, d\tau \tag{6.68}$$

where the magnetic energy density w_m is given by eqn. 6.63, $d\tau$ is an element of volume, and the integral is taken throughout the field.

6.9 Inductors in the presence of linear magnetic material

The discussion of a system of inductors in Sections 5.8, 5.9 and 5.10 applies when the surrounding region is free space, but it is not restricted to that situation. The argument only depends on the linearity of eqns. 5.70 relating the magnetic fluxes to the electric currents. Taking the electric currents $I_1, I_2, \ldots$ to be the free currents round the circuits, the results apply no matter what distribution of magnetic material may exist in the surrounding region so long as the material behaves linearly. If the material is non-isotropic or non-homogeneous it may be difficult to calculate the coefficients of self and mutual inductance, but when once they are calculated (or measured), all equations in Sections 5.8, 5.9 and 5.10 are applicable so long as non-linear material is excluded.

In particular, the magnetic energy stored in an inductor of inductance L whose winding carries a free electric current I causing a right-handed threading magnetic flux Φ is

$$W_m = \frac{1}{2}I\Phi = \frac{1}{2}LI^2 = \frac{1}{2L}\Phi^2 \tag{6.69}$$

These equations apply no matter what magnetic material may be present so long as it behaves linearly and so long as magnetic energy is defined as described in the previous section.

6.10   Magnetic stress in the presence of linear magnetic materials

We are now in a position to discuss magnetic stress in linear magnetic materials. The argument follows the same lines as those used for the magnetic field in free space (Section 5.11) and the same lines as those used for the electric field in free space (Section 4.7) and for the electric field in linear dielectric material (Section 4.9). Only the results will be stated.

Let us first consider linear isotropic material. Tubes of magnetic flux are in tension, and the tension per unit cross-sectional area is (c.f. eqns. 4.60 and 5.88)

$$T_m = \frac{1}{2}H'B = \frac{1}{2}\mu H'^2 = \frac{1}{2\mu}B^2 \tag{6.70}$$

Tubes of magnetic flux also exert a sideways pressure given by (c.f. eqns. 4.61 and 5.89)

$$p_m = \frac{1}{2}H'B = \frac{1}{2}\mu H'^2 = \frac{1}{2\mu}B^2 \tag{6.71}$$

If the x axis is directed along a tube of magnetic flux at a particular location in the material, the magnetic stress tensor is (c.f. exprs. 4.62 and 4.63; 5.91 and 5.92)

$$\begin{pmatrix} H'B & 0 & 0 \\ 0 & 0 & 0 \\ 0 & 0 & 0 \end{pmatrix} \tag{6.72}$$

plus

$$\begin{pmatrix} -\frac{1}{2}H'B & 0 & 0 \\ 0 & -\frac{1}{2}H'B & 0 \\ 0 & 0 & -\frac{1}{2}H'B \end{pmatrix} \tag{6.73}$$

If the x axis does not have this special orientation, the tensor (6.72) becomes (c.f. the tensors 4.64 and 5.93)

$$\begin{pmatrix} H'_x B_x & H'_y B_x & H'_z B_x \\ H'_x B_y & H'_y B_y & H'_z B_y \\ H'_x B_z & H'_y B_z & H'_z B_z \end{pmatrix} \tag{6.74}$$

For linear non-isotropic magnetic material, the tensor (6.74) still applies, but the pressure tensor (6.73) must be written (c.f. the tensor 4.65)

$$\begin{pmatrix} -\tfrac{1}{2} H' \cdot B & 0 & 0 \\ & -\tfrac{1}{2} H' \cdot B & 0 \\ 0 & 0 & -\tfrac{1}{2} H' \cdot B \end{pmatrix} \tag{6.75}$$

so that the total magnetic stress tensor is then the sum of the tensors (6.74) and (6.75). For linear non-isotropic magnetic material, the relation between B and H' is that given by eqns. 6.64.

Because the definition of magnetic energy adopted in Section 6.8 omits the organised electronic spin-energy of quantum-mechanical origin, it follows that the associated quantum-mechanical actions within and between atoms are also omitted from the magnetic stress tensor. Discussion of these interactions is appropriate to a text on quantum mechanics. The same applies to the interactions encountered in classical mechanics. Stress in magnetic materials due to geometrical strain is omitted; it should be studied in a test dealing with the strength of materials.

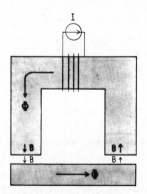

Fig. 6.3.   *Illustrating attraction of an armature by an electromagnet*

The fact that mechanical stress is not included in the magnetic stress tensor may be illustrated with the aid of Fig. 6.3, which depicts the lifting of an armature by an electromagnet. Let us suppose that the magnetic material is linear and isotropic (permeability μ/μ_0) while the surrounding air may be regarded as non-magnetic. As a result of a steady electric current I round the winding, let a steady magnetic flux Φ pass through the electromagnet and the armature. We suppose that, except where it crosses the gaps, the magnetic flux is confined to the magnetic material. If S is the area of a pole-face, the magnetic flux density in a gap is (fringing neglected)

$$B = \Phi/S \tag{6.76}$$

and this same magnetic flux density exists in the material of the electromagnet adjacent to either pole-face. The tension per unit cross-sectional area of the tubes where they cross the gaps is (eqns. 5.88)

$$\frac{1}{2\mu_0} B^2 \tag{6.77}$$

and this provides the force that lifts the armature. But the tension of the tubes per unit cross-sectional area in the electromagnet is (eqns. 6.70)

$$\frac{1}{2\mu} B^2 \tag{6.78}$$

If the permeability of the magnetic material is 10^3, the tension of the tubes of magnetic flux in the material is only 0·1% of what it is in the gaps. The remaining 99·9% of the tension is taken up by the mechanical stress in the material of which the electromagnet is constructed.

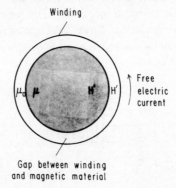

Fig. 6.4.   *Illustrating a normal cross-section of an inductor nearly filled with magnetic material of permeability μ/μ_0*

Again, let us consider a toroidal inductor such as that illustrated in Fig. 6.1. Let the inductor be nearly filled with linear isotropic magnetic material of permeability μ/μ_0, with a gap between the material and the winding as illustrated cross-sectionally in Fig. 6.4. Outside the magnetic material we suppose that the permeability is unity. Let the free electric current applied to each unit length of the inductor by means of the winding be H', so that the magnetic flux density in the gap between the winding and the magnetic material is

$$\mu_0 H' \tag{6.79}$$

Since there is no free electric current on the surface of the magnetic material, there is no discontinuity of H' in passing from the gap into the material (eqns. 2.18). Hence the magnetic flux density in the magnetic material is

$$\mu H' \tag{6.80}$$

If the permeability is 10^3, the magnetic flux density is a thousand times bigger in the material than it is in the gap because of the effect of electron spin. The pressure exerted on the winding in Fig. 6.4 by the magnetic field in the free space constituting the gap is (eqns. 5.89)

$$\tfrac{1}{2}\mu_0 H'^2 \tag{6.81}$$

For the winding to be in equilibrium this pressure has to be offset; it is balanced in practice by mechanical stress in the winding or in its support. But the magnetic pressure in the magnetic material is (eqns. 6.71)

$$\tfrac{1}{2}\mu H'^2 \tag{6.82}$$

and, if the permeability is 10^3, the pressure given by expr. 6.82 is a thousand times bigger than the pressure in the gap (expr. 6.81). This difference of pressure is balanced by non-magnetic stress in the magnetic material. In practice, this stress would cause distortion of the material and this we have neglected. Elastic deformation of magnetic material by a magnetic field is what is known as magnetostriction.

The magnetic stress tensor formed by summing the tensors (6.74) and (6.75) implies that that magnetic stress is exerted across any surface drawn in a magnetic field. However, the existence of this stress can only be verified experimentally in certain circumstances. For example, a static magnetic field may be dissected along a surface composed of lines of magnetic flux. The magnetic field on one side of the dissection surface may be abolished and that on the other side may be supported by electric current on the dissection surface. The pressure exerted by the magnetic field on this electric current may then be studied, as was done for the winding in Fig. 6.4. For a surface not composed of lines of magnetic flux, however, dissection would intercept tubes of magnetic flux, and this would require a distribution of magnetic charge over the dissection surface as well as a distribution of electric current (Section 1.14). However, even though perfect dissection of a magnetic field by means of a surface perpendicular to the field is not possible, nevertheless the concept of tension in tubes of magnetic flux may be quite well demonstrated experimentally as described in connection with Fig. 6.3.

6.11 Non-linear ferromagnetic material

Except for small changes in the applied magnetic field, ferromagnetic material in fact behaves non-linearly. This may be demonstrated experimentally by means of a toroidal inductor filled with ferromagnetic material in the manner illustrated in Fig. 6.1. With a narrow gap in the material as indicated in the diagram, the magnetic flux density B may be measured where the tube of magnetic flux crosses the gap. This may then be compared with the free electric current per unit length H' that must be applied by means of the winding in order to create the magnetic flux.

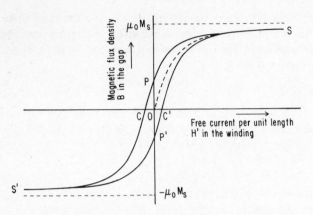

Fig. 6.5.   *Illustrating magnetisation curves for ferromagnetic material*

For simplicity, let us suppose initially that the ferromagnetic material, although non-linear, is isotropic. The direction of the magnetic moment per unit volume M is then the same as that of magnetic flux density vector B. The magnetic flux density B in the toroidal inductor of Fig. 6.1 varies with the free electric current per unit length H' applied via the winding in a manner such as that indicated in Fig. 6.5. The origin 0 in this diagram corresponds to a situation in which there is (i) no free electric current round the winding, (ii) no magnetic moment per unit volume in the material, and (iii) no magnetic flux threading the inductor. If, starting from this condition, the free current per unit length H' round the winding is increased, the material becomes magnetised, and the magnetic flux density in the inductor increases in a manner such as that indicated by the broken curve in Fig. 6.5. The magnetic flux density quickly becomes far greater than the value $\mu_0 H'$ directly created by the free electric current in the winding. This is because the magnetic moments of the various small domains of the material cease to be more-or-less randomly oriented and become somewhat aligned in the direction of the magnetic field applied by the winding. It is these partially aligned magnetic domains that produce most of the magnetic flux round the inductor even though it is the free electric current in the winding that controls the extent to which they do so.

When all the electron spins are aligned in the same direction, the material has its saturation magnetic moment per unit volume M_s, and the magnetic flux density created by the spins is $\mu_0 M_s$. With increase in the free electric current per unit length H' on the winding, the magnetic flux density saturates as shown by the broken curve in Fig. 6.5. After the spin contribution to the magnetic flux density saturates at the value $\mu_0 M_s$, the magnetic flux density continues to increase slowly because of the small contribution $\mu_0 H'$ produced directly by the free electric current in the winding. Hence the asymptotes in Fig. 6.5 that look horizontal in fact have a slight upward slope.

If, after the electric current in the winding has been increased sufficiently to saturate the magnetic moment per unit volume, the winding-current is now decreased again, then the magnetic flux density in the inductor follows a curve such as SP in Fig. 6.5. When the electric current in the winding has been reduced to zero, there is still magnetic flux round the inductor with a density corresponding to the point P. This magnetic flux is now produced entirely by electron spin in the material which, in these circumstances, is described as a 'permanent' magnet.

If the current in the winding now becomes negative, the value of the magnetic flux density round the inductor follows a curve such as PCS' in Fig. 6.5. At the point C there is no magnetic flux round the inductor because that created by electron spin in the material is balanced by that created directly by the free electric current in the winding.

If the winding-current is made sufficiently negative, the magnetic moment per unit volume in the material saturates in the negative direction. If we now pro-ceed to reduce the negative electric current in the winding, the value of the mag-netic flux density in the inductor follows a curve such as S'P', and at P' the material is again a 'permanent' magnet. If we now make the winding-current positive, the magnetic flux density follows a curve such as P'C'S until saturation in the positive direction again occurs. Cycling the winding-current back and forth between a large positive value and a large negative value causes the point in Fig. 6.5 corresponding to the value of the magnetic flux density in the inductor to cycle round the loop SPCS'P'C'S, known as a hysteresis loop.

Theoretical discussion of the precise shape of the magnetisation curves of ferro-magnetic material involves the quantum mechanics of electron spin. Nevertheless, curves such as those shown in Fig. 6.5 are easily measured with an arrangement such as that shown in Fig. 6.1. The magnetisation curves for many ferromagnetic materials were known long before the quantum-mechanical concept of electron spin was invented. Magnetisation curves have been used for design of electromag-nets since the early years of electrical machinery.

Let us calculate how much work is done by the generator connected into the winding of Fig. 6.1 in the process of creating the magnetic flux density round the inductor, assuming that the magnetisation behaviour for the ferromagnetic material filling the inductor is known (c.f. Fig. 6.5). If the winding is pictured as providing continuously distributed free electric current per unit length H' over the entire perimeter l of the toroidal inductor, the total free electric current flowing solen-oidally round the curved surface is

$$I = H'l \tag{6.83}$$

If the cross-sectional area of the toroid is S, the magnetic flux threading the inductor is

$$\Phi = BS \tag{6.84}$$

where B is deduced from H' in accordance with the appropriate magnetisation curve. If B is changed to $B + dB$, the corresponding increment in the threading magnetic flux is

$$d\Phi = S dB \tag{6.85}$$

In accordance with expr. 6.2, the work done by the generator connected into the winding (disregarding any losses due to ohmic resistance of the winding) is

$$Id\Phi = (H'dB)(lS) \tag{6.86}$$

or, measured per unit volume of the ferromagnetic material,

$$H'dB \tag{6.87}$$

The work done on the ferromagnetic material via the winding in changing B from B_1 to B_2 is therefore, per unit volume,

$$\int_{B_1}^{B_2} H' dB \tag{6.88}$$

Evaluation of the integral (6.88) has to be performed using the appropriate magnetisation curve relating H' to B. Moreover, a different curve is in general involved in changing the magnetic flux density from B_2 to B_1 from that involved in changing it from B_1 to B_2. In particular, the work done on the ferromagnetic material per unit volume if the free electric current in the winding is cycled back and forth between a large positive value and a large negative value is given, per cycle, by the area of the hysteris loop SPCS'P'C'S in Fig. 6.5. This work is not used to create organised spin-energy in the ferromagnetic material. The latter is loaned quantum-mechanically from the reservoir of energy associated with the aggregate of atomic particles that constitutes the material and is repayable to that reservoir. The work represented by the area of a hysteresis loop is work done in rearranging the small domains of uniform electron spin in the material in the process of cycling the state of magnetisation. It is work that is dissipated in the material, and constitutes what is known as hysteresis loss.

If the ferromagnetic material is non-isotropic as well as non-linear, the direction of magnetisation in Fig. 6.1 is at an angle to the lines of magnetic flux, and the same is true for the direction of the magnetic vector H'. The appropriate component of H' therefore has to be used in calculating expr. 6.86. In consequence, $H'dB$ in exprs. 6.86, 6.87 and 6.88 is replaced by $H' \cdot dB$. The work done on the non-isotropic non-linear ferromagnetic material in the inductor via the winding in the process of changing B from B_1 to B_2 is therefore, per unit volume,

$$\int_{B_1}^{B_2} H' \cdot dB \tag{6.89}$$

For linear isotropic behaviour of ferromagnetic material, expr. 6.89 reduces to expr. 6.88 and may be integrated without difficulty. In Fig. 6.1 let the material in the inductor be initially unmagnetized and subject to no free electric current round the winding. Let us use only small winding-currents so that only locations close to the origin in Fig. 6.5 are involved. Furthermore, let us take the magnetisation curve near the origin to be given by the linear relation

$$B = \mu H' \tag{6.90}$$

so that no hysteresis loop is involved and no hysteresis loss occurs. We are then dealing with linear isotropic magnetic material of permeability μ/μ_0. Substitution from eqn. 6.90 into expr. 6.88 then permits evaluation of the integral and leads to eqns. 6.61 for the magnetic energy stored, per unit volume, in linear isotropic ferromagnetic material.

6.12   The concept of magnetopotential energy

Magnetic charge does not exist: magnetic fields are produced by electric charge in motion. This is true even in the case of a permanent magnet, although the moving charge then takes the form of quantum-mechanical electron spin. Nevertheless, magnetic charge is a useful concept and often provides the simplest way of performing a particular calculation. For example, a thin solenoid conveying magnetic flux Φ from a point P to a point Q in free space produces an external magnetic field that is most simply calculated as though there were a source of magnetic flux Φ at Q and a sink $- \Phi$ at P, that is, as though there were a positive magnetic charge at Q and an equal and opposite magnetic charge at P (Fig. 1.9). It is, of course, essential to remember that this procedure gives only the external magnetic field of the solenoid. There is a strong internal magnetic field for which the magnetic flux density is obtained by dividing the threading magnetic flux Φ by the small cross-sectional area of the solenoid.

Let us suppose that such a thin solenoid conveying magnetic flux Φ from P to Q in free space is placed in an external magnetic field. This field exerts forces on the moving charges the constitute the electric current round the solenoid. So also does the magnetic field of the solenoid itself. The latter tends to distort the solenoid because of the magnetic stress in the field, particularly because of the sideways pressure of tubes of magnetic flux (c.f. Fig. 6.4). Let us suppose that such distortion is completely prevented by mechanical stress in the winding, so that the winding may be treated as a structure that is incapable of geometrical distortion. The external magnetic field still exerts forces on the charge moving in the winding. These forces tend to rotate the rigid solenoid and, unless the external field is uniform, they also tend to give the rigid solenoid a motion of translation towards locations of stronger magnetic field (Problems 5.1–5.5, 6.4–6.6). The forces exerted by the external magnetic field on the charges moving in the winding have a resultant that may be described as a combination of a force and a torque. The resultant force and the resultant torque can be correctly calculated if one imagines that there is (i) a magnetic charge of strength $- \Phi$ at

P subject to a force per unit charge equal to the external magnetic vector H_P at P, and (ii) a magnetic charge of strength $+ \Phi$ at Q subject to a force per unit charge equal to the external magnetic vector H_Q at Q (Problems 1.2 and 1.3). These two forces at P and Q give the same resultant force and torque as the actual forces applied to the moving charges in the winding. That the forces are actually applied to the moving charges in the winding and not to magnetic charges at P and Q could be made evident if the winding of the solenoid were capable of geometrical distortion. But if the solenoid is rigid, the total mechanical action may be correctly calculated from the magnetic charges, and moreover this is the easiest way to arrive at the resultant mechanical action.

It is appropriate to ask what use there may be for the concept of magnetic charge in discussing the energy of an electromagnetic system. Energy is used to calculate mechanical action, and the mechanical action on a rigid magnetic dipole in a magnetic field in free space is correctly calculated if it is regarded as a pair of separated magnetic charges. The mechanical action on magnetised matter is therefore correctly calculated if it is regarded as an aggregate of rigid magnetic dipoles in free space. Moreover, the negative poles may be regarded as a distribution of negative magnetic charge and the positive poles as a slightly displaced distribution of positive magnetic charge. The mechanical action on a distribution of magnetic moment per unit volume in space may therefore be correctly calculated using methods analogous to those employed in Sections 4.8–4.10 for a distribution of electric moment per unit volume. However, if this were done, the additional energy due to presence of the magnetic material would be presented as potential energy. We could call this magnetopotential energy. Moreover, if we were also to call the energy associated with the movement of electric charge round electric currents magnetokinetic energy, we could then say that magnetic energy is the sum of magnetokinetic energy and magnetopotential energy.

Although this procedure results in correct calculations of mechanical action on magnetic materials due to electromagnetic fields, it is important not to forget that magnetic charge does not exist and that what has just been called magnetopotential energy is not in reality potential energy at all. Consider diamagnetic material. In diamagnetic material the magnetic moments of the atoms are turned towards the direction opposite to the applied magnetic field, and this gives each atom increased magnetopotential energy. But what the atom really possesses is increased motional energy as a result of the Larmor precession of its electronic structure (Section 6.5). Again, consider ferromagnetic material. Suppose that the inductor illustrated in Fig. 6.1 is filled with saturated ferromagnetic material. The atoms have magnetic moments that are associated with electron spin. These magnetic moments are pointing in the direction of the applied magnetic field, and this gives each atom reduced magnetopotential energy. What has really happened, however, is that reduced motional energy has been supplied via the winding in accordance with eqns. 6.54 and 6.56. This is because part of the motional energy (eqn. 6.57), although controlled by the winding-current, actually came from the quantum-mechanical reservoir of energy associated with the atomic structure of the material and will have to be repaid to that reservoir when the material is demagnetised.

Nevertheless, the concept of magnetopotential energy is often convenient when dealing with permanent magnets. If the pole-strengths may be treated as constant when the magnets are moved relative to each other, it is convenient to apply eqn. 5.80 when calculating mechanical action. If the energy W_m in this equation is then regarded as potential energy, no practical difficulty ensues. Most problems about the mechanical action in a system of permanently magnetised needles are conveniently handled in terms of the mutual magnetopotential energy of a system of magnetic point charges of prescribed magnitudes, using the counterpart of the electrostatic methods developed in Chapters 3 and 4. However, arguments based on magnetic poles, while quite useful, never tell the whole truth. So far as energy is concerned, arguments based on magnetic poles correctly calculate the magnetic energy associated with bound magnetic moment per unit volume in magnetic material, but they incorrectly present it as potential energy when in reality it is motional energy.

Problems

6.1   A magnetostatic field in free space is created by a system of fixed permanent magnets having prescribed pole-strengths. At a point in the field where the magnetic vector is H and the magnetic-flux-density vector is B, there exists a small thin straight solenoid of length l and radius a ($a \ll l$). The solenoid has n closely wound turns of wire, and its terminals are short-circuited. The solenoid is cooled to near the absolute zero of temperature, so that the wire is superconducting and its resistance may be neglected. Because of the superconducting behaviour, no battery is needed in the circuit to prevent current from decaying. With the axis of the solenoid at right angles to the magnetic vector H, an electric current is started in the solenoid so as to produce a magnetic flux Φ threading the solenoid. The vector length of the solenoid in the direction of the threading magnetic flux is l. In this condition the magnetic energy of the entire system is W_0. The solenoid is now slowly turned so that l is no longer perpendicular to H. Explain why, as the solenoid turns, the electric current round it adjusts so as to maintain the threading magnetic flux at the initial value Φ. Use the result in Problem 1.3 to deduce that the magnetic energy of the system is now

$$W_m = -\Phi l \cdot H + W_0$$

and may be written

$$W_m = -m \cdot B + W_0$$

where m is the magnetic dipole moment $\Phi l / \mu_0$ of the solenoid.

6.2   A long thin non-conducting circular cylinder in free space has length l and radius a ($a \ll l$). The cylinder carries a total charge Q uniformly distributed over its surface, and the mass of the cylinder is negligible. The cylinder can spin on its axis without friction. It is mounted in a uniform static external magnetic field of field strength H with the spin axis perpendicular to the lines of magnetic flux. The cylinder is given an initial angular velocity of spin ω_0, and it is then slowly turned so that the axial direction that is related by the right-hand-screw rule to the direction of spin makes an angle θ with the magnetic vector of the external field. Explain why, as the spin axis turns, the rate of spin adjusts so as to preserve the value of the magnetic flux threading the cylinder. Calculate the angular velocity of spin as a function of θ, and show that it can reverse if $\omega_0 < 2\pi H l/Q$.

6.3   In the previous problem, the constant magnetic flux threading the cylinder is denoted by Φ. Show that, when the axis of the thin spinning cylinder is in the direction θ, the electric current round the cylinder is approximately

$$\frac{\Phi - \pi a^2 \mu H \cos\theta}{\mu_0 \pi a^2/l}$$

and the energy of rotation is approximately

$$\frac{1}{2}\frac{\Phi^2}{\mu_0 \pi a^2/l} - \Phi l H \cos\theta + \tfrac{1}{2}\mu_0 (H\cos\theta)^2 \pi a^2 l$$

Identify the first term in this expression as the energy of rotation when $\theta = \tfrac{1}{2}\pi$. Identify the last term as energy that was stored in the volume now occupied by the cylinder even when the cylinder was in the position $\theta = \tfrac{1}{2}\pi$. Deduce that the reduction in the energy of the system caused by turning it from position $\tfrac{1}{2}\pi$ to position θ is $\Phi l H \cos\theta$, and hence verify the result obtained in Problem 6.1.

6.4   A magnetostatic field in free space is created by a system of fixed permanent magnets that have prescribed pole-strengths. At a point in the field where the magnetic vector is H and the magnetic-flux-density vector is B, there exists a small magnetic dipole having prescribed pole-strengths $\pm\Phi$ at a prescribed distance l apart. The vector length l points from the negative pole to the positive pole, and $\Phi l/\mu_0$ is the vector magnetic dipole moment m. Initially l is perpendicular to H, and the total energy of the system is W_0. The dipole is now turned so that l is no longer perpendicular to H, and all pole-strengths are kept constant in the process. Treating magnetic poles as though they were magnetic charges, calculate the new magnetic energy of the system, and show that it has the value calculated in Problem 6.1.

6.5   In the previous problem, the length l of the magnetic dipole makes an angle θ with the magnetic vector H. By expressing the energy $-\Phi l \cdot H + W_0$ in terms of θ and differentiating with respect to θ, show that the torque T tending to increase θ is $-\Phi l H \sin\theta$ and that it may be expressed vectorially either as

$$T = \Phi l \times H$$

or as

$$T = m \times B$$

6.6   In Problem 6.4, the dipole is held so that there is a fixed angle between the vectors l and H, but the dipole is capable of displacement in the direction of a linear co-ordinate x. In this direction the rate of increase of the magnetic vector is $\partial H/\partial x$. Use the expression $-\Phi l \cdot H + W_0$ for magnetic energy to deduce that the force tending to move the dipole in the direction x increasing is $\Phi l \cdot \partial H/\partial x$ or $m \cdot \partial B/\partial x$.

6.7   In free space there exist two small magnetic dipoles whose pole strengths have prescribed values Φ_1 and Φ_2. For the first dipole the position vector of the positive pole with respect to the negative pole is l_1, and for the second dipole it is l_2. The lengths l_1 and l_2 have prescribed values, and the energy necessary to construct the dipoles when remotely separated from each other is W_0. The position vector of the dipole of length l_2 with respect to the dipole of length l_1 ir r, and r is large compared with l_1 and l_2. The three vectors l_1, l_2, and r are coplanar. The angle from the direction of l_1 to the direction of r is θ_1, and the angle from the direction of r to the direction of l_2 is θ_2. Show that the magnetic energy of the magnetic dipoles is

$$- \frac{l_1 l_2}{4\pi\mu_0 r^3} \Phi_1 \Phi_2 \left(2 \cos\theta_1 \cos\theta_2 + \sin\theta_1 \sin\theta_2\right) + W_0$$

6.8   Use the expression for energy derived in the previous problem to calculate the mechanical action between the dipoles, and compare the results with those obtained for a pair of small loops of current in Problem 5.9. If the magnetic dipoles are ferromagnetic needles, explain why the method used in Problems 5.8 and 5.9 is closer to the physics of the phenomena than the method used in this problem and the previous problem.

6.9   There exists in space a distribution of magnetic material that is linear but not necessarily either isotropic or homogeneous. Using version 2 of the electromagnetic equations in materials (Table 2.1), a comparison is made between two situations. In the first the material is subject to a magnetostatic field (H_1', B_1) and in the second to a magnetostatic field (H_2', B_2). By considering a situation in which the material is subject to a magnetostatic field $[(1-x)H_1' + xH_2', (1-x)B_1 + xB_2]$ and allowing x to go from 0 to 1, show that the work done per unit volume in converting field 1 into field 2 is (c.f. Problem 4.1).

$$H_1' \cdot (B_2 - B_1) + \tfrac{1}{2}(H_2' - H_1') \cdot (B_2 - B_1)$$

By equating this to the increase in the energy per unit volume, deduce that

$$H_1' \cdot B_2 = H_2' \cdot B_1$$

Verify the truth of this relation for isotropic material of permeability μ/μ_0.

6.10   In the previous problem, the linear material is non-isotropic and has a permeability tensor μ. Express the result $H_1' \cdot B_2 = H_2' \cdot B_1$ in terms of the elements in the permeability tensor and the cartesian components of H_1' and H_2'. By taking $H_1' = (1, 0, 0)$ and $H_2' = (0, 1, 0)$, deduce that $\mu_{xy} = \mu_{yx}$. Hence establish the truth of all of eqns. 6.65.

6.11   A long straight solenoidal inductor is filled with homogeneous diamagnetic material of the type described in Section 6.5. Show that production of a magnetic flux density B in the inductor requires a current per unit length round the curved surface equal to

$$\frac{B}{\mu_0} \left(1 + \mu_0 \frac{Ne^2a^2}{2m}\right)$$

Deduce that the magnetic energy that is stored within the inductor per unit volume in association with the magnetic field applied to the atoms is

$$\frac{B^2}{2\mu_0} \left(1 + \mu_0 \frac{Ne^2a^2}{2m}\right)$$

Use eqn. 5.77 with $d\Phi = 0$ to show that the electric current involved in Larmor rotation of the electronic structure of an atom (expr. 6.21) reduces the stored magnetic energy by

$$\frac{1}{2} \left(\frac{e^2}{2\pi m} B\right) (\pi a^2 B)$$

Deduce that the total magnetic energy stored within the inductor per unit volume in the presence of the diamagnetic material is $B^2/(2\mu_0)$. Hence verify that the total motional energy per unit volume within the inductor in the presence of the diamagnetic material exceeds $B^2/(2\mu_0)$ by the kinetic energy per unit volume derived in expr. 6.29.

6.12   A magnetoplasma has N free electrons per unit volume (charge e, mass m), and is subject to a steady magnetic field of flux density B. The electrons are in thermic motion at temperature T, so that the rms velocity perpendicular to the magnetic field is v where

$$\tfrac{1}{2}mv^2 \;=\; KT$$

and K is Boltzmann's constant $(1 \cdot 380 \times 10^{-23}$ J/deg). Show that the radius of a typical electronic orbit round the magnetic field is

$$(2KT/m)^{1/2}(eB/m)^{-1}$$

and that the electric current round the orbit is

$$e(eB/m)/(2\pi)$$

Deduce the magnetic moment of the orbit, and show that the free magnetic moment per unit volume is

$$M \;=\; -(NKT/B^2)B$$

Verify that a magnetoplasma is diamagnetic, and derive the magnetic susceptibility.

Chapter 7

Flow of energy in an electromagnetic field

7.1 Introduction

In Chapters 3 and 4 consideration was given to the storage of potential energy in an electric field. In Chapters 5 and 6, consideration was given to the storage of motional energy in a magnetic field. We now consider the interaction between electric and magnetic energy in systems where both forms of energy are involved. Such systems may be compared with mechanical systems in which both potential and kinetic energy are stored.

7.2 A simple electromagnetic oscillatory system

In mechanics a simple form of oscillatory system is a pendulum in which energy oscillates between the potential form when the bob has its maximum height and the kinetic form when the bob has its maximum velocity. Another well known mechanical oscillatory system is the balance-wheel and hair-spring of a clock. Here energy oscillates between the rotatory kinetic energy of the wheel and the potential energy of stress in the spring. In an oscillatory electromagnetic system motional energy associated with mass is replaced by motional energy associated with charge and takes the form of magnetic energy stored in an inductor as discussed in connection with eqns. 6.69. Potential energy, on the other hand, takes the form of electric energy stored in a capacitor as discussed in connection with eqns. 3.84. A simple electromagnetic oscillatory system therefore takes the form of an inductor connected to a capacitor in such a way that energy can oscillate back and forth between the two.

An example of an oscillatory electromagnetic system is shown in Fig. 7.1. An inductor in the form of a hollow metal cylinder of radius a and length l in free space is connected to a parallel-plate capacitor having plates of length l, width d and separation b. When electric current flows round the inductor it charges the capacitor; when the capacitor discharges, it causes electric current round the inductor. For quasi-static behaviour of the electric field in the capacitor and of the magnetic field in the inductor the size-limit summarised in Table 4.1 needs to be satisfied. At the lower frequencies, multiturn inductors and multiplate capacitors are normally used to prevent the systems from being inconveniently large. Such systems are drawn schematically as shown in Fig. 7.2 and often have a physical structure not radically different from the schematic circuit diagram. Such systems constitute resonant oscillatory circuits.

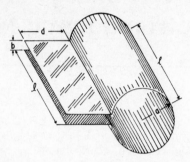

Fig. 7.1   *Simple oscillatory electromagnetic system consisting of an inductor connected in parallel with a capacitor*

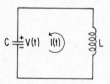

Fig. 7.2   *Schematic diagram of an oscillatory circuit*

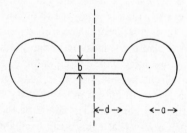

Fig. 7.3   *Axial cross-section of an electromagnetic oscillatory system consisting of an inductor in the form of a metal toroid [radius a, perimeter 2 (d + a)] connected across a parallel-plate capacitor with circular plates (radius d, separation b)*

Let us consider an arrangement that is of type shown in Fig. 7.1 but that is constructed in an axially symmetrical form as shown in Fig. 7.3. There is then no radiation from the system assuming that the metal has a thickness large compared with the skin depth at the frequency of interest. Let us assume that the metal is a perfect conductor and that the interior of the resonator is a vacuum. There are then no losses at all and, assuming that $b \ll a \ll d$, the inductance and capacitance are, respectively, (Problem 2.1)

$$L = \tfrac{1}{2}\mu_0 a^2/(a + d), \quad C = \pi\epsilon_0 d^2/b \tag{7.1}$$

At time t, let the current round the inductor be $I(t)$ and the voltage of the capacitor be $V(t)$. As shown in Fig. 7.2, let the positive direction of the current round the inductor be directed towards the positive plate of the capacitor, so that the electric field in the capacitor points round the circuit in the same direction as the electric current. At time t, the energies stored in the inductor and the capacitor are respectively (eqns. 5.64 and 3.43)

$$W_m = \tfrac{1}{2}LI^2, \quad W_e = \tfrac{1}{2}CV^2 \tag{7.2}$$

and their time rates of change are

$$\frac{dW_m}{dt} = LI\frac{dI}{dt}, \quad \frac{dW_e}{dt} = CV\frac{dV}{dt} \tag{7.3}$$

From Faraday's law of induction and the principle of conservation of charge we have

$$V = -L\frac{dI}{dt}, \quad I = C\frac{dV}{dt} \tag{7.4}$$

Eqns. 7.3 may therefore be written

$$\frac{dW_m}{dt} = -VI, \quad \frac{dW_e}{dt} = VI \tag{7.5}$$

Addition of this pair of equations then gives

$$\frac{d}{dt}(W_m + W_e) = 0 \tag{7.6}$$

so that

$$W_m + W_e = W_0 \tag{7.7}$$

where W_0 is the constant total electromagnetic energy of the system. Eqn. 7.7 is the energy equation for the system. It states that the magnetic energy stored in the inductor decreases as the electric energy in the capacitor increases, and vice versa. The total energy of the system remains constant because we are neglecting losses (radiation, conduction and hysteresis).

Eqns. 7.5 show that the product VI is the rate of increase of the energy stored in the capacitor and the rate of decrease of the energy stored in the inductor. The product VI is the rate of flow of energy from the inductor to the capacitor. If the agreed positive direction for either V or I were reversed in Fig. 7.2, the product of the capacitor voltage and the inductor current would give the rate of flow of energy from the capacitor to the inductor.

If we eliminate either V or I from eqns. 7.4 we obtain the equation of simple harmonic vibration:

$$\frac{d^2I}{dt^2} + \omega^2 I = 0, \quad \frac{d^2V}{dt^2} + \omega^2 V = 0 \tag{7.8}$$

where

$$\omega = (LC)^{-1/2} \tag{7.9}$$

This is the resonant angular frequency of the system. If time is measured from the instant when the voltage V in Fig. 7.2 has its maximum positive value V_0, the solution of eqns. 7.4 is

$$I = I_0 \cos(\omega t + \tfrac{1}{2}\pi), \quad V = V_0 \cos \omega t \tag{7.10}$$

where

$$\frac{V_0}{I_0} = \omega L = \frac{1}{\omega C} = \left(\frac{L}{C}\right)^{1/2} \tag{7.11}$$

Substitution from eqns. 7.10 into eqns. 7.2 gives

$$W_m = W_0 \sin^2 \omega t, \quad W_e = W_0 \cos^2 \omega t \tag{7.12}$$

where the total energy in the system is

$$W_0 = \tfrac{1}{2} L I_0^2 = \tfrac{1}{2} C V_0^2 \tag{7.13}$$

Eqns. 7.12 may be written

$$W_m = \tfrac{1}{2} W_0 (1 - \cos 2\omega t), \quad W_e = \tfrac{1}{2} W_0 (1 + \cos 2\omega t) \tag{7.14}$$

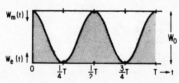

Fig. 7.4 *Illustrating oscillation of energy between the capacitor and the inductor of a resonant oscillatory circuit*

They clearly satisfy the energy equation (7.7). The complementary variations of W_m and W_e with t are illustrated in Fig. 7.4. The electric energy stored in the capacitor is plotted vertically upward from the bottom of the diagram, while the magnetic energy stored in the inductor is plotted vertically downwards from the top of the diagram. The quantity T in Fig. 7.4 is the period $2\pi/\omega$ of the oscillation; in accordance with eqn. 7.9, it is given by

$$T = 2\pi (LC)^{1/2} \tag{7.15}$$

Notice that the energy oscillation repeats in each time-interval $\tfrac{1}{2}T$ because the sign of the voltage (current) is irrelevant for storage of energy in the capacitor (inductor). Eqns. 7.14 show that, on the average, half the energy of the system is stored in the inductor and half in the capacitor:

$$\bar{W}_m = \bar{W}_e = \tfrac{1}{2} W_0 \tag{7.16}$$

Substitution from eqns. 7.14 into eqns. 7.5 shows that

$$VI = -\omega W_0 \sin 2\omega t = \omega W_0 \cos (2\omega t + \tfrac{1}{2}\pi) \qquad (7.17)$$

This quantity is plotted as a function of time in Fig. 7.5. It gives the rate of flow of energy from the inductor to the capacitor; when negative, the magnitude gives the rate of flow of energy from the capacitor to the inductor.

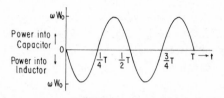

Fig. 7.5   *Illustrating, as a function of time, the rate of flow of energy between the inductor and the capacitor of a resonant oscillatory circuit*

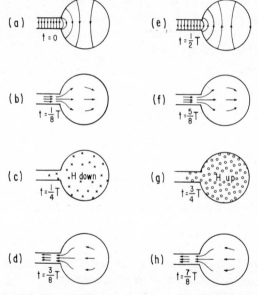

Fig. 7.6   *Illustrating, for a single azimuthal cross-section, the process of oscillation in the resonant oscillatory system of Fig. 7.3*

The second of eqns. 7.10 shows that the voltage of the capacitor is a maximum at time zero. Consequently the electric field in the capacitor is a maximum at time zero as indicated at Fig. 7.6a. A quarter of a cycle later ($t = \tfrac{1}{4}T$) the electric field in the capacitor vanishes, while at $t = \tfrac{1}{2}T$ this electric field is again a maximum, but with the opposite polarity as shown in Fig. 7.6e. The first of eqns. 7.10 shows that the electric current round the inductor, and consequently the magnetic field threading the inductor, is a maximum at $t = \frac{1}{4}T$, the direction being as indicated in Fig. 7.6c. The magnetic field in the inductor again maximises at $t = \frac{3}{4}T$, but it has th opposite direction as indicated in Fig. 7.6g. The magnetic field vanishes at $t = 0$ and $t = \frac{1}{2}T$. The electric energy stored in the capacitor at $t = 0$ flows into the inductor during the first quarter-cycle as shown in Fig. 7.6b. The magnetic energy stored in the inductor at $t = \frac{1}{4}T$ flows back into the capacitor during the second quarter-cycle. The process is repeated during the third and fourth quarter cycles but with the electromagnetic field oppositely directed. There is a flow of energy back and forth between the capacitor and the inductor; it is described by Fig. 7.5 and takes place in the manner indicated in Fig. 7.6b, d, f and h. We need to understand the character of such a flow of energy.

7.3 A simple electromagnetic transmission system

A simple arrangement for generating energy in one location and using it in another is a direct-current transmission line connecting a battery (voltage V) to a resistor (resistance R) as shown in Fig. 7.7. If the system is supposed for simplicity to be loss free except for the resistive load, there exists round the circuit a steady free electric current I given by Ohm's law:

$$I = V/R \tag{7.18}$$

Fig. 7.7   *Illustrating a direct-current transmission system*

Electromagnetic energy is being generated in the battery at a rate per unit time equal to

$$VI \tag{7.19}$$

It is being dissipated in the resistor at the same rate. Energy is therefore being transmitted steadily along the line at the rate VI per unit time.

Let us suppose that the line connecting the source to the sink is a strip transmission line consisting of a pair of perfectly conducting strips each of length l, width b and separation a. For simplicity, let us assume that fringing of the electromagnetic field is avoided, so that the electric and magnetic fields between the strips may be assumed uniform. Let the strips be perfectly conducting, and let the material between them be linear homogeneous material of dielectric constant ϵ/ϵ_0, permeability μ/μ_0 and conductivity σ. Then the line has an inductance per unit length L, a capacitance per unit length C and a conductance per unit length G given by (Problem 2.2)

$$L = \mu a/b, \quad C = \epsilon b/a, \quad G = \sigma b/a \tag{7.20}$$

Using version 2 of the electromagnetic equations for linear isotropic materials (Table 2.2), the electric field strength E and the magnetic field strength H' between the strips are

$$E = V/a, \quad H' = I/b \tag{7.21}$$

so that H' is the free electric current per unit width on the strips, omitting the bound current on the surfaces of the material in contact with the strips. The electric and magnetic flux densities between the strips are

$$D' = \epsilon V/a, \quad B = \mu I/b \tag{7.22}$$

where $\pm D'$ are the free charges per unit area on the strips, omitting the bound charges on the surface of the material in contact with the strips. From eqns. 7.21 and 7.22, the electric and magnetic energies stored per unit volume between the strips are (eqns. 4.54 and 6.61)

$$\tfrac{1}{2}ED' = \tfrac{1}{2}\epsilon(V/a)^2, \quad \tfrac{1}{2}H'B = \tfrac{1}{2}\mu(I/b)^2 \tag{7.23}$$

These expressions also give the tensions in the tubes of electric and magnetic flux, and the sideways pressures that they exert (eqns. 4.60, 4.61, 6.70 and 6.71). The tension in the tubes of electric flux tends to pull the strips together, while the sideways pressure in the tubes of magnetic flux tends to push them apart.

An arrangement for avoiding non-uniformity of the electric and magnetic field between the strips could take the form illustrated in Fig. 7.8. The sideways pressure of the tubes of electric flux at the edge of the dielectric material between the strips is then largely balanced by mechanical stress in this material. Likewise the tension in the tubes of magnetic flux between the strips is largely balanced by mechanical stress in the material used to complete the magnetic circuit.

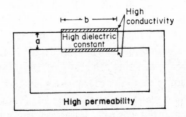

Fig. 7.8   *Illustrating a normal cross-section of a perfectly conducting strip transmission line possessing (i) material of high dielectric constant between the strips to keep the electric field uniform, and (ii) a non-conducting magnetic circuit of high permeability to keep the magnetic field between the strips uniform*

Let us assume that the material between the strips is non-conducting except for a short length τ of the line near the end as shown in Fig. 7.9, and that this terminating section constitutes the load resistor. If the terminating load is made of material of conductivity σ, then the terminating conductance and resistance are (Problem 2.1)

$$G = \sigma\tau b/a, \quad R = a/(\sigma\tau b) \tag{7.24}$$

It is often convenient to regard the load as a resistive sheet whose surface conductance is g and surface resistance is r, defined by

$$g = \sigma\tau, \quad r = 1/(\sigma\tau) \tag{7.25}$$

These quantities are the conductance and resistance between opposite edges of a square of the resistive sheet. In terms of g and r the conductance and resistance of the load may be written (eqns. 7.24)

$$G = gb/a, \quad R = ra/b \tag{7.26}$$

Battery   Resistive sheet

Fig. 7.9 *Illustrating the flow of energy from a battery to a resistor between the conductors of a strip transmission line*

The region between the strips may be regarded as a tank of electromagnetic energy in which the volume densities of electric and magnetic energy are given by eqns. 7.23. At the resistor end of the line, energy 'drips' out of the tank as each free electron in the resistor is accelerated between one collision and the next, where it is converted into the random motion of heat. At the battery end of the line, energy drips into the tank each time that an electron in the electrolyte adjacent to an electrode is separated from a positive ion by the chemical process involved in the functioning of the battery. As energy drips into the tank at the battery end and drips out at the resistor end, there is a flow of energy through the tank as indicated in Fig. 7.9. If account were taken of the finite conductivity of the conductors of the transmission line, some of the energy flowing through the tank would be diverted sideways to maintain the flow of the electric current in the line.

The electric field between the strips of the transmission line is an electrostatic field for which the equipotential surfaces are planes parallel to the perfectly conducting strips. Any such plane can be replaced by a thin perfectly conducting sheet at the appropriate potential without upsetting the electric field on either side. Because the sheet does not intercept tubes of magnetic flux, the magnetic field is also undisturbed. Hence the space between the strips in Fig. 7.9 may be dissected into numerous similar strip transmission lines for each of which the separation between the strips is small. In particular, a transmission line could be provided for each of the cells of which the battery is composed, so that each cell feeds its own line and dissipates energy in its own portion of the terminating resistive sheet.

Let us dissect the electromagnetic field between the strips in Fig. 7.9 by means of closely spaced geometrical planes perpendicular to the electric field and closely spaced geometrical planes perpendicular to the magnetic field. In this way, the region in which energy is flowing is subdivided into channels having small rectangular cross-sections. Let dx be the separation between a pair of adjacent planes perpendicular to the electric field, and dy be the separation between a pair of adjacent planes perpendicular to the magnetic field. Then the electromagnetic field in a channel whose cross-sectional dimensions are (dx, dy) may be regarded as that for a strip transmission line having separation dx and voltage Edx, and having width dy and free electric current $H'dy$. For a typical centrally located channel, fringing does not need to be taken into account because the electromagnetic fields in the adjacent channels prevent it. The product of the voltage and current for the channel of cross-sectional area $dxdy$ is

$$(Edx)(H'dy) = (EH')(dxdy) \tag{7.27}$$

and this is the energy flowing down this channel per unit time. Hence energy is flowing down the channels at the rate

$$EH' \tag{7.28}$$

per unit cross-sectional area per unit time. For the entire space between the strips, the rate of flow of energy per unit time is, neglecting fringing,

$$(EH')(ab) \tag{7.29}$$

and, using eqns. 7.21, we verify that this agrees with expr. 7.19.

The energy flowing down each channel is dissipated in the portion of the terminating resistive sheet corresponding to this channel. For a unit square of the resistive sheet with edges parallel to the electromagnetic field, the voltage across it is E and the free current through it is H', while the conductance and resistance across it are g and r, given by eqns. 7.25. Consequently, the energy dissipated in the terminating resistive sheet per unit area per unit time is (eqns. 3.90)

$$EH' = gE^2 = rH'^2 \tag{7.30}$$

If the terminating resistor does not take the form of a uniform resistive sheet, the electromagnetic field near the end of the transmission line is distorted, and with it the flow of energy. At the battery end of the line non-uniformity of the energy flow also occurs because the actual source of the electromagnetic energy lies in the chemical reaction taking place at the surfaces of contact between the electrodes and the electrolyte.

We have described the flow of energy along a strip transmission line neglecting fringing, but a similar description can be given taking fringing into account. This can also be done for a coaxial transmission line (Fig. 7.10) and for a twin-wire transmission line (Fig. 7.11). A plane perpendicular to the length of the line may be dissected into small squares by means of mutually perpendicular lines of electric force and magnetic flux. Corresponding to each small square, there is a channel along the line for which the voltage per unit separation is the electric field strength

E in the channel and the free electric current per unit width is the magnetic field strength H', so that the flow of energy down the line per unit cross-sectional area per unit time is EH'. Fringing is not involved because each channel is guarded by the channels adjacent to it.

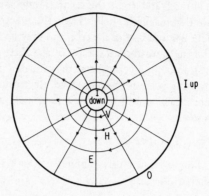

Fig. 7.10   *Illustrating the electromagnetic field in a normal cross-section of a coaxial transmission line*

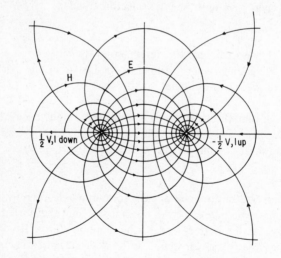

Fig. 7.11   *Illustrating the electromagnetic field in a normal cross-section of a twin-wire transmission line*

However, there is now a variation of the power per unit area over a plane perpendicular to the length of the line. Consider a perfectly conducting coaxial transmission line (inner radius a and outer radius b) with uniform linear non-conducting material between the conductors. If V is the excess of the voltage of the inner conductor over that of the outer conductor and I is the free electric current on the inner conductor directed from the generator towards the load, then the electromagnetic field between the conductors at distance r from the axis may be written (Problem 2.4)

$$E = \frac{V}{\ln (b/a)} \frac{1}{r}, \quad H' = \frac{I}{2\pi r}$$ (7.31)

the directions being as shown in Fig. 7.10. The power down the line per unit cross-sectional area at distance r from the axis is therefore

$$EH' = \frac{VI}{2\pi \ln (b/a)} \frac{1}{r^2}$$ (7.32)

It is therefore largest close to the inner conductor and smallest close to the outer conductor. Integration of the expression on the right-hand side of eqn. 7.32 over the entire cross-sectional area of the transmission line gives VI as it should. A similar calculation may be performed for a twin-wire transmission line.

7.4 The concept of the Poynting vector

In Sections 5.2, 5.3 and 5.4 we studied how energy is stored in a stationary isolated tube of magnetic flux in free space by electric current flowing solenoidally over the surface. We pictured the current as created by a charge q per unit area moving solenoidally over the surface with velocity v, so that the electric current per unit length is (eqn. 5.37)

$$H = qv$$ (7.33)

Expressed vectorially, this surface density of electric current is (eqn. 5.38)

$$i = n \times H = qv$$ (7.34)

where n is a unit normal to the surface drawn into the tube.

If the solenoidal flow is accelerated, the magnetic flux threading the tube is increasing with time. In accordance with Faraday's law of induction, an electric field is brought into play that is illustrated in Fig. 5.1 for a straight tube of circular cross-section. To maintain the acceleration, the effect of this electric field on the moving charge must be offset. Consequently, a force F must be applied to the charged surface given, per unit area, by (eqn. 5.39)

$$F = -qE$$ (7.35)

The rate at which this force is doing work per unit area of the surface of the tube is (eqn. 5.40)

$$F \cdot v = -qv \cdot E$$ (7.36)

Substitution for qv from the second of eqns. 7.34 into eqn. 7.36 shows that the rate at which work is being done on the tube per unit area of its surface is

$$F \cdot v = -(n \times H) \cdot E = n \cdot (E \times H)$$ (7.37)

It follows that, at a point on the surface of a stationary isolated tube of magnetic flux in free space where the electric vector is E and the magnetic vector is H, energy must be supplied at a rate per unit time per unit area equal to

$$n \cdot (E \times H) \qquad (7.38)$$

This is the inward normal component of the vector

$$f = E \times H \qquad (7.39)$$

which is known as the Poynting vector.

If in the above argument the tube were filled with magnetic material, the force per unit area F would be applied to the free electric current on the surface of the tube, exclusive of bound electric current associated with the magnetic moment per unit volume. Consequently, H would be replaced by H' as used in version 2 of the equations for the electromagnetic field in materials (Table 2.1). The energy supplied to the surface of the tube per unit time per unit area is then (c.f. expr. 7.38)

$$n \cdot (E \times H') \qquad (7.40)$$

which is the inward normal component of the Poynting vector

$$f = E \times H' \qquad (7.41)$$

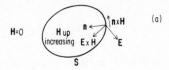

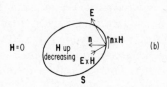

Fig. 7.12   *Illustrating the rate, per unit area per unit time, at which energy is provided by or to the electric current flowing solenoidally over the surface of a stationary tube of magnetic flux in which the flux is changing with time*

The result at which we have arrived is illustrated in Fig. 7.12. A stationary isolated tube of magnetic flux in free space is bounded by a surface S, and the diagrams show a cross-section of the tube perpendicular to the magnetic flux. Normal to this cross-section there is a magnetic vector H inside S, but no magnetic field outside S. In consequence, there is a surface current of density $n \times H$ over the surface of the tube, where n is a unit normal drawn into the tube. If the magnetic flux threading the tube is increasing steadily with time, an electric field is brought into play in accordance with Faraday's law of induction. If E is the electric vector at the point of S where the inward unit normal is n, then the current per unit width $n \times H$ has to be driven against the electric field E, and this requires energy per unit time per unit area of S equal to the inward normal component of $E \times H$ (Fig. 7.12a). On the other hand, if the magnetic flux threading the tube is decreasing steadily with time, then the associated electric vector E is in the opposite direction, and energy is being made available to the bounding current at a rate per unit time per unit area equal to the outward normal component of $E \times H$ (Fig. 7.12b). If dS is a vector element of area of S directed normally inward, then the power made available to the tube at dS is

$$(E \times H) \cdot dS \qquad (7.42)$$

if this quantity is positive. If it is negative, then power is being made available by the tube at the element of area dS. The total power made available to the tube by the forces applied to the surface current is

$$\int_S (E \times H) \cdot dS \qquad (7.43)$$

taken over the entire surface S of the tube. If the tube is filled with magnetic material, expr. 7.42 becomes

$$(E \times H') \cdot dS \qquad (7.44)$$

and expr. 7.43 becomes

$$\int_S (E \times H') \cdot dS \qquad (7.45)$$

7.5 The physical significance of the Poynting vector

The tube of magnetic flux in free space considered in Fig. 7.12a may be subdivided into two tubes by means of an internal surface S$'$ as shown in Fig. 7.13. Surface densities of electric current $\pm n \times H$ may be introduced on S$'$, one of which bounds the tube within S$'$ and the other bounds the tube between S$'$ and S. The normal component of $E \times H$ at S$'$, measured into the tube existing within S$'$, gives the energy that must be supplied per unit time per unit area to S$'$ to create the rate of increase of magnetic flux within S$'$. The same component of $E \times H$ gives the energy that is made available per unit time per unit area at S$'$ by the tube existing between S$'$ and S. The increase of magnetic energy between S$'$ and S is the excess of that made available by forces applied at S over that used at S$'$ to supply energy to the tube within S$'$, and this applies to any subdivision of the large tube into smaller tubes. The entire process is conveniently described by saying that there is a flow of energy in the tube at each point given, in magnitude and direction, by $E \times H$ per unit time per unit area.

For the circular tube of magnetic flux illustrated in Fig. 5.1, the flow of energy is radially inwards as shown in Fig. 7.14. The rate EH per unit time per unit area decreases linearly with r from a maximum at the surface $(r = a)$ to zero at the centre $(r = 0)$. It may be verified from the two diagrams at the bottom of Fig. 7.14 that the excess of the rate of inflow of energy per unit time per unit length into a tube having any selected radius, over that for a coaxial tube having any smaller radius, accounts for the rate of accumulation of magnetic energy in the intervening volume.

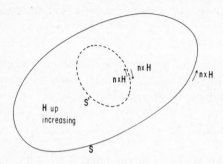

Fig. 7.13   *Illustrating subdivision of a tube of magnetic flux into two tubes by means of an internal surface S'*

As described in Section 5.2, let us think of the electric current round the circular tube of magnetic flux in Fig. 7.14 as created by rotation, with steadily increasing angular velocity ω, of a circular cylinder of radius a carrying a charge Q per unit length, uniformly distributed. The tube may be subdivided into coaxial thin-walled elementary tubes of which a typical one has inner radius r and outer radius $r + dr$ as shown in Fig. 7.15. Each such hollow tube can be maintained by charges $+Q$ on its outer surface and $-Q$ on its inner surface if both surfaces rotate in the same direction with the same steadily increasing angular velocity ω. Each elementary tube of wall-thickness dr may then be regarded as driven by the one that envelops it; in its turn it drives the one that it envolops. The Poynting vector describes the process of inward transmission of energy.

The Poynting vector also describes steady flow of energy along (i) a strip transmission line (Fig. 7.9), (ii) a coaxial transmission line (Fig. 7.10) and (iii) a twin-wire transmission line (Fig. 7.11). The Poynting vector $E \times H'$ is parallel to the length of the line and is directed from the generator-end towards the load-end. The directions of E and H' are perpendicular, and the magnitude of the Poynting vector is EH'. For a coaxial transmission line, the magnitude of the Poynting vector is calculated in eqn. 7.32.

Likewise, the Poynting vector describes the flow of energy involved in a resonant oscillatory circuit. As illustrated in Fig. 7.6, the electric field of the capacitor fringes into the inductor and the magnetic field of the inductor fringes into the capacitor. Particularly near the join between the capacitor and the inductor, the electric and magnetic fields coexist, leading to a flow of energy described by the Poynting vector. This flow is indicated in Fig. 7.6*b*, *d*, *f* and *h*.

For a resonant oscillatory circuit consisting of a multiturn inductor connected in parallel with a multiplate capacitor by means of a short piece of twin-wire transmission line, the line constitutes the means where by the electric field of the capacitor is fringed into the magnetic field of the inductor and the magnetic field of the inductor is fringed into the electric field of the capacitor. In the overlapping electric and magnetic fields thereby created, flow of energy takes place along the line back and forth between the inductor and the capacitor in accordance with the Poynting vector.

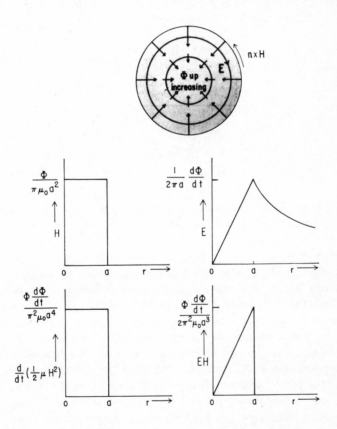

Fig. 7.14   *Illustrating flow of energy in a straight tube of magnetic flux which has circular cross-section and in which the flux is increasing steadily with time*

The physical significance of the Poynting vector may be described more generally as follows. Consider any electromagnetic field in free space. At a particular time t, let us draw in the field a surface S that is composed of lines of magnetic flux, and that divides the volume occupied by the field into two regions V_1 and V_2.

Let n be a unit normal to S at any point P, directed from V_1 to V_2, and let E and H be the electric and magnetic vectors at P. Consider a situation in which the electromagnetic field in V_1 adjacent to S remains unchanged, but the tangential magnetic field in V_2 adjacent to S is reduced to zero. This may be done by introducing on S an electric current whose surface density is $-n \times H$. The electric field E then does work on the introduced electric current (c.f. eqns. 7.37) at a rate per unit area per unit time equal to $-(n \times H) \cdot E$ or $n \cdot (E \times H)$. The electric current introduced on the surface S absorbs energy at a rate per unit area per unit time equal to the normal component of the Poynting vector $E \times H$ directed into S from the region where the electromagnetic field is undisturbed. If magnetic material is present we use free electric current on S, so that H is replaced by H' and $E \times H$ by $E \times H'$.

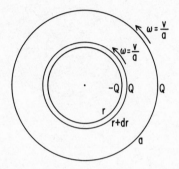

Fig. 7.15   *Illustrating inward transmission of energy in a rotating hollow charged circular cylinder for which the angular velocity is steadily increasing*

The Poynting vector gives a prescription for intercepting the flow of energy in an electromagnetic field by a surface S composed of lines of magnetic flux, and thereby measuring the flow. Let us consider how this works for a transmission line. Suppose that energy is flowing down the line, and that we wish to measure how much energy per unit time is passing a given plane S perpendicular to the length of the line. We could do this by disconnecting the portion of the line beyond S, thereby reducing to zero the electric current on the line beyond S, and simultaneously connecting across the remainder of the line at S a terminating resistor in the form of a resistive sheet (c.f. Fig. 7.9) that leaves the field in this portion of the line undisturbed. We have then intercepted the flow of energy by the terminating resistor and can thereby measure the flow.

To intercept the flow of energy in an electromagnetic field, the procedure is to reduce the tangential magnetic field just beyond the interception surface S to zero but leave the electromagnetic field on the opposite side of S unaffected. The tangential magnetic field at the interception surface is thereby completely converted into electric current on the interception surface, and the energy absorbed by the interception surface per unit area per unit time is the inward normal component of the Poynting vector.

Consider a situation in which crossed uniform static electric and magnetic fields exist in free space, intersecting at an angle ϕ. Let the electric vector be E and the magnetic vector be H, so that there is a uniform flow of energy through space at the rate $E \times H$ per unit area per unit time. What does this flow mean?

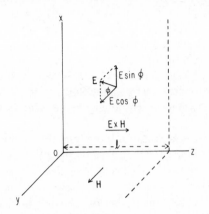

Fig. 7.16 *Illustrating crossed uniform static electric and magnetic fields in free space, causing a uniform flow of energy E X H per unit area per unit time*

Let us use a system of cartesian co-ordinates for which the positive z axis is in the direction $E \times H$, so that E and H, although intersecting at angle ϕ, are both parallel to the planes $z = $ constant. Let H be in the direction of the positive y axis as shown in Fig. 7.16. Then

$$E = (E \sin \phi, E \cos \phi, 0) \tag{7.46}$$

$$H = (0, H, 0) \tag{7.47}$$

so that

$$E \times H = (0, 0, EH \sin \phi) \tag{7.48}$$

To examine the meaning of this flow of energy let us intercept it at the plane $z = l$. We reduce the magnetic field to zero in the region $z > l$ by introducing on the plane $z = l$ a steady electric current of surface density

$$i = (H, 0, 0) \tag{7.49}$$

Eqns. 7.46 and 7.49 show that each unit square of the plane $z = l$ with edges parallel to the x and y axes carries a steady electric current H, across which the drop in voltage is $E \sin \phi$. It therefore absorbs energy at the rate $EH \sin \phi$ per unit time in agreement with the Poynting vector given in eqn. 7.48.

If in Fig. 7.16 the magnetic field is abolished not only for $z > l$ but also for $z < 0$, then electric current has to be driven on the plane $z = 0$. Energy is then being supplied to the region $0 < z < l$ at the plane $z = 0$ at the rate $EH \sin \phi$ per unit area per unit time and is being removed at the plane $z = l$ at the same rate. This is what the Poynting vector in eqn. 7.48 means.

If the interception plane $z = l$ in Fig. 7.16 were turned round the y direction through an angle θ, the energy absorbed per unit area per unit time would be multiplied by $\cos\theta$. If the interception plane were turned round the x direction, the same would be true. But it would not then be possible completely to abolish the magnetic field to the right of this plane because the tubes of magnetic flux would then be intercepted by the plane. To stop them abruptly at this plane would require the availability of magnetic charge. However, it would still be possible to abolish only the component of magnetic field tangential to the interception plane. Alternatively, it would be possible to fill the region beyond the interception plane with non-conducting material of high permeability. This would reduce the H' field in this region to a low value, so that nearly all the flow of energy on the opposite side of the interception plane would be absorbed by the free electric current in the interception plane.

The simplest theoretical procedure for visualising the meaning of the Poynting vector is to imagine that magnetic charge and magnetic current are available in addition to electric charge and electric current. Any electromagnetic field may then be dissected along a surface S and the electromagnetic field on one side of S abolished (Section 1.14). This requires the introduction of electric and magnetic charges and currents on S given by eqns. 1.92 and 1.91. The rate at which work is being done on the surface S at a point A per unit area per unit time by the undisturbed electromagnetic field is then the inward normal component at A of the Poynting vector. Theoretically, any electromagnetic field may be terminated at any surface. The flow of energy described by the Poynting vector is then absorbed or emitted by the terminating surface. This can only be tested experimentally for situations such that no magnetic charge or current is needed, but these are the situations of interest in practice.

That a flow of electromagnetic energy can exist in space is a fact of far-reaching importance. It is this phenomenon that accounts for much of the energy transmitted from the Sun to the Earth. It is this phenomenon that permits energy to pass from radio and television stations to distant receivers. Even for telephone communication using a transmission line, the flow of energy is between the conductors: the conductors guide the energy, but the flow is primarily between them. The same is true for the flow of energy from generating stations along power lines to consumers (c.f. Fig. 7.9).

Chapter 8

Conservation of energy and momentum in electromagnetic fields

8.1 Introduction

Now that we understand the storage and flow of energy in electromagnetic fields we are in a position to study the electromagnetic energy budget and to examine whether it balances. We begin by considering electromagnetic fields in free space.

8.2 Poynting's energy theorem for electromagnetic fields in free space

Although the presence of polarisable and magnetisable atoms and molecules is ruled out for the time being, the presence of free charged particles is permitted. However, the particles are assumed initially to be massless. No kinetic energy and no gravitational potential energy is then involved.

We do, however, contemplate the possibility that the free charged particles may be acted on by a possibly non-electromagnetic force capable of doing work on them and thereby increasing the energy of the system. We also contemplate the possibility that the free charged particles may be acted on by a force capable of doing work against them and thereby removing energy from the system. Such a force arises when the free electrons in a conductor collide with atomic or molecular obstacles. Electromagnetic energy is then being dissipated in the system by the collisional process discussed in Section 4.12, or generated by the process discussed in Section 4.13.

In Sections 4.12 and 4.13 consideration was given to a situation in which the free charged particles possessed a steady drift-velocity. We found that, if at a particular location and time the electric vector is E and the volume-current-density vector is J, then electromagnetic energy is being dissipated per unit volume per unit time at the rate (eqn. 4.80)

$$p = E \cdot J \tag{8.1}$$

and if this is negative electromagnetic energy is being generated per unit volume. We are now concerned with a situation in which the drift-velocity of the particles is not necessarily steady. The particles would therefore be gaining or losing kinetic energy were it not for the fact that we are neglecting the mass of the particles for the time being. Because we are neglecting mass, eqn. 8.1 still gives the electromagnetic energy being dissipated or generated even if the electromagnetic field varies with time. In a volume V, the net excess of the electromagnetic energy dissipated per unit time over that generated is

$$P = \int_V p \, d\tau \tag{8.2}$$

where p is given by eqn. 8.1.

Electromagnetic energy is also being lost or gained by the system in the volume V as a result of the flow of electromagnetic energy across the bounding surface S. At a location in free space where the electric vector is E and the magnetic vector is H, the flow of energy is given, per unit area per unit time, by the Poynting vector (eqn. 7.39)

$$f = E \times H \tag{8.3}$$

If dS is a vector element of area of S directed normally outwards, the net electromagnetic energy being lost by the volume V as a result of flow across S is

$$F = \int_S f \cdot dS \tag{8.4}$$

where f is given by eqn. 8.3.

The electromagnetic field is also storing electric and magnetic energy in the volume V. If, at a particular location and time, the electric vector is E and the magnetic vector is H, then the electric and magnetic energies stored per unit volume are (eqns. 4.6 and 5.55)

$$w_e = \tfrac{1}{2}\epsilon_0 E^2, \qquad w_m = \tfrac{1}{2}\mu_0 H^2 \tag{8.5}$$

If $d\tau$ is an element of volume in a region V, the total electric and magnetic energies stored in V are

$$W_e = \int_V w_e \, d\tau, \qquad W_m = \int_V w_m \, d\tau \tag{8.6}$$

where w_e and w_m are given by eqns. 8.5.

Since only electromagnetic energy is involved in the system, we expect the net rate of loss of electromagnetic energy from a volume V to be equal to the rate of decrease of the electromagnetic energy stored in V. This means that

$$P + F = -\frac{d}{dt}(W_e + W_m) \tag{8.7}$$

where P, F, W_e and W_m are given by eqns. 8.2, 8.4 and 8.6. Let us verify the energy equation (8.7) using a volume V whose surface S is fixed relative to the observer. Eqn. 8.7 may then be written

$$\frac{\partial}{\partial t}(W_e + W_m) + P + F = 0 \tag{8.8}$$

or (eqns. 8.1–8.6)

$$\frac{\partial}{\partial t}\int_V (\tfrac{1}{2}\epsilon_0 E^2 + \tfrac{1}{2}\mu_0 H^2)\, d\tau + \int_V E \cdot J\, d\tau + \int_S (E \times H) \cdot dS = 0 \tag{8.9}$$

This is Poynting's energy theorem for an electromagnetic system in free space. It is proved as follows.

8.3   Proof of Poynting's energy theorem for electromagnetic fields in free space

Use of the divergence theorem (Identity 17, Appendix 2), followed by Identity 9, gives

$$\int_S (E \times H) \cdot dS = \int_V \operatorname{div}(E \times H)\, d\tau \tag{8.10}$$

$$= \int_V (-E \cdot \operatorname{curl} H + H \cdot \operatorname{curl} E)\, d\tau \tag{8.11}$$

Replacing curl E by $-\partial B/\partial t$ and curl H by $J + \partial D/\partial t$ in accordance with Maxwell's curl equations, we then obtain

$$\int_S (E \times H) \cdot dS = -\int_V \{E \cdot (J + \partial D/\partial t) + H \cdot \partial B/\partial t\}\, d\tau \tag{8.12}$$

or

$$\int_V (E \cdot \partial D/\partial t + H \cdot \partial B/\partial t)\, d\tau + \int_V E \cdot J\, d\tau + \int_S (E \times H) \cdot dS = 0 \tag{8.13}$$

Putting

$$D = \epsilon_0 E, \qquad B = \mu_0 H \tag{8.14}$$

in eqn. 8.13 yields eqn. 8.9. Poynting's energy theory for electromagnetic systems in free space is thereby established.

> It may be noted that it is quite possible for the volume V under consideration to be the region exterior to a closed surface S. Some care is then necessary with the convergence of the integrals at infinity. This may be handled by assuming that the sources did not exist prior to some instant of time. The resulting electromagnetic field then occupies a finite region of space determined by the velocity of light. However, it is often convenient to assume instead that all radiation is absorbed somewhere. On this basis a slight exponential decrease of the electromagnetic field with distance may be introduced to make the electromagnetic field decay at large distances.

8.4   The electromagnetic energy budget in the presence of linear materials

In practical materials, mechanical stress is often of great importance. Moreover, discussion of electromagnetic stress inevitably involves mechanical stress. As mentioned in connection with Fig. 6.3, when a weight is lifted by a magnet, the tension in the tubes of magnetic flux in the air is largely taken up by mechanical tension in the magnet: mechanical potential energy is thereby stored in the magnetic material. Likewise kinetic energy can be stored in material if the strain is varying with time, for example, if an elastic wave is passing through the material. Materials constitute electromechanical systems, and full discussion of the energy budget of the system involves mechanical potential energy, electrical potential energy, motional energy associated with mass and motional energy associated with charge.

In this text we are disconnecting the electromagnetic properties of materials from the mechanical properties of materials by assuming that, although mechanical stress is involved, it does not cause strain. With no geometrical distortion of the material, the mechanical stress does no work and consequently does not store energy. In these circumstances the electromagnetic energy budget should balance by itself. Let us see whether this is true.

We now use version 2 of the electromagnetic equations in materials (Table 2.1), so that the fields D' and H' appear in place of D and H, while J and ρ are the volume densities of free electric current and charge. Eqn. 8.13 now becomes

$$\int_V (E \cdot \partial D'/\partial t + H' \cdot \partial B/\partial t)\, d\tau + \int_V E \cdot J\, d\tau + \int_S (E \times H') \cdot dS = 0$$

(8.15)

In accordance with eqn. 7.41, the third integral in this equation is the excess of the rate of flow of electromagnetic energy out of the volume V over that into V. The second integral is the excess of the rate of dissipation of electromagnetic energy in V over the rate of generation in V. We therefore need to verify that the first integral constitutes the time-rate of increase of the electric and magnetic energy stored in V.

The electric energy stored in material is discussed in Section 4.8, and the magnetic energy in Section 6.8. For linear isotropic material of dielectric constant ϵ/ϵ_0 and permeability μ/μ_0 we have (eqns. 4.54 and 6.61)

$$w_e = \tfrac{1}{2}\epsilon E^2, \qquad w_m = \tfrac{1}{2}\mu H'^2 \tag{8.16}$$

and (eqns. 2.25)

$$D' = \epsilon E, \qquad B = \mu H' \tag{8.17}$$

It follows that

$$E \cdot \partial D'/\partial t = \partial w_e/\partial t, \qquad H' \cdot \partial B/\partial t = \partial w_m/\partial t \tag{8.18}$$

For isotropic material this proves that the first integral in eqns. 8.15 is the time-rate of increase of the electromagnetic energy stored in the volume V.

For non-isotropic material, eqns. 8.17 have to be replaced by eqns. 4.55

and 6.64, while eqns. 8.16 have to be replaced by eqns. 4.56 and 6.66. Nevertheless, eqns. 8.18 are still true so that, even for linear non-isotropic material, eqn. 8.15 may be written in the form given in eqn. 8.8.

This verifies that there is a balance in the electromagnetic energy budget for any system involving linear materials, so long as no mechanical distortion of the materials occurs.

8.5   Electromagnetic work done in non-linear dielectric and magnetic materials

For non-linear materials statements must be made, not in terms of stored energy, but in terms of the work done in a specific operation. This is because different operations connecting the same initial and final states of polarisation and magnetisation involve different amounts of work. In a magnetisation diagram of the type shown in Fig. 6.5, or in a corresponding polarisation diagram for a non-linear dielectric, a specific operation involves a curve joining an initial point corresponding to the state of the material at time t_1 to a final point corresponding to the state of the material at time t_2. The curves joining the initial and final points specify (i) E and D' as functions of the time t between t_1 and t_2, and (ii) H' and B as functions of t between t_1 and t_2. For material that is electrically and magnetically non-linear, the work done per unit volume in performing the operation is (exprs. 4.74 and 6.89)

$$\int_{t_1}^{t_2} (E \cdot \partial D'/\partial t + H' \cdot \partial B/\partial t)\, dt \tag{8.19}$$

If eqn. 8.15 is integrated with regard to t from t_1 to t_2, the first integral is identical with what expr. 8.19 becomes when integrated with regard to volume throughout V. The resulting equation states that the electromagnetic budget for work done on non-linear materials is balanced so long as no mechanical distortion of the material occurs.

8.6   The energy budget in a plasma

In the preceeding discussion of the energy budget no account was taken of the mass of the free electrons in conducting material. However, for the free electrons in an ionised gas, or plasma, their kinetic energy plays an important role. We need to include the kinetic energy of free charged particles in discussing the energy budget of conducting materials.

In order to average out the random thermic motions of the free electrons, we think in terms of the average velocity of a group of electrons. This is the drift-velocity v employed in Sections 2.12 and 4.12. Consider a location in the conducting material where the drift-velocity is v, the electric vector is E and the magnetic flux-density vector is B. The equation of drift-motion of an electron of charge e and mass m is (c.f. eqn. 2.59)

$$m \frac{dv}{dt} = e(E + v \times B) - F \tag{8.20}$$

when F is the average drag due to collisions of the electrons with atoms. In Section 4.12 we were concerned with a steady drift-velocity, so that dv/dt in eqn. 8.20 vanished and the average collisional drag per electron could be taken as (eqn. 4.77)

$$F = e(E + v \times B) \tag{8.21}$$

leading to eqn. 4.80 for the rate of dissipation of energy per unit volume per unit time. In Section 8.2, dv/dt in eqn. 8.20 did not vanish, but m was taken as zero. Consequently, eqn. 8.21 was still true, leading once again to eqn. 8.1. We now need to consider what happens when the rate of change of momentum $m\, dv/dt$ in eqn. 8.20 cannot be neglected.

Multiplying eqn. 8.20 scalarly by v we obtain

$$\frac{d}{dt}(\tfrac{1}{2}mv^2) = eE \cdot v - F \cdot v \tag{8.22}$$

This equation states that the kinetic energy of an electron due to the drift-motion increases at a rate equal to the excess of (i) the rate at which energy is supplied to it by the electromagnetic field over (ii) the rate at which energy is converted by collisions into the random motion of heat. Multiplying eqn. 8.22 by the number N of free electrons per unit volume we obtain

$$\frac{d}{dt}(\tfrac{1}{2}Nmv^2) = NeE \cdot v - NF \cdot v \tag{8.23}$$

and this may be written

where

$$dw_k/dt = E \cdot J - p \tag{8.24}$$

$$\begin{cases} w_k = \tfrac{1}{2}Nmv^2 & \text{(8.25)} \\ J = Nev & \text{(8.26)} \\ p = NF \cdot v & \text{(8.27)} \end{cases}$$

The quantity w_k is the ordered kinetic energy stored by the free electrons per unit volume, the vector J gives the volume current density in the conductor, and p is the power dissipated per unit volume as a result of collisions. If p is negative, power is being generated per unit volume by an agency that is applying force to the electrons. If we need to take into account not only electrons but also ions of several species, then summation signs are required in eqns. 8.23, 8.25, 8.26 and 8.27, but eqn. 8.24 is unmodified.

Eqn. 8.24 shows that, if the masses of the charge-carriers are taken into account, eqn. 8.1 is replaced by

$$p = E \cdot J - dw_k/dt \tag{8.28}$$

and eqn. 8.2 is replaced by

$$P = \int_V (E \cdot J)\, d\tau - dW_k/dt \qquad (8.29)$$

where W_k is the total kinetic energy stored in the volume V by the ordered motion of the charge-carriers, namely

$$W_k = \int_V w_k\, d\tau \qquad (8.30)$$

Even when the mass of the charge-carriers in the conducting material is taken in account, eqn. 8.13 still follows from Maxwell's equations, and therefore eqn. 8.9 is still true. But the first integral now no longer represents the total rate of increase of the stored energy in the volume V because it does not include the kinetic energy given by eqn. 8.30. Neither does the second integral in eqn. 8.9 represent the rate of dissipation of energy because this is now given by eqn. 8.29, where allowance is made for the fact that energy stored in ordered kinetic form is not energy dissipated. For a volume V whose boundary is fixed relative to the observer, eqn. 8.29 may be written

$$\int_V E \cdot J\, d\tau = P + \partial W_k/\partial t \qquad (8.31)$$

and substitution into eqn. 8.9 then yields, instead of eqn. 8.8,

$$\frac{\partial}{\partial t}(W_e + W_m + W_k) + P + F = 0 \qquad (8.32)$$

Hence it is the rate of increase of the total energy in the volume V (electric, magnetic and kinetic) that is balanced partly by the excess of the rate of production of energy in V over the rate of dissipation, and partly by the excess of the rate of flow of energy out of the bounding surface S over the rate of inflow.

For an electric circuit, the terms in eqns. 8.31 and 8.32 involving the kinetic energy of the moving free electrons are almost always numerically unimportant. For an ionised gas, however, they play a decisive role.

8.7   Use of free electric moment per unit volume

When dealing with a plasma, it is frequently convenient to employ version 3 of the electromagnetic equations (Table 2.1). In version 3 use is made of the electric-flux-density vector D'' and the magnetic vector H'' defined by (eqns. 2.19)

$$D'' = \epsilon_0 E + P_b + P_f, \qquad H'' = \frac{1}{\mu_0}B - M_b - M_f \qquad (8.33)$$

in place of the electric-flux-density vector D' and the magnetic vector H' of version 2 defined by (eqns. 2.16)

$$D' = \epsilon_0 E + P_b, \qquad H' = \frac{1}{\mu_0} B - M_b \tag{8.34}$$

Here P_b and M_b are electric and magnetic moments per unit volume associated with electrons bound in atoms, whereas P_f and M_f are the electric and magnetic moments per unit volume associated with the free charges.

In a plasma or other conducting material, the free magnetic moment per unit volume is usually unimportant, but the free electric moment per unit volume plays a decisive role. It is from P_f that the free electric-current-density vector J and the free electric charge density ρ are deduced. In accordance with eqns. 2.1, we have

$$J = \partial P_f / \partial t, \qquad \rho = - \operatorname{div} P_f \tag{8.35}$$

From eqns. 8.33 and 8.34, neglecting M_f, we deduce that

$$D'' = D' + P_f, \qquad H'' = H' \tag{8.36}$$

Differentiation of the first of these equations with respect to time and substitution from the first of eqns. 8.35 gives

$$\partial D'' / \partial t = \partial D' / \partial t + J \tag{8.37}$$

Taking the scalar product with the electric vector E, we obtain

$$E \cdot \partial D'' / \partial t = E \cdot \partial D' / \partial t + E \cdot J \tag{8.38}$$

and application of eqns. 8.18 and 8.28 then gives

$$E \cdot \partial D'' / \partial t = \frac{\partial}{\partial t} (w_e + w_k) + p \tag{8.39}$$

If use is made of version 3 of Maxwell's equations instead of version 2 (Table 2.1), eqn. 8.15 becomes

$$\int_V (E \cdot \partial D'' / \partial t + H'' \cdot \partial B / \partial t) \, d\tau + \int_S (E \times H'') \cdot dS = 0 \tag{8.40}$$

Substitution for $E \cdot \partial D'' / \partial t$ into this equation from eqn. 8.39 combined with use of the second of eqns. 8.36 and of eqns. 8.18, then reproduces the energy relation

$$\frac{\partial}{\partial t} (W_e + W_m + W_k) + P + F = 0 \tag{8.41}$$

derived in eqn. 8.32.

It should be noted, however, that the term $E \cdot \partial D'' / \partial t$ in eqn. 8.40 is not simply the time-rate of increase of the electric energy of the plasma per unit volume. In accordance with eqn. 8.39, $E \cdot \partial D'' / \partial t$ includes p, the collisional dissipation of energy in the plasma per unit volume per unit time. Moreover, eqn. 8.39 also shows that the term $E \cdot \partial D''/\partial t$ includes the time-rate of increase of part of the motional energy of the plasma per unit volume, namely, the kinetic energy density w_k of the free charges caused by their ordered motion under the influence of the electric field. The only motional energy per unit volume included in the term $H'' \cdot \partial B/\partial t$ in eqn. 8.40 is that associated with the free-space magnetic field, plus the normally negligible contributions from the bound and free magnetic moments per unit volume.

8.8   The concept of electromagnetic momentum per unit volume

When a gun fires a projectile along its barrel, the gun recoils in accordance with the principle of conservation of mechanical momentum. When a radar fires a pulse along its beam, the radar too recoils, but so slightly that the effect is usually overlooked. When a generator connected to one end of a transmission line sends an electromagnetic pulse down the line, the generator recoils slightly. Moreover, a terminating resistor in which the energy of the pulse is absorbed reacts like a target that has been impacted by a projectile of low momentum. These effects are described by saying that the pulse not only possesses electromagnetic energy but also electromagnetic momentum.

> It may be wondered why the recoil of a radar when it fires a pulse is so easily overlooked whereas that of a gun when it fires a projectile seems quite obvious. It should be noted that the total energy conveyed by the projectile is not simply its kinetic energy. The Einstein rest energy is also conveyed, so that a projectile of mass m is conveying energy mc^2 (Appendix C). Calculated in this way, a projectile conveying a total energy of one joule would have a mass of only about 10^{-17} k. One joule is typical for the energy of a pulse fired by a radar but it is ridiculously small for the total energy of a projectile fired by a gun.

For simplicity, consider a strip transmission line with perfectly conducting strips of width b and separation a in free space. Let a be small compared with b, and let fringing be neglected. Using cartesian co-ordinates, let the x axis be in the positive direction of the electric field, the y axis in the positive direction of the magnetic field, and the z axis along the transmission line in the direction of propagation. Let the origin be centrally located between the strips, so that the strips lie in the planes $x = \pm \frac{1}{2}a$ and the open edges of the line lie in the planes $y = \pm \frac{1}{2}b$. At location z along the line, let the voltage between the strips be V and the currents along the strips be I. The electric field strength E and the magnetic field strength H between the strips are then

$$E = V/a, \qquad H = I/b \tag{8.42}$$

and the electric and magnetic flux densities are

$$D = \epsilon_0 V/a, \qquad B = \mu_0 I/b \tag{8.43}$$

The magnetic vector potential between the strips has no x and y components, but its z component is (Problem 8.1)

$$A = Bx \qquad (8.44)$$

Hence the charge on the strip $x = \frac{1}{2}a$ has an electromagnetic momentum in the z direction given, per unit charge, by (eqns. 5.20)

$$A_1 = \frac{1}{2}Ba \qquad (8.45)$$

and that on the strip $x = -\frac{1}{2}a$ has

$$A_2 = -\frac{1}{2}Ba \qquad (8.46)$$

The charge per unit length on the strip $x = \frac{1}{2}a$ is

$$Q_1 = Db \qquad (8.47)$$

and that on the strip $x = -\frac{1}{2}a$ is

$$Q_2 = -Db \qquad (8.48)$$

Hence the electromagnetic momentum per unit length of the transmission line is

$$Q_1 A_1 + Q_2 A_2 = DBab \qquad (8.49)$$

which corresponds to an electromagnetic momentum per unit volume between the strips given by

$$g = DB \qquad (8.50)$$

This momentum is directed along the transmission line from the generator end to the load end. It has to be created by the generator and absorbed by the load.

The concept of electromagnetic momentum is closely related to the concept of electromagnetic energy flow. For the strip transmission line just considered there exists between the strips a flow of electromagnetic energy from the generator end to the load end. It is given, per unit cross-sectional area per unit time, by

$$f = EH \qquad (8.51)$$

In terms of D and B, this may be written

$$f = \frac{DB}{\epsilon_0 \mu_0} \qquad (8.52)$$

Comparison of eqns. 8.50 and 8.52 shows that

$$f = c^2 g \qquad (8.53)$$

where c is the velocity of light in free space. Moreover, both the flow of energy and the electromagnetic momentum are directed along the transmission line from the generator end to the load end, so that eqn. 8.53 may be written vectorially as

$$f = c^2 g \qquad (8.54)$$

Eqn. 8.54 illustrates a result that is in fact true for any electromagnetic field in free space. Consider a point in an electromagnetic field in free space where the electric vector is E, the magnetic vector is H, the electric-flux-density vector is D and the magnetic-flux-density vector is B. We already know (eqn. 7.39) that the flow of electromagnetic energy at this location per unit area per unit time is, in magnitude and direction,

$$f = E \times H \tag{8.55}$$

In agreement with eqn. 8.50, we shall find in the next section that, at this location, the electromagnetic momentum per unit volume is, in magnitude and direction,

$$g = D \times B \tag{8.56}$$

The two vectors f and g in eqns. 8.55 and 8.56 are related in accordance with eqn. 8.54.

A similar relation is obtained between energy flow and momentum for a mechanical system provided that the energy associated with the mass is taken into account. Consider a gas consisting of N particles per unit volume each of mass m moving with velocity v relative to the observer. Each particle is conveying an amount of energy mc^2 (Appendix C), and the flow of energy per unit area per unit time is

$$f = Nmc^2v \tag{8.57}$$

But the momentum per unit volume is

$$g = Nmv \tag{8.58}$$

Eqns. 8.57 and 8.58 show that f and g are related by eqn. 8.54.

For an acoustic wave travelling down an acoustic transmission line, the gas between two adjacent normal cross-sections at locations z and $z + dz$ has mechanical momentum parallel to the direction of propagation, and its time-rate of increase is created by the excess of the pressure at location z over the pressure at location $z + dz$. For an electromagnetic wave travelling down an electromagnetic transmission line in free space, the field between two adjacent normal cross-sections at locations z and $z + dz$ has electromagnetic momentum per unit volume g given by eqn. 8.50, and this momentum in general varies with time as a disturbance is propagated down the line. There is also a difference of the electromagnetic pressure p at the locations z and $z + dz$ that can be calculated from the electromagnetic stress tensors discussed in Chapters 4 and 5. Let us ask whether the time-rate of increase of the electro-magnetic momentum in the z direction existing between locations z and $z + dz$ is equal to the excess of the electromagnetic pressure at location z over that at location $z + dz$. In other words, we ask whether, as in an acoustic transmission line,

$$\frac{\partial g}{\partial t} = -\frac{\partial p}{\partial z} \tag{8.59}$$

For the strip transmission line discussed earlier in this section, let the generator be located in the plane $z = 0$ and the load in the plane $z = l$. Propagation of a disturbance along the line is then expressed by the equations (Problems 8.2–8.6)

$$E = \epsilon_0^{-1/2}F(t-z/c), \qquad D = \epsilon_0^{1/2}F(t-z/c) \left.\right) $$
$$H = \mu_0^{-1/2}F(t-z/c), \qquad B = \mu_0^{1/2}F(t-z/c) \left.\right\}$$
$$\tag{8.60}$$

where the function $F(t)$ describes the time-variation created at the location $z = 0$ by the generator. Examples of such a function are the gaussian pulse

$$F(t) = A \exp\{-\tfrac{1}{2}(t/T)^2\} \tag{8.61}$$

and the sinusoidal fluctuation

$$F(t) = A \cos \omega t \tag{8.62}$$

Both the flow of energy per unit area per unit time and the electromagnetic momentum are directed along the line in the z direction, and their magnitudes are (eqns. 8.51 and 8.50)

$$f = cF^2(t-z/c), \qquad g = c^{-1}F^2(t-z/c) \tag{8.63}$$

The electric and magnetic energy densities are (eqns. 4.6 and 5.55)

$$\tfrac{1}{2}\epsilon_0 E^2 = \tfrac{1}{2}F^2(t-z/c), \qquad \tfrac{1}{2}\mu_0 H^2 = \tfrac{1}{2}F^2(t-z/c) \tag{8.64}$$

and these expressions also give the tensions and sideways pressures in the tubes of electric and magnetic flux (eqns. 4.22, 4.28, 5.88 and 5.89). The total pressure exerted by the electromagnetic field on one side of a normal cross-section of the transmission line on the electromagnetic field on the other side is therefore

$$p = \tfrac{1}{2}\epsilon_0 E^2 + \tfrac{1}{2}\mu_0 H^2 = F^2(t-z/c) \tag{8.65}$$

and the negative gradient of this pressure in the direction of propagation is

$$-\partial p/\partial z = 2c^{-1}F(t-z/c)F'(t-z/c) \tag{8.66}$$

But differentiation of the second of eqns. 8.63 with regard to time gives

$$\partial g/\partial t = 2c^{-1}F(t-z/c)F'(t-z/c) \tag{8.67}$$

It follows from eqns. 8.66 and 8.67 that eqn. 8.59 is in fact true.

This result means that the difference of electromagnetic pressure between a pair of adjacent normal cross-sections of the transmission line is what is required to create the time rate of increase of the electromagnetic momentum in the intervening portion of the field. This illustrates with a simple example (i) the idea that a portion of an electromagnetic field has momentum, and (ii) the fact that the time rate of increase of this electromagnetic momentum is equal to the net force exerted on that portion of the field as a result of electromagnetic stress.

8.9   The momentum theorem for electromagnetic fields

To establish the principle of conservation of momentum for any electromagnetic field in free space, we start with Maxwell's equations in the form

$$\nabla \times E = -\partial B/\partial t \qquad \nabla \cdot B = 0 \left.\right\}$$
$$\nabla \times H = J + \partial D/\partial t \qquad \nabla \cdot D = \rho \left.\right\}$$
(8.68)

We multiply the electric curl equation vectorially by D, the magnetic curl equation vectorially by B, and add to obtain

$$(\nabla \times E) \times D + (\nabla \times H) \times B = \frac{\partial}{\partial t}(D \times B) + J \times B$$
(8.69)

Making use of the Maxwell divergence equations, this result may be written

$$\{(\nabla \times E) \times D + (\nabla \cdot D)E\} + \{(\nabla \times H) \times B + (\nabla \cdot B)H\}$$

$$= \frac{\partial}{\partial t}(D \times B) + \rho E + J \times B$$
(8.70)

The x component of the electric term on the left-hand side of eqn. 8.70 is

$$\left(\frac{\partial E_x}{\partial z} - \frac{\partial E_z}{\partial x}\right)D_z - \left(\frac{\partial E_y}{\partial x} - \frac{\partial E_x}{\partial y}\right)D_y + \left(\frac{\partial D_x}{\partial x} + \frac{\partial D_y}{\partial y} + \frac{\partial D_z}{\partial z}\right)E_x$$

Using the fact that $D = \epsilon_0 E$, this expression may be written

$$\frac{\partial}{\partial x}(E_x D_x - \tfrac{1}{2}E \cdot D) + \frac{\partial}{\partial y}(E_x D_y) + \frac{\partial}{\partial z}(E_x D_z)$$

or

$$\text{div}\,\{T_{xx}^{(e)}, T_{yx}^{(e)}, T_{zx}^{(e)}\}$$
(8.71)

where $\{T_{xx}^{(e)}, T_{yx}^{(e)}, T_{zx}^{(e)}\}$ is the vector corresponding to the first column of the electric stress tensor $T^{(e)}$ given by the sum of the tensors (4.34) and (4.35). In the same way the x component of the magnetic term on the left-hand side of eqn. 8.70 is

$$\text{div}\,\{T_{xx}^{(m)}, T_{yx}^{(m)}, T_{zx}^{(m)}\}$$
(8.72)

where $T^{(m)}$ is the magnetic stress tensor given by the sum of the tensors (5.92) and (5.93). It follows that, if T is the total electromagnetic stress tensor given by

$$T = T^{(e)} + T^{(m)}$$
(8.73)

then the left-hand side of eqn. 8.70 is the vector whose cartesian components are

$$[\text{div}\,(T_{xx}, T_{yx}, T_{zx}), \text{div}\,(T_{xy}, T_{yy}, T_{zy}), \text{div}\,(T_{xz}, T_{yz}, T_{zz})]$$
(8.74)

Let us now replace the left-hand side of eqn. 8.70 by expr. 8.74, integrate throughout a fixed volume V whose bounding surface is S, and apply the divergence theorem (Identity 17, Appendix B). If dS is a vector element of area of S directed normally outward, and $d\tau$ is an element of volume of V, we obtain

$$\int_S T \cdot dS = \frac{\partial}{\partial t} \int_V g \, d\tau + \int_V (\rho E + J \times B) \, d\tau \tag{8.75}$$

where g is given by eqn. 8.56 and the left-hand side of eqn. 8.75 denotes the vector whose cartesian components are

$$\int_S (T_{xx}, T_{yx}, T_{zx}) \cdot dS, \int_S (T_{xy}, T_{yy}, T_{zy}) \cdot dS, \int_S (T_{xz}, T_{yz}, T_{zz}) \cdot dS \tag{8.76}$$

Expr. 8.76 is the force exerted on the electromagnetic system within the volume V in consequence of the electromagnetic stress T acting across the bounding surface S. Eqn. 8.75 states that this force provides (i) the time-rate of increase of the electromagnetic momentum within V, and (ii) the forces exerted by the electromagnetic field on the charges and currents within V. This constitutes the momentum theorem for an electromagnetic field in free space.

If homogeneous linear isotropic materials are present, it is convenient to use version 2 of the electromagnetic equations in materials (Table 2.2). The vectors H and D involved in eqns. 8.68–8.76 are then replaced by H' and D', while J and ρ are the volume densities of free electric current and charge. It follows that the electromagnetic momentum per unit volume in the presence of homogeneous linear isotropic material is

$$g = D' \times B \tag{8.77}$$

Discussion of electromagnetic momentum in non-homogeneous, non-isotropic and non-linear materials is more difficult. Account has to be taken of both mechanical and electromagnetic momentum. As mentioned in Sections 4.10 and 6.10, geometrical distortion of the material in general occurs. When this varies with time, mechanical momentum is generated and must be taken into account. Even for an isotropic plasma the mechanical momentum of the charge-carriers must not be overlooked. A term corresponding to the time-rate of change of this mechanical momentum then appears in the second integral on the right-hand side of eqn. 8.85, and it must be transferred to the first integral (c.f. Section 8.6).

Problems

8.1 In cartesian co-ordinates (x, y, z) in free space, steady electric current flows in the plane $x = \frac{1}{2}a$ in the direction of the positive z axis and returns along the plane $x = -\frac{1}{2}a$ in the direction of the negative z axis. The surface density of the electric current in both planes is uniform and has magnitude i. Show that the magnetic-flux-density vector is given by

$$B = \begin{cases} 0, & x > \frac{1}{2}a \\ -(\mu_0 i/b)\hat{y}, & -\frac{1}{2}a < x < \frac{1}{2}a \\ 0, & x < -\frac{1}{2}a \end{cases}$$

and the magnetic vector potential by

$$
A = \begin{cases}
(\tfrac{1}{2}Ba)\hat{z}, & x \geqslant \tfrac{1}{2}a \\
Bx\hat{z}, & -\tfrac{1}{2}a \leqslant x \leqslant \tfrac{1}{2}a \\
-(\tfrac{1}{2}Ba)\hat{z}, & x \leqslant -\tfrac{1}{2}a
\end{cases}
$$

Describe qualitatively how the A and B fields are modified if the strips have a finite width b in the y direction $(b \gg a)$, showing that the magnitude of the magnetic vector potential then maximises on the current sheets.

8.2   A pair of perfect conductors (e.g. parallel wires) constitutes an infinite transmission line in free space having uniform inductance L per unit length and uniform capacitance C per unit length. At position z measured along the line, the voltage between the conductors is V and the current along the conductors is I, the positive conductor being the one on which the current is taken as positive in the direction z increasing. By dissecting out the portion of the electromagnetic field between normal cross-sections at positions z and $z + dz$, show that it constitutes an inductor of inductance $L\, dz$ and a capacitor of capacitance $C\, dz$, and that in consequence

$$
-\frac{\partial V}{\partial z} = L\frac{\partial I}{\partial t}, \qquad -\frac{\partial I}{\partial t} = C\frac{\partial V}{\partial t}
$$

where t denotes time. Deduce that both V and I satisfy the one-dimensional wave equation and that the velocity of propagation of waves along the line is

$$
v = (LC)^{-1/2}
$$

Show that the voltage may be written

$$
V = F(t - z/v) + G(t + z/v)
$$

where F and G are arbitrary differentiable functions. Deduce that, with $G = 0$, the solution represents a wave travelling in the positive z direction, and that for this wave

$$
\frac{V}{I} = \left(\frac{L}{C}\right)^{1/2}
$$

The quantity $(L/C)^{1/2}$ is known as the characteristic resistance of the transmission line.

8.3   By making use of the expressions for inductance and capacitance per unit length derived in Problems 2.3 and 2.5, deduce from the previous problem the velocity of propagation along a perfectly conducting coaxial transmission line and a perfectly conducting twin-wire transmission line in free space. Show that in both cases the velocity of propagation is the velocity of light in free space.

8.4   A strip transmission line consists of a pair of parallel infinitely long horizontal perfectly conducting strips of width b and separation a in free space, arranged one vertically above the other. It may be assumed that a is small compared with b and that, in consequence, the electromagnetic field between the strips is large compared with the external electromagnetic field, except close to the open edges. From the inductance and capacitance per unit length of the line derived in Problem 2.2, verify that the velocity of propagation of waves along the line is the velocity of light in free space. If E and H are the electric and magnetic field strengths between the strips at position z along the line and fringing is neglected, show that the voltage and current for the line at position z are

$$V = Ea, \qquad I = Hb$$

Hence show that, for the strip transmission line, the differential equations for the voltage and current in Problem 8.2 may be rewritten

$$-\frac{\partial E}{\partial z} = \mu_0 \frac{\partial H}{\partial t}, \qquad -\frac{\partial H}{\partial z} = \epsilon_0 \frac{\partial E}{\partial t}$$

8.5   In the previous problem, cartesian co-ordinates (x, y, z) are chosen between the strips such that z is measured along the line, x is measured in the positive direction of the electric field (from the positive strip to the negative strip) and y is measured in the positive direction of the magnetic field (right-handed about the positive directions of the currents in the strips). The dimension b is now allowed to tend to infinity, and then the dimension a. Show that the electromagnetic field takes the form

$$E = \{E(t, z), 0, 0\}, \qquad H = \{0, H(t, z), 0\}$$

where E and H are functions of t and z only. Write down Maxwell's equations for such a field in free space, and show that they are the differential equations for E and H derived in the previous problem.

8.6   For an electromagnetic field of the form

$$E = \{E(t, z), 0, 0\}, \qquad H = \{0, H(t, z), 0\}$$

in free space, where E and H are independent of the cartesian co-ordinates x and y, obtain the solution of Maxwell's equations that takes the form

$$E = F(t - z/c) + G(t + z/c)$$

where F and G are arbitrary differentiable functions. Show that, with $G = 0$, the solution represents a plane wave travelling in the positive z direction with the velocity c of light in free space, and that for this wave

$$\frac{E}{H} = \left(\frac{\mu_0}{\epsilon_0}\right)^{1/2}.$$

Show that this quantity is a resistance whose numerical magnitude is 377Ω approximately. Show that the direction of the electric vector, the direction of the magnetic vector and the direction of propagation form a right-handed system of directions both for a wave travelling in the $+z$ direction ($G = 0$) and for a wave travelling in the $-z$ direction ($F = 0$).

Chapter 9

Energy in oscillatory electric circuits

9.1 Introduction

In Section 7.2 we discussed the behaviour of a simple oscillatory electromagentic system consisting of an inductor of inductance L connected in parallel with a capacitor of capacitance C in circumstances where losses could be neglected. The system oscillates with angular frequency (eqn. 7.9)

$$\omega = (LC)^{-1/2} \tag{9.1}$$

in the manner indicated in Fig. 7.6. A total energy W_0 existing in the capacitor at time zero oscillates back and forth between the capacitor and the inductor with angular frequency 2ω. At time t the energies stored in the inductor and the capacitor, respectively, are (eqns. 7.14)

$$W_m = \tfrac{1}{2}W_0(1 - \cos 2\omega t), \quad W_e = \tfrac{1}{2}W_0(1 + \cos 2\omega t) \tag{9.2}$$

On the average, half the energy of the system is stored in the inductor and half in the capacitor (eqns. 7.16):

$$\bar{W}_m = \bar{W}_e = \tfrac{1}{2}W_0 \tag{9.3}$$

The variations with time of the magnetic energy W_m stored in the inductor and of the electric energy W_e stored in the capacitor are illustrated in Fig. 7.4. This oscillation of energy may also be represented with the aid of the vector diagram shown in Fig. 9.1. In this diagram $OA = AM = AE = W_0$ and the diameter MAE rotates counterclockwise with angular velocity 2ω. The reference components OM' and OE' of the vector $\overrightarrow{OM}$ and $\overrightarrow{OE}$ are then given by the right-hand sides of eqns. 9.2. As the diameter MAE rotates, therefore, the reference components of the vectors $\overrightarrow{OM}$ and $\overrightarrow{OE}$ describe the time variations of the magnetic energy stored in the inductor and of the electric energy stored in the capacitor.

The rate of flow of energy per unit time from the inductor to the capacitor is (eqn. 7.17)

$$\omega W_0 \cos (2\omega t + \tfrac{1}{2}\pi) \tag{9.4}$$

and when this quantity is negative its magnitude gives the rate of flow of energy from the capacitor to the inductor. Expr. 9.4 is plotted as a function of time in Fig. 7.5. This quantity may be illustrated with the aid of a vector diagram as shown in Fig. 9.2. The vector is of length ωW_0 and at time t it makes a counterclockwise

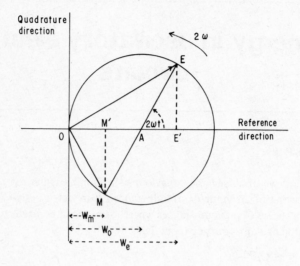

Fig. 9.1   *Vector representation of the energy stored in the capacitor and the inductor of a resonant oscillatory circuit*

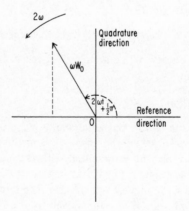

Fig. 9.2   *Vector representation of the reactive flow of energy between the capacitor and the inductor of a resonant oscillatory circuit*

angle $2\omega t + \frac{1}{2}\pi$ with the reference direction, so that its reference component is expr. 9.4. As the vector rotates in the counterclockwise direction with angular velocity 2ω, its reference component gives the rate of flow of energy from the inductor to the capacitor, negative values indicating flow in the reverse direction. Such an oscillatory flow of energy is known as a reactive flow of energy.

By contrast, a resistive flow of energy is unidirectional but not necessarily independent of time. Consider the arrangement shown in Fig. 9.3. A load of resistance R is connected across a current generator delivering a sinusoidal current of angular frequency ω and amplitude I_0 given by

$$I = I_0 \cos \omega t \tag{9.5}$$

The rate of flow of energy from the generator to the resistive load is then

$$RI^2 = RI_0^2 \cos^2 \omega t \tag{9.6}$$

and this may be written

$$\tfrac{1}{2}RI_0^2(1 + \cos 2\omega t) \tag{9.7}$$

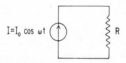

Fig. 9.3 *Resistive load connected across an oscillatory current generator*

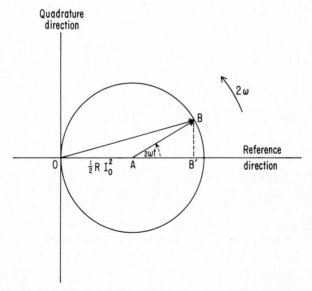

Fig. 9.4 *Vector representation of the resistive flow of energy from an oscillatory generator to a resistive load*

This quantity may be represented in a vector diagram as shown in Fig. 9.4. The radius of the circle is $\tfrac{1}{2}RI_0^2$, and the radius AB rotates in the counter-clockwise direction with angular velocity 2ω. The reference component OB$'$ of the vector $\overrightarrow{\text{OB}}$ is equal to expr. 9.6. We see that the rate of flow of energy from the generator to the resistor varies between a maximum value of RI_0^2 and a minimum value of zero, the mean value being $\frac{1}{2}RI_0^2$. The rate of flow of energy from the generator to the resistor varies with time, but there is never any backflow of energy from the resistor to the generator. Absence of backflow is the characteristic feature of a resistive flow of energy.

9.2 The concepts of resistive and reactive power

Oscillatory behaviour of networks of inductors, capacitors and resistors is of enormous importance in electrical engineering. Such networks involve the concepts of resistive and reactive flow of energy that are important in studying all oscillatory electromagnetic fields. It will be convenient to gain experience with these concepts by studying the simple system shown in Fig. 9.5. This consists of a resistor of resistance R, an inductor of inductance L and a capacitor of capacitance C connected in series across a pair of terminals. Any resistance possessed by the inductor is supposed to be included with the resistor. We suppose that the circuit is excited by a source of energy that is connected across the terminals and that is supplying alternating current of given angular frequency ω. We assume that no non-linear behaviour is involved.

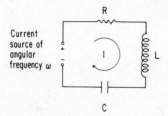

Fig. 9.5   *Illustrating a series resonant circuit driven by an oscillatory source of energy applied at the terminals*

For relating the amplitudes and phases of the various voltages and currents in an oscillatory electric network of angular frequency ω it is convenient to represent the voltages and currents as the reference components of vectors rotating in the complex plane with angular velocity ω (c.f. Section 2.9). For the circuit illustrated in Fig. 9.5, let the current have amplitude $|I|$ and be represented in the complex plane as the reference component (real part) of the counterclockwise rotating vector

$$I = |I| \exp(j\omega t). \tag{9.8}$$

We may assume that ω has an indefinitely small imaginary part (quadrature component) so that the current round the circuit and the voltage across the capacitor were zero at $t = -\infty$. Then, at time t, the voltages across the resistor, the inductor and the capacitor, respectively, are the reference components of the rotating vectors

$$RI, \quad L\frac{dI}{dt}, \quad \frac{1}{C}\int_{-\infty}^{t} I \, dt \tag{9.9}$$

or (eqns. 2.39 and 2.40)

$$RI, \quad j\omega LI, \quad \frac{1}{j\omega C}I \tag{9.10}$$

The voltage across the terminals in Fig. 9.5 is therefore the reference component of the rotating vector

$$V = RI + L\frac{dI}{dt} + \frac{1}{C}\int_{-\infty}^{t} I \, dt \tag{9.11}$$

$$= \left(R + j\omega L + \frac{1}{j\omega C}\right)I \tag{9.12}$$

$$= ZI \tag{9.13}$$

where

$$Z = R + j\left(\omega L - \frac{1}{\omega C}\right) \tag{9.14}$$

The reference component of eqn. 9.11 is Faraday's law of induction applied to the circuit. Notice that the sign conventions are those shown in Fig. 9.5; positive current enters the circuit at the positive terminal and leaves at the negative terminal.

The relation between the rotating vectors (9.10) is illustrated in Fig. 9.6. The relation between the rotating vectors V and I is illustrated in Fig. 9.7a. In accordance with eqn. 9.13, the ratio of the lengths of these vectors and the angle between them are given by the absolute value and the argument of the vector Z illustrated in Fig. 9.7b. This vector is known as the impedance looking into the terminals of the circuit shown in Fig. 9.5.

For calculating the time-variation of energy storage and flow, use must be made, not of the rotating vectors representing the voltages and currents, but of the actual oscillatory voltages and currents themselves. Squares and products of the voltages and currents are required, and it has to be remembered that the real part of the product of a pair of complex numbers is not in general equal to the product of the real parts. Let us therefore examine the actual voltages and currents associated with the circuit illustrated in Fig. 9.5.

The actual current round the circuit is the reference component of the rotating vector (9.8) and is therefore

$$|I| \cos \omega t \tag{9.15}$$

Likewise the actual voltages across the resistor, the inductor and the capacitor respectively are the reference components of the rotating vectors (9.10) illustrated in Fig. 9.6, and are therefore

$$R|I| \cos \omega t, \quad \omega L \, |I| : \cos (\omega t + \tfrac{1}{2}\pi), \quad \frac{1}{\omega C} |I| \cos (\omega t - \tfrac{1}{2}\pi) \tag{9.16}$$

By addition we see that the actual voltage across the terminals of the circuit in Fig. 9.5 is

$$R \, |I| \cos \omega t + \left(\omega L - \frac{1}{\omega C} \right) |I| \cos (\omega t + \tfrac{1}{2}\pi) \tag{9.17}$$

By forming the product of the voltage (9.17) and the current (9.15), we see that the rate per unit time at which energy is being supplied to the circuit by the source of energy connected to the terminals is

$$P |= R|I|^2 \cos^2 \omega t + \left(\omega L - \frac{1}{\omega C} \right) |I|^2 \cos \omega t \cos (\omega t + \tfrac{1}{2}\pi)$$

which may be rewritten

$$P = \tfrac{1}{2} R|I|^2 (1 + \cos 2\omega t) + \tfrac{1}{2} \left(\omega L - \frac{1}{\omega C} \right) |I|^2 \cos (2\omega t + \tfrac{1}{2}\pi) \tag{9.18}$$

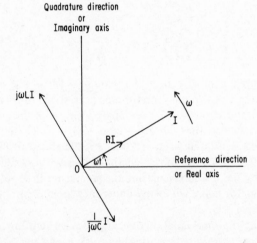

Fig. 9.6 *Illustrating the rotating vectors whose reference components give the voltage across the resistor, the inductor and the capacitor in Fig. 9.5*

The time-variation of this power may be illustrated by means of reference components in the vector diagram shown in Fig. 9.8. Here the vector $\overrightarrow{OA}$ represents the constant term on the right-hand side of eqn. 9.18, namely

$$\tfrac{1}{2} R|I|^2 \tag{9.19}$$

The vector $\overrightarrow{AB}$ is represented in the complex plane as

$$\tfrac{1}{2} R|I|^2 \exp (j2\omega t) \tag{9.20}$$

and its reference component is the second term on the right-hand side of eqn. 9.18, namely

$$\tfrac{1}{2}R|I|^2\cos 2\omega t \tag{9.21}$$

The vector $\overrightarrow{BC}$ in Fig. 9.8 is represented in the complex plane as

$$\frac{1}{2}\left(\omega L - \frac{1}{\omega C}\right)|I|^2\exp\left\{j\left(2\omega t + \tfrac{1}{2}\pi\right)\right\} \tag{9.22}$$

and its reference component is the remaining term in eqn. 9.18, namely

$$\frac{1}{2}\left(\omega L - \frac{1}{\omega C}\right)|I|^2\cos\left(2\omega t + \tfrac{1}{2}\pi\right) \tag{9.23}$$

The vectors $\overrightarrow{AB}$ and $\overrightarrow{BC}$ are mutually perpendicular and they rotate solidly about the point A with angular velocity 2ω. In accordance with eqn. 9.18, the reference component of the resultant vector $\overrightarrow{OC}$ gives the time-variation of the rate of supply of energy to the circuit.

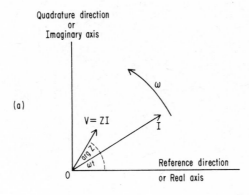

(a)

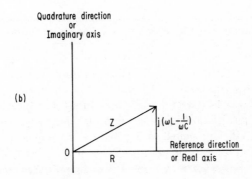

(b)

Fig. 9.7 *Illustrating the relation between the rotating vectors whose reference components give the voltage and current at the terminals of the circuit shown in Fig. 9.5*

Because the circle traced out by the point C in Fig. 9.8 in general envelopes the origin it follows that, although energy is being supplied to the circuit over a major portion of each half cycle of the oscillation of angular frequency ω, nevertheless there is a minor portion of the half cycle during which energy is being repaid to the source by the circuit. If there were only resistance in the circuit we would only have the vectors $\overrightarrow{OA}$ and $\overrightarrow{AB}$; the vector $\overrightarrow{BC}$ would be missing. The resultant vector would then be $\overrightarrow{OB}$ and its tip would move round the broken-line circle shown in Fig. 9.8. There would be no back-flow of energy. It is the existence of energy storage in the inductor and the capacitor that causes the back-flow. On the other hand, if there were no resistance in the circuit, only the rotating vector $\overrightarrow{BC}$ would appear in Fig. 9.8. This would correspond to an oscillation of energy into and out of the circuit given by expr. 9.23. Energy would then be loaned by the source to the circuit and repaid by the circuit to the source in alternate quarter-periods of the oscillation of angular frequency ω.

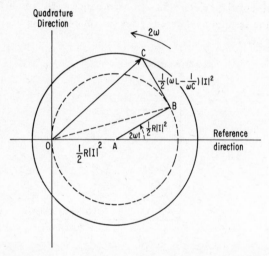

Fig. 9.8 *Vector representation of the time-variation of the power supplied to the circuit in Fig. 9.5 by the oscillatory source of energy applied at the terminals*

The oscillation of energy into and out of the circuit given by expr. 9.23 and represented in Fig. 9.8 by the vector $\overrightarrow{BC}$ constitutes a reactive rate of flow of energy, that is, reactive power (c.f. Fig. 9.2). It is the presence of reactive power that causes back-flow of energy to the source during part of each half-period of the oscillation. The rate of flow of energy into the circuit given by the sum of expr. 9.19 and 9.21, and represented in Fig. 9.8 by the reference component of the vector $\overrightarrow{OB}$, constitutes resistive power (c.f. Fig. 9.4). Resistive power is the sum of (i) mean resistive power given by expr. 9.19 and represented by the vector $\overrightarrow{OA}$ and (ii) oscillatory resistive power given by expr. 9.21 and represented by the reference component of the vector $\overrightarrow{AB}$.

Notice that the concept of reactive power is not identical with the concept of oscillatory power. In Fig. 9.8, the vector $\overrightarrow{OA}$ corresponds to the mean power and the reference component of the sum of the vectors $\overrightarrow{AB}$ and $\overrightarrow{BC}$ gives the oscillatory power. But it is the reference component of the sum of the vectors $\overrightarrow{OA}$ and $\overrightarrow{AB}$ that gives the resistive power, and the reference component of only the vector $\overrightarrow{BC}$ that gives the reactive power. The presence of reactive power connotes back-flow of energy to the source during part of each half-cycle of the oscillation of angular frequency ω.

9.3 The importance of reactive power

Reactive power is important because, although it constitutes only a loan-business in energy, nevertheless in practice it leads to loss of energy and frequently to unnecessary loss of energy. This happens when the source of energy, instead of being connected directly to a circuit such as that illustrated in Fig. 9.5, is instead at some distance, the two being connected by an intervening transmission line. An example would be an alternating current generating station connected to a consumer via a power line. Due to energy losses in the interconnecting transmission line, a purely loan-business in energy at the load-end results in actual expenditure of effort at the source-end, because the source has to provide a little more energy than the load borrows, and likewise receives back a little less than the load repays. Efficiency therefore requires that no unnecessary loan-business in energy be transacted at the load-end. It is consequently important, not only to be able to recognize when reactive oscillation of energy is taking place, but also to recognize what action is needed to eliminate it.

In this connection the sign of the factor

$$\omega L - \frac{1}{\omega C} \tag{9.24}$$

in the reactive oscillation given by expr. 9.23 is important. To eliminate the reactive power we must make expr. 9.24 vanish so that the full-line circle in Fig. 9.8 becomes identical with the broken-line circle. From expr. 9.24 this requires that

$$\omega = (LC)^{-1/2} \tag{9.25}$$

From eqn. 9.1 we see that this means that the inductance L and the capacitance C in Fig. 9.5 must be adjusted so as to form a resonant circuit at the operating angular frequency ω. If this condition is not satisfied, then the sign of expr. 9.24 tells us whether the reactive oscillation arises from too large an inductive effect or too large a capacitive effect.

The relative importance of the inductance and capacitance is conveniently studied in terms of the energy that they store. From expr. 9.15, the magnetic energy stored by the inductor is

$$W_m = \tfrac{1}{2}L|I|^2 \cos^2 \omega t = \tfrac{1}{4}L|I|^2(1 + \cos 2\omega t) \tag{9.26}$$

From the last of exprs. 9.16, the electric energy stored in the capacitor is

$$W_e = \frac{1}{2}\frac{1}{\omega^2 C}|I|^2 \sin^2 \omega t = \frac{1}{4}\frac{1}{\omega^2 C}|I|^2(1 - \cos 2\omega t) \tag{9.27}$$

Adding eqns. 9.26 and 9.27 we see that the total electromagnetic energy stored in the circuit illustrated in Fig. 9.5 is

$$W = \frac{1}{4}\left(\omega L + \frac{1}{\omega C}\right)\frac{1}{\omega}|I|^2 + \frac{1}{4}\left(\omega L - \frac{1}{\omega C}\right)\frac{1}{\omega}|I|^2 \cos 2\omega t \qquad (9.28)$$

The time rate of change of the total electromagnetic energy stored by the circuit is therefore

$$\frac{dW}{dt} = \frac{1}{2}\left(\omega L - \frac{1}{\omega C}\right)|I|^2 \cos\left(2\omega t + \tfrac{1}{2}\pi\right) \qquad (9.29)$$

This is identical with expr. 9.23 for the reactive power, verifying that the reactive component of the rate of supply of energy to the circuit is used to store electromagnetic energy in the circuit.

When the condition (9.25) is satisfied, there is no reactive oscillation of energy at the terminals of the circuit illustrated in Fig. 9.5, and the total electromagnetic energy stored in the circuit is constant. But this does not mean that there is no reactive oscillation of energy associated with the circuit. With the total energy W constant and equal to W_0, eqns. 9.26 and 9.27 describe a reactive oscillation of energy between the inductor and the capacitor as discussed in connection with Fig. 9.2. However, this does not lead to a reactive oscillation of energy at the terminals of the circuit. When condition (9.25) is satisfied, we may describe the activity as follows. Starting from the instant of maximum energy storage in the inductor, this energy is emptied into the capacitor where it becomes electric energy that reaches its maximum value a quarter of a cycle later. This energy is then emptied back into the inductor. Although there is reactive oscillation of energy between the inductor and the capacitor, the source connected to the terminals does not engage in an energy loan-business with the circuit. There is no reactive oscillation of energy between the source and the circuit even though there is between the inductor and the capacitor.

Now suppose that condition (9.25) is not satisfied and that, in particular, expr. 9.24 is positive. The total electromagnetic energy W stored in the circuit then has an oscillatory portion as described by eqn. 9.28. The total stored energy is a maximum at $t = 0$. At this time the magnetic energy W_m stored in the inductor is a maximum (eqn. 9.26) and the electric energy stored in the capacitor is zero (eqn. 9.27). What ensues is illustrated in Fig. 9.9a. During the first quarter-cycle of the oscillation of angular frequency ω the inductor empties its energy into the capacitor, which cannot however hold all of it, and the balance has to be repaid to the source. During the following quarter-cycle the capacitor empties its energy back into the inductor, which however needs more, and the balance has to be loaned back from the source. On the other hand, if expr. 9.24 is negative, it is the capacitor that has too much energy on the average, and the sequence of events is shown in Fig. 9.9b.

We see that reactive oscillation of energy at the terminals of the circuit arises from lack of balance between the amounts of electric and magnetic energy stored by the circuit. We can verify this from the mean values of the stored magnetic and electric energies given, in accordance with eqns. 9.26 and 9.27, by

$$\bar{W}_m = \tfrac{1}{4}L|I|^2 \tag{9.30}$$

$$\bar{W}_e = \frac{1}{4}\frac{1}{\omega^2 C}|I|^2 \tag{9.31}$$

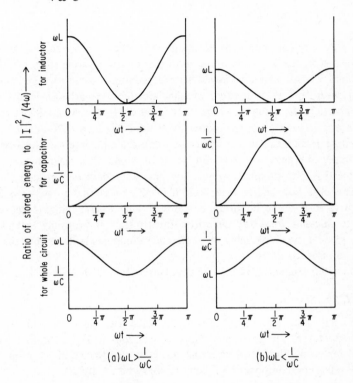

Fig. 9.9   *Illustrating, for the circuit shown in Fig. 9.5, the time-variation of the energy stored in the inductor, the energy stored in the capacitor and the energy stored in both*

We see that the excess of the mean stored magnetic energy over the mean stored electric energy is

$$\bar{W}_m - \bar{W}_e = \frac{1}{4}\left(\omega L - \frac{1}{\omega C}\right)\frac{1}{\omega}|I|^2 \tag{9.32}$$

Comparison between eqns. 9.29 and 9.32 shows that the time rate of change of the total electromagnetic energy stored by the circuit may be written

$$\frac{dW}{dt} = 2\omega(\bar{W}_m - \bar{W}_e)\cos(2\omega t + \tfrac{1}{2}\pi) \tag{9.33}$$

In accordance with expr. 9.23 and eqn. 9.32, the expression on the right-hand side of eqn. 9.33 is also the reactive oscillation in the rate of supply of energy to the circuit at its terminals, and we see that the amplitude of this oscillation is proportional to the difference between $\bar{W}_m$ and $\bar{W}_e$. It is therefore lack of balance between the mean values of the electric and magnetic energies stored by the system that leads to a reactive oscillation of energy at the terminals.

9.4 The concept of complex power

It will be noticed that, while the discussion in the preceding two sections of the circuit illustrated in Fig. 9.5 was commenced in eqn. 9.8 in terms of vectors rotating in the complex plane with angular velocity ω, the rotating vectors were abandoned beginning with eqn. 9.15 in order to calculate the time-variation in the rate of delivery of energy to the circuit (eqn. 9.18), and the time-variation in the amounts of magnetic and electric energy stored by the circuit (eqns. 9.26 and 9.27). This abandonment is essential if one wishes to study the complete time-history of the rate of supply of energy to the circuit and of the amounts of magnetic and electric energy stored by the circuit. However, now that we understand the general character of the energy flow and storage, we shall be able to see that a few key quantities are adequate to describe important features of what is happening. Moreover, it is possible to calculate these key quantities directly from the rotating vectors in the complex plane instead of using the reference components that give the actual vibrating voltages and currents.

The complex conjugate of the complex current round the circuit is, from eqn. 9.8,

$$I^* = |I| \exp(-j\omega t) \tag{9.34}$$

This is a vector that rotates in the complex plane with angular velocity ω in the clockwise direction. The reference component of the clockwise rotating vector in eqn. 9.34 represents the actual vibrating current round the circuit as completely as does the reference component of counterclockwise rotating vector in eqn. 9.8. Moreover, the complex product of the two counter-rotating vectors is independent of time:

$$II^* = |I|^2 \tag{9.35}$$

Hence eqn. 9.30 for the mean value of the magnetic energy stored in the inductor may be written

$$\bar{W}_m = \tfrac{1}{4}LII^* \tag{9.36}$$

In this way the mean magnetic energy stored in the inductor is expressed directly in terms of the complex current I associated with the inductor. In accordance with eqn. 9.26, the instantaneous value of the magnetic energy stored in the inductor varies between zero and a maximum value that is double expr. 9.36. If we want to know the epoch of maximum storage of energy in the inductor, eqn. 9.36 does not tell us. But eqn. 9.36 does tell us everything else that we need to know about storage of magnetic energy in the inductor.

Likewise it follows from the third of exprs. 9.10 that the voltage across the capacitor is the reference component of the rotating vector

$$V_C = \frac{1}{j\omega C}I \tag{9.37}$$

and the complex conjugate of this is

$$V_C^* = -\frac{1}{j\omega C}I^* \tag{9.38}$$

Hence

$$V_C V_C^* = \frac{1}{\omega^2 C^2}|I|^2 \tag{9.39}$$

and eqn. 9.31 may be written

$$\bar{W}_e = \tfrac{1}{4}C V_C V_C^* \tag{9.40}$$

Again we see from eqns. 9.27 and 9.39 that the instantaneous value of the electric energy stored in the capacitor varies between zero and a maximum value that is double expr. 9.40. Unless we have to know the epoch of maximum storage, eqn. 9.40 gives us the key information on storage of electric energy in the capacitor directly in terms of the complex voltage V_C across the capacitor.

To handle the rate of flow of energy from the source to the circuit directly in terms of rotating vectors in the complex plane, we combine the complex voltage across the terminals with the conjugate of the complex current at the terminals. We define a complex quantity $\tilde{P}$, known as the complex power from the source to the circuit at the terminals, by the equation

$$\tilde{P} = \tfrac{1}{2}VI^* \tag{9.41}$$

where V is the complex voltage across the terminals given by eqn. 9.12 and I^* is the conjugate of the complex current at the terminals given by eqn. 9.34. From eqns. 9.8 and 9.12, we see that

$$V = \left\{ R + j\left(\omega L - \frac{1}{\omega C}\right) \right\} |I| \exp(j\omega t) \tag{9.42}$$

Substitution from eqns. 9.34 and 9.42 into eqn. 9.41 then gives for the complex power

$$\tilde{P} = \frac{1}{2}\left\{ R + j\left(\omega L - \frac{1}{\omega C}\right) \right\}|I|^2 \tag{9.43}$$

Notice that the complex power is independent of time. It is a complex quantity whose real part gives the length of the vectors $\overrightarrow{OA}$ and $\overrightarrow{AB}$ in Fig. 9.8, and whose imaginary part gives the length of the vector $\overrightarrow{BC}$. As the mutually perpendicular vectors $\overrightarrow{AB}$ and $\overrightarrow{BC}$ in Fig. 9.8 rotate solidly with angular velocity 2ω, the instantaneous value of the resistive power varies between zero and a maximum value that is double the real part of the complex power $\tilde{P}$, while the reactive power oscillates in accordance with expr. 9.23 with an amplitude equal to the magnitude of the imaginary part of the complex power $\tilde{P}$. Moreover, the sign of the imaginary part of the complex power is the same as that of expr. 9.24 and tells us whether the reactive oscillation of energy at the terminals of the circuit arises for an excess of mean stored magnetic energy in the circuit or an excess of mean stored electric energy. Hence the complex power defined in eqn. 9.41 gives us the key information about rate of transfer of energy between the source and the circuit at the terminals, and does so directly in terms of the complex voltage and the complex current at the terminals. What the complex power fails to tell us is the orientation of the rotating perpendicular vectors $\overrightarrow{AB}$ and $\overrightarrow{BC}$ in Fig. 9.8 at any particular instant of time, but this we frequently do not need to know.

Complex power is a fixed vector in the complex plane, and its real and imaginary parts are called the resistive and reactive components of the complex power. Positive reactive components are said to be inductively reactive because they correspond to a system that is rich in mean stored magnetic energy; negative reactive components are said to be capacitively reactive because they correspond to a system that is rich in mean stored electric energy. This may be verified by using eqn. 9.32 to write eqn. 9.43 in the form

$$\tilde{P} = \tfrac{1}{2}R|I|^2 + j2\omega(\bar{W}_m - \bar{W}_e) \tag{9.44}$$

The resistive component of the complex power gives the mean rate of dissipation of energy in the circuit, corresponding to expr. 9.19 and to the vector $\overrightarrow{OA}$ in Fig. 9.8. The resistive component of the complex power also gives the amplitude of the oscillatory component of the resistive power, corresponding to expr. 9.21 and to the vector $\overrightarrow{AB}$ in Fig. 9.8. On the other hand, the reactive component of the complex power in eqn. 9.43 has a magnitude equal to the amplitude of the reactive oscillation of energy between the source and the circuit, corresponding to expr. 9.23 and to the vector $\overrightarrow{BC}$ in Fig. 9.8. In addition, the sign of the reactive component of the complex power tells us how to readjust the relative levels of mean stored magnetic and electric energies if we wish to eliminate the reactive oscillation of energy between the circuit and the source.

9.5 Use of root-mean-square voltages and currents

In handling storage and flow of energy in an oscillatory system it is often convenient to use, instead of the amplitudes of the various vibrations, their root-mean-square (RMS) values. The current through the inductor in Fig. 9.5 is given by expr. 9.15, and its RMS value is

$$\bar{I} = \frac{1}{\sqrt{2}}|I| \tag{9.45}$$

Hence, in the formulas of this chapter, $|I|^2$ may, if desired, be replaced by $2\bar{I}^2$. Likewise, the voltage across the terminals in Fig. 9.5 has amplitude $|V|$ derivable from eqn. 9.13 and, in terms of the RMS voltage $\bar{V}$, $|V|$ may be replaced by $2\bar{V}^2$.

From eqn. 9.37 the voltage across the capacitor in Fig. 9.5 has amplitude

$$|V_C| = \frac{1}{\omega C}|I| \tag{9.46}$$

and its RMS value is therefore

$$\bar{V}_C = \frac{1}{\omega C}\bar{I} \tag{9.47}$$

In terms of RMS voltages and currents, the mean energies stored in the inductor and the capacitor may be written, respectively, as (eqns. 9.30 and 9.31)

$$\bar{W}_m = \tfrac{1}{2}L\bar{I}^2, \quad \bar{W}_e = \tfrac{1}{2}C\bar{V}_C^2 \tag{9.48}$$

and the instantaneous values as (eqns. 9.26 and 9.27)

$$W_m = \tfrac{1}{2}L\bar{I}^2(1 + \cos 2\omega t), \quad W_e = \tfrac{1}{2}C\bar{V}_C^2(1 - \cos \omega t) \tag{9.49}$$

Likewise, the mean power entering the circuit at the terminals is (eqn. 9.19)

$$\bar{P} = R\bar{I}^2 \tag{9.50}$$

and the instantaneous value is (eqn. 9.18)

$$P = R\bar{I}^2(1 + \cos 2\omega t) + \left(\omega L - \frac{1}{\omega C}\right)\bar{I}^2 \cos (2\omega t + \tfrac{1}{2}\pi) \tag{9.51}$$

In Fig. 9.8, the lengths of the vectors $\overrightarrow{OA}$, $\overrightarrow{AB}$ and $\overrightarrow{BC}$, if expressed in terms of the RMS current round the circuit, are, respectively,

$$R\bar{I}^2, \quad R\bar{I}^2, \quad \left(\omega L - \frac{1}{\omega C}\right)\bar{I}^2 \tag{9.52}$$

The complex power entering the circuit at the terminals becomes (eqn. 9.43)

$$\tilde{P} = R\bar{I}^2 + j\left(\omega L - \frac{1}{\omega C}\right)\bar{I}^2 \tag{9.53}$$

or (eqn. 9.44)

$$\tilde{P} = R\bar{I}^2 + j2\omega(\bar{W}_m - \bar{W}_e) \tag{9.54}$$

The key information about the instantaneous power in eqn. 9.51 is contained in the complex power in eqn. 9.53. As shown in Fig. 9.8, backflow of energy at the terminals is avoided by eliminating the reactive power, and eqn. 9.54 shows that this involves balancing the mean magnetic and electric energies stored in the system.

9.6 The concept of impedance

For the oscillatory system shown in Fig. 9.5, the complex power from the source to the circuit at its terminals may be written (eqn. 9.53)

$$\tilde{P} = \left\{R + j\left(\omega L - \frac{1}{\omega C}\right)\right\}\bar{I}^2 \tag{9.55}$$

and from eqn. 9.14 we then see that the coefficient of $\overline{I}^2$ in eqn. 9.55 is the impedance Z looking into the terminals. The impedance looking into the terminals is thus the complex power from the source to the system measured per unit mean square current at the terminals. The impedance therefore provides the information on complex power in a form that is normalised to unit mean square current at the terminals. Moreover, impedance is a vector in the complex plane that is directly and simply calculated from the rotating vectors V and I in the complex plane in accordance with eqn. 9.13; to obtain Z we divide the rotating vector V whose reference component is the oscillatory voltage across the terminals by the rotating vector I whose reference component is the oscillatory current at the terminals. The impedance Z thus calculated is the complex power from the source to the system, measured per unit mean square current at the terminals.

We may express these results as follows. By definition

$$Z = \frac{V}{I} \tag{9.56}$$

and

$$\widetilde{P} = \tfrac{1}{2}VI^* \tag{9.57}$$

Substitution for V from eqn. 9.56 into eqn. 9.57 gives

$$\widetilde{P} = \tfrac{1}{2}ZII^* = \tfrac{1}{2}Z|I|^2 = Z\overline{I}^2 \tag{9.58}$$

Hence, Z is $\widetilde{P}$ measured per unit mean square current.

Impedance depends on how the system is constructed but is independent of the strength of excitation of the system. If we put

$$Z = R + jX \tag{9.59}$$

where R and X are real then, in accordance with eqn. 9.54, the sign of X tells us whether the circuit is rich in magnetic energy ($X > 0$) or rich in electric energy ($X < 0$). Moreover, in accordance with eqns. 9.51 and 9.53, $|X|$ is the amplitude of the reactive oscillation in the rate of flow of energy at the terminals, measured per unit mean square current, while R is the mean rate at which energy is being transferred permanently from the source to the system, also measured per unit mean square current at the terminals.

The impedance looking into a system at a pair of terminals is a key parameter of the system. It is readily calculated in terms of the complex voltage and current at the terminals and is represented by a vector in the complex plane as shown in Fig. 9.10*a*. The positive real axis in such an impedance diagram is said to be in the resistive direction, the positive imaginary axis in the inductively reactive direction, and the negative imaginary axis in the capacitively reactive direction.

It is often convenient to make use of the reciprocal of the impedance. This is known as the admittance looking into the system at the terminals, and it may be written

$$Y = G + jS \tag{9.60}$$

where G and S are real. Since

$$Y = \frac{I}{V} \tag{9.61}$$

it follows that

$$\tilde{P} = \tfrac{1}{2}Y^*VV^* = \tfrac{1}{2}Y^*|V|^2 = Y^*\bar{V}^2 \tag{9.62}$$

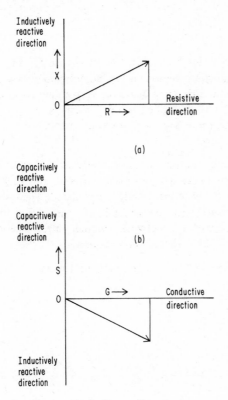

Fig. 9.10 *Illustrating impedance and admittance diagrams*

where $\bar{V}$ is the RMS voltage across the terminals. Hence the admittance looking into the terminals is the conjugate of the complex power from the source to the system, measured per unit mean square voltage across the terminals. The admittance is calculated by taking, in the complex plane, the quotient of the rotating vector whose reference component is the oscillatory current at the terminals by the rotating vector whose reference component is the oscillatory voltage across the terminals. Admittance is represented as a vector in the complex plane as shown in Fig. 9.10b. Because, in accordance with eqns. 9.56 and 9.61, the vectors $R + jX$ and $G + jS$ are the reciprocals of each other in the complex plane, the real parts R and G have the same sign but the imaginary parts X and S have opposite signs. The quantities R, G, X and S are known, respectively, as the resistance, conductance, reactance and susceptance associated with the impedance or admittance.

9.7 Electric network theory

Fig. 9.5 illustrates a one mesh example of an electric network that can store electric and magnetic energy, that can absorb electromagnetic energy resistively from a source and that can engage in a reactive loan-business in energy with a source. Multi-mesh networks of this type are of great practical importance, and their study is the subject of electric network theory. There are two main methods for handling the oscillatory behaviour of such networks:

(i) In the mesh current method a complex clockwise current is assigned to each mesh, so that elements common to two adjacent meshes are associated with a current equal to the difference of the corresponding mesh currents. Faraday's law of induction is then applied to each mesh as in eqns. 9.9–9.14. Enough linear algebraic equations are thereby obtained to solve for the complex mesh currents.

(ii) In the node potential method a complex electric scalar potential is assigned to each node of the network and, by means of equations such as (9.12), each of the complex currents entering a node is calculated. Their sum is then equated to zero in accordance with the principle of conservation of charge. By doing this for each node of the network, enough linear algebraic equations are obtained to solve for the complex node potentials.

There are many excellent texts on electric network theory.

Chapter 10

Energy storage in oscillatory electromagnetic fields

10.1 Introduction

The concepts of energy storage and flow employed in the previous chapter for an oscillatory electric circuit are capable of being applied to any electromagnetic field for which all components of all electromagnetic vectors everywhere vibrate with the same angular frequency ω. We shall discuss energy storage in oscillatory electromagnetic fields in this chapter and energy flow in the next.

All non-linear behaviour is ruled out in such a study. A sinusoidal variation in the electric current on the winding of an inductor with a ferromagnetic core does not give a sinusoidal variation in the threading magnetic flux unless the amplitude of the current oscillation is so small that the ferromagnetic material may be regarded as behaving linearly. Only if we are dealing with linear material is it possible for all cartesian components of the electromagnetic vectors to vary sinusoidally with time.

10.2 Two- and three-dimensional vector algebra

In the process of discussing electromagnetic vectors for which each cartesian component is oscillating with a specified angular frequency ω, we shall be involved in using, in three-dimensional space, vectors for which each cartesian component is a rotating vector in the complex plane (see Section 2.9). It will be necessary to distinguish a vector such as w in the complex plane having reference and quadrature components u, v and written

$$w = u + jv \tag{10.1}$$

from a vector such as A in three dimensional space having cartesian components A_x, A_y, A_z and written

$$A = (A_x, A_y, A_z) \tag{10.2}$$

It will be particularly important to distinguish between the magnitude of a vector in the complex plane and the magnitude of a vector in three-dimensional space. The magnitude of the two-dimensional vector (10.1) will be written $|w|$ where

$$|w|^2 = u^2 + v^2 \tag{10.3}$$

The magnitude of the three-dimensional vector (10.2) will be written A where

$$A^2 = A_x^2 + A_y^2 + A_z^2 \tag{10.4}$$

Notice that each cartesian component of the three-dimensional vector (10.2) may be a complex number, that is, a vector in the complex plane of the type (10.1). Thus each of the quantities A_x, A_y and A_z in eqn. 10.4 is in general complex, with the result that A is also complex. Hence the quantity A in eqn. 10.4, although it is the magnitude of the three dimensional vector (A_x, A_y, A_z), is nevertheless a vector in the complex plane of magnitude $|A|$ where

$$|A|^2 = |A_x^2 + A_y^2 + A_z^2| \tag{10.5}$$

Notice that the expression on the right-hand side of eqn. 10.5 is not the same thing as $|A_x|^2 + |A_y|^2 + |A_z|^2$. The latter quantity is the sum of the squares of the magnitudes of the complex numbers A_x, A_y and A_z constituting the cartesian components of the three-dimensional vector A, whereas for $|A_x^2 + A_y^2 + A_z^2|$ the complex numbers A_x, A_y and A_z are squared and added in the complex plane before the magnitude of this complex number is evaluated.

A convenient expression for the quantity $|A_x|^2 + |A_y|^2 + |A_z|^2$, when required, may be obtained as follows. The complex conjugate of the three-dimensional vector (10.2) is

$$A^* = (A_x^*, A_y^*, A_z^*) \tag{10.6}$$

and if we form the scalar product of the vectors (10.2) and (10.6) we obtain

or
$$A \cdot A^* = A_x A_x^* + A_y A_y^* + A_z A_z^*$$
$$A \cdot A^* = |A_x|^2 + |A_y|^2 + |A_z|^2 \tag{10.7}$$

From statements (10.5) and (10.7) it follows that

$$|A|^2 \neq A \cdot A^* \tag{10.8}$$

When handling oscillatory electromagnetic fields each cartesian component of each oscillatory electromagnetic vector is represented in the complex plane as the reference component of a rotating vector. Care is necessary not to confuse the three-dimensional vector algebra involving scalar and vector products with the two-dimensional vector algebra in the complex plane involving complex products and quotients.

10.3   The concept of elliptical polarisation

The electromagnetic fields associated with a circuit of the type discussed in the previous chapter are such that the tip of each electromagnetic vector at each point oscillates sinusoidally back and forth along a line. Such an electromagnetic field is said to be linearly polarised. In general, however, electromagnetic fields are more complicated than this, even if they have a specified angular frequency of oscillation ω. Each electromagnetic vector in general has three cartesian components all vibrating sinusoidally with the same angular frequency ω. But the amplitudes and phases of the three cartesian components of vibration are, in general, different and we need to ask what motion is executed by the tip of the resultant vector.

Consider an electromagnetic field of angular frequency ω and let the cartesian components of the complex electric and magnetic-flux-density vectors at a particular location be (Section 2.9)

$$E = (E_x^0, E_y^0, E_z^0) \exp{(j\omega t)} \tag{10.9}$$

$$B = (B_x^0, B_y^0, B_z^0) \exp{(j\omega t)} \tag{10.10}$$

The quantities E_x^0, E_y^0, E_z^0 and B_x^0, B_y^0, B_z^0 are the cartesian components of the three-dimensional vectors E and B at time zero; they are the complex amplitudes of the vibrations executed by the cartesian components of the electric and magnetic-flux-density vectors. For simplicity let us assume that ω is real, although the discussion can be adapted to cover a complex ω if desired.

The x component of the electric vector is a sinusoidal oscillation of phase $\arg E_x^0$ and amplitude $|E_x^0|$; the amplitude may also be written $|E_x|$. Similar statements apply to the y and z components, and also to the three components of the magnetic-flux-density vector. In terms of these amplitudes and phases, we may write eqns. 10.9 and 10.10 as

$$E = [|E_x| \exp{\{j(\omega t + \arg E_x^0)\}}, |E_y| \exp{\{j(\omega t + \arg E_y^0)\}},$$

$$|E_z| \exp{\{j(\omega t + \arg E_z^0)\}}] \tag{10.11}$$

$$B = [|B_x| \exp{\{j(\omega t + \arg B_x^0)\}}, |B_y| \exp{\{j(\omega t + \arg B_y^0)\}},$$

$$|B_z| \exp{\{j(\omega t + \arg B_z^0)\}}] \tag{10.12}$$

Taking the real parts, we see that the cartesian components of the oscillatory electric vector are

$$[|E_x| \cos{(\omega t + \arg E_x^0)}, |E_y| \cos{(\omega t + \arg E_y^0)}, |E_z| \cos{(\omega t + \arg E_z^0)}] \tag{10.13}$$

and those of the oscillatory magnetic-flux-density vector are

$$[|B_x| \cos{(\omega t + \arg B_x^0)}, |B_y| \cos{(\omega t + \arg B_y^0)}, |B_z| \cos{(\omega t + \arg B_z^0)}] \tag{10.14}$$

Any pair of cartesian components of either the electric or magnetic-flux-density vector constitute a pair of mutually perpendicular sinusoidal vibrations of the same frequency. For example, the component of the electric vector in the xy plane is a combination of a sinusoidal vibration of amplitude $|E_x|$ and phase

$\arg E_x^0$ along the x axis, and a sinusoidal vibration of amplitude $|E_y|$ and phase $\arg E_y^0$ along the y axis. These two sinusoidal vibrations of the same frequency cause the tip of the component of the electric vector in the xy plane to follow an ellipse whose major and minor axes, as shown in Fig. 10.1, do not in general coincide with the co-ordinate axes. The locus of a point from which the tangents to the ellipse are mutually perpendicular is a circle that is known as the director circle and is shown broken in Fig. 10.1. The radius of the circle is

$$\{|E_x|^2 + |E_y|^2\}^{1/2} \tag{10.15}$$

so that this expression is independent of the orientation of the co-ordinate axes. The square of the magnitude of the component of the electric vector in the xy plane at time t is

$$|E_x|^2 \cos^2(\omega t + \arg E_x^0) + |E_y|^2 \cos^2(\omega t + \arg E_y^0) \tag{10.16}$$

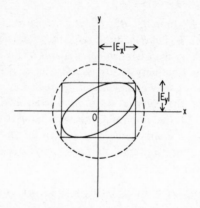

Fig. 10.1.   *Illustrating elliptical polarisation of the electric vector in the (x, y) plane*

Since the average value of each squared cosine over a period of the vibration is one-half, the time-average value of expr. 10.16 is

$$\frac{1}{2}\{|E_x|^2 + |E_y|^2\} \tag{10.17}$$

Hence the root-mean-square magnitude of the component of the electric vector in the xy plane is

$$\frac{1}{\sqrt{2}}\{|E_x|^2 + |E_y|^2\}^{1/2} \tag{10.18}$$

which is $1/\sqrt{2}$ times the radius of the director circle in Fig. 10.1. Similar statements apply to the other two co-ordinate planes for the electric vector, and in the three co-ordinate planes for the magnetic-flux-density vector.

The tip of the electric vector in three dimensions traces a curve whose projections onto the three co-ordinate planes are the ellipses just discussed. The three elliptic cylinders of projection intersect in an ellipse whose plane does not in general coincide with a co-ordinate plane. This is the ellipse traced out by the tip of the electric vector in three dimensional space. The tip of the magnetic-flux-density vector likewise traces out an ellipse whose plane does not in general coincide with a co-ordinate plane or with the plane of the electric ellipse.

In simple situations, these ellipses collapse to straight lines. When the electric vector at a particular point in an oscillatory electromagnetic field is linearly polarised, the tip of the vector vibrates harmonically back and forth along a fixed straight line. This occurs when the three cartesian components of the vibrating vector in expr. 10.13 have the same phase. Taking this common phase to be α, the complex electric vector in eqn. 10.11 then becomes

$$E = (|E_x|, |E_y|, |E_z|), \exp\{j(\omega t + \alpha)\}. \tag{10.19}$$

Hence the vector

$$(|E_x|, |E_y|, |E_z|), \tag{10.20}$$

defines the line along which the tip of the linearly polarised electric vector vibrates.

For a linearly polarised oscillatory electromagnetic field there is, at each point, a unique direction in space associated with each electromagnetic vector, namely the direction of the straight line along which this electromagnetic vector is vibrating (c.f. expr. 10.20). On the other hand, for elliptical polarisation, a pair of directions are involved for each electromagnetic vector at each point of space. Consider the complex electric vector in eqn. 10.11 under conditions where $\arg E_x^0$, $\arg E_y^0$ and $\arg E_z^0$ are not all equal. We may then separate E_x, E_y and E_z into their real and imaginary parts:

$$\left.\begin{aligned} E_x &= E_x^{(r)} + jE_x^{(i)} \\ E_y &= E_y^{(r)} + jE_y^{(i)} \\ E_z &= E_z^{(r)} + jE_z^{(i)} \end{aligned}\right\} \tag{10.21}$$

and the two vectors

$$(E_x^{(r)}, E_y^{(r)}, E_z^{(r)}) \quad \text{and} \quad (E_x^{(i)}, E_y^{(i)}, E_z^{(i)}) \tag{10.22}$$

then point in different directions. It is only for linear polarisation that these two vectors have the same direction. This direction is then also that of the vector (10.20).

The quantities

$$\left\{\begin{aligned} \bar{E} &= \frac{1}{\sqrt{2}}\{|E_x|^2 + |E_y|^2 + |E_z|^2\}^{1/2} \end{aligned}\right. \tag{10.23}$$

$$\left. \bar{B} = \frac{1}{\sqrt{2}}\{|B_x|^2 + |B_y|^2 + |B_z|^2\}^{1/2} \right. \tag{10.24}$$

are of interest. The quantities in the brackets are the sums of the squares of the amplitudes of the vibrations along the three co-ordinate axes for the electric vector and for the magnetic-flux-density vector, respectively. From exprs. 10.13 and 10.14 we may see (c.f. exprs. 10.16–10.18) that $\bar{E}$ and $\bar{B}$, are respectively, the root-mean-square magnitudes of the electric vector and of the magnetic-flux-density vector over a period of the vibration.

The quantities $\bar{E}$ and $\bar{B}$ defined in eqns. 10.23 and 10.24 are independent of the orientation of the co-ordinates axes and may be expressed in terms of scalar products. The complex conjugates of the vectors (10.9) and (10.10) are

$$\begin{cases} E^* = (E_x^{0*}, E_y^{0*}, E_z^{0*}) \exp\left(-j\omega t\right) & (10.25) \\ B^* = (B_x^{0*}, B_y^{0*}, B_z^{0*}) \exp\left(-j\omega t\right) & (10.26) \end{cases}$$

and each cartesian component in these equations is a vector in the complex plane that rotates in the clockwise direction with angular velocity ω. From eqns. 10.9, 10.10, 10.25 and 10.26, it follows that

$$\begin{cases} E \cdot E^* = E_x^0 E_x^{0*} + E_y^0 E_y^{0*} + E_z^0 E_z^{0*} & (10.27) \\ B \cdot B^* = B_x^0 B_x^{0*} + B_y^0 B_y^{0*} + B_z^0 B_z^{0*} & (10.28) \end{cases}$$

or

$$\begin{cases} E \cdot E^* = |E_x|^2 + |E_y|^2 + |E_z|^2 & (10.29) \\ B \cdot B^* = |B_x|^2 + |B_y|^2 + |B_z|^2 & (10.30) \end{cases}$$

Hence eqns. 10.23 and 10.24 may be written

$$\begin{cases} \bar{E} = \dfrac{1}{\sqrt{2}} (E \cdot E^*)^{1/2} & (10.31) \\[4mm] \bar{B} = \dfrac{1}{\sqrt{2}} (B \cdot B^*)^{1/2} & (10.32) \end{cases}$$

Similar statements may be made for all other electromagnetic vectors in an electromagnetic field that oscillates sinusoidally.

10.4   The time-variation of energy density in oscillatory electromagnetic fields

Let us consider what is involved in obtaining the complete information on the time-variation of energy density in an oscillatory electromagnetic field of angular frequency ω. We assume that we are located in material having both dielectric and magnetic properties, and we use version 2 of the electromagnetic equations in materials (Table 2.1). The material is linear, but it may be non-isotropic. The electric energy density is given by eqn. 4.56 and the magnetic energy density by eqn. 6.66. The oscillatory electric vector and magnetic-flux-density vector are given by eqns. 10.13 and 10.14 with similar expressions for the electric-flux-density vector and the magnetic vector.

The electric energy density is

$$w_e = \tfrac{1}{2}|E_xD'_x|\cos(\omega t + \arg E^0_x)\cos(\omega t + \arg D'^0_x)$$
$$+ \tfrac{1}{2}|E_yD'_y|\cos(\omega t + \arg E^0_y)\cos(\omega t + \arg D'^0_y)$$
$$+ \tfrac{1}{2}|E_zD'_z|\cos(\omega t + \arg E^0_z)\cos(\omega t + \arg D'^0_z) \qquad (10.33)$$

and using appropriate trigonometric formulas this may be written

$$w_e = \tfrac{1}{4}|E_xD'_x|\cos(\arg E^0_x - \arg D'^0_x) + \cos(2\omega t + \arg E^0_x + \arg D'^0_x)\}$$
$$+ \tfrac{1}{4}|E_yD'_x|\cos(\arg E^0_y - \arg D'^0_y) + \cos(2\omega t + \arg E^0_y + \arg D'^0_y)\}$$
$$+ \tfrac{1}{4}|E_zD'_z|\cos(\arg E^0_z - \arg D'^0_z) + \cos(2\omega t + \arg E^0_z + \arg D'^0_z)\}$$
$$(10.34)$$

of which the mean value is

$$\bar{w}_e = \tfrac{1}{4}\{|E_xD'_x|\cos(\arg E^0_x - \arg D'^0_x)$$
$$+ |E_yD'_y|\cos(\arg E^0_y - \arg D'^0_y) + |E_zD'_z|\cos(\arg E^0_z - \arg D'^0_z)$$
$$(10.35)$$

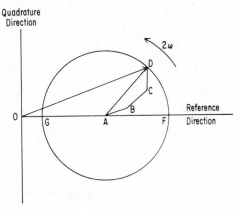

Fig. 10.2   *Vector diagram representing the time-variation of energy density*

The time-variation of expr. 10.34 may be described with the aid of the vector diagram shown in Fig. 10.2. Here $\overrightarrow{OA}$ is a fixed vector in the reference direction whose length is $\bar{w}_e$ given by eqn. 10.35. The vectors $\overrightarrow{AB}$, $\overrightarrow{BC}$ and $\overrightarrow{CD}$, together with their resultant $\overrightarrow{AD}$, form a quadrilateral of constant shape that rotates about the point A in the counterclockwise direction with angular velocity 2ω. We define

$$\overrightarrow{AB} = \tfrac{1}{4}|E_xD'_x|\exp\{j(2\omega t + \arg E^0_x + \arg D'^0_x)\}$$
$$\overrightarrow{BC} = \tfrac{1}{4}|E_yD'_y|\exp\{j(2\omega t + \arg E^0_y + \arg D'^0_y)\} \qquad (10.36)$$
$$\overrightarrow{CD} = \tfrac{1}{4}|E_zD'_z|\exp\{j(2\omega t + \arg E^0_z + \arg D'^0_z)\}$$

so that the reference components of these vectors are the three oscillations of angular frequency 2ω appearing in eqn. 10.34. The vector $\overrightarrow{OD}$ is then the vector sum of $\overrightarrow{OA}$, $\overrightarrow{AB}$, $\overrightarrow{BC}$, $\overrightarrow{CD}$, so that the reference component of $\overrightarrow{OD}$ is the electric energy density w_e in eqn. 10.34. Consequently the variation of the reference component of the vector $\overrightarrow{AD}$ in Fig. 10.2 as the point D moves round the circle with angular velocity 2ω gives the time-variation of the electric energy density.

The electric energy density oscillates between a maximum value OA + AD corresponding to the point F and a minimum value OA–AD corresponding to the point G. The point G cannot be to the left of the origin because the electric energy density cannot be negative. But G can coincide with 0. This occurs when both the electric vector and the electric-flux-density vector vibrate linearly in the same phase:

$$\left. \begin{aligned} \arg E_x^0 &= \arg E_y^0 = \arg E_z^0 \\ &= \arg D_x'^0 = \arg D_y'^0 = \arg D_z'^0 \end{aligned} \right\} \tag{10.37}$$

On the other hand, the circle in Fig. 10.2 can shrink to a point at A, and there is then no time-variation in the electric energy density. This occurs when the field is circularly polarised, for example, when

$$\left. \begin{aligned} |E_x D_x'| &= |E_y D_y'|; \quad |E_z D_z'| = 0 \\ \arg E_x^0 &= \arg D_x'^0 = 0 \\ \arg E_y^0 &= \arg D_y'^0 = \tfrac{1}{2}\pi \end{aligned} \right\} \tag{10.38}$$

Similar statements apply for the magnetic energy density, the mean value of which is given by (c.f. eqn. 10.35).

$$\bar{w}_m = \tfrac{1}{4}\{|H_x' B_x| \cos{(\arg H_x'^0 - \arg B_x^0)} + |H_y' B_y| \cos{(\arg H_y'^0 - \arg B_y^0)}$$
$$+ |H_z' B_z| \cos{(\arg H_z'^0 - \arg B_z^0)}\} \tag{10.39}$$

After calculating the vectors $\overrightarrow{OD}$ in Fig. 10.2 for both the electric and magnetic energy densities, we can form the resultant of the two vectors. The reference component of this resultant then gives the time-variation of the total electromagnetic energy density $w_e + w_m$.

10.5   Mean energy density expressed in terms of the complex field vectors

The most important quantities connected with electric and magnetic energy density in an oscillatory electromagnetic field are their mean values $\bar{w}_e$ and $\bar{w}_m$ given by eqns. 10.35 and 10.39. We can show that these quantities are directly expressible in terms of the complex electromagnetic field (E, D', H', B) by means of the relations

$$\bar{w}_e = \tfrac{1}{4} E \cdot D'^*, \qquad \bar{w}_m = \tfrac{1}{4} H'^* \cdot B \tag{10.40}$$

where D'^* and H'^* are the complex conjugates of D' and H'.

The first of these equations involves the scalar product of the complex electric vector E in eqn. 10.11 with the vector

$$D'^* = [|D'_x| \exp\{-j(\omega t + \arg D'^0_x)\}, |D'_y| \exp\{-j(\omega t + \arg D'^0_y)\},$$

$$|D'_z| \exp\{-j(\omega t + \arg D'^0_z)\}] \tag{10.41}$$

so that

$$\begin{aligned}\tfrac{1}{4} E \cdot D'^* = \tfrac{1}{4}[&|E_x D'_x| \exp\{j(\arg E^0_x - \arg D'^0_x)\} \\ &+ |E_y D'_y| \exp\{j(\arg E^0_y - \arg D'^0_y)\} \\ &+ |E_z D'_z| \exp\{j(\arg E^0_z - \arg D'^0_z)\}]\end{aligned} \tag{10.42}$$

Similarly,

$$\begin{aligned}\tfrac{1}{4} H'^* \cdot B = \tfrac{1}{4}[&|H'_x B_x| \exp\{-j(\arg H'^0_x - \arg B^0_x)\} \\ &+ |H'_y B_y| \exp\{-j(\arg H'^0_y - \arg B^0_y)\} \\ &+ |H'_z B_z| \exp\{-j(\arg H'^0_z - \arg B^0_z)\}]\end{aligned} \tag{10.43}$$

It is immediately clear that the real parts of the expressions on the right-hand sides of eqns. 10.42 and 10.43 are the expressions for the mean stored electric and magnetic energy densities $\bar{w}_e$ and $\bar{w}_m$ given in eqns. 10.35 and 10.39. However, to establish the truth of eqns. 10.40 we also have to show that the imaginary parts of $\tfrac{1}{4} E \cdot D'^*$ and $\tfrac{1}{4} H'^* \cdot B$ vanish even when the electromagnetic field is elliptically polarised.

First let us consider isotropic material. For linear isotropic material of dielectric constant ϵ/ϵ_0 and permeability μ/μ_0 we have (eqns. 2.25)

$$D' = \epsilon E, \qquad H' = \frac{1}{\mu} B \tag{10.44}$$

For each three-dimensional cartesian component, the phase involved in the vibration of the electric flux density is then the same as the phase involved in the vibration of the electric field, so that

$$\arg E^0_x = \arg D'^0_x, \quad \arg E^0_y = \arg D'^0_y, \quad \arg E^0_z = \arg D'^0_z \tag{10.45}$$

Substitution from eqns. 10.45 into eqn. 10.42 gives

$$\tfrac{1}{4} E \cdot D'^* = \tfrac{1}{4}(|E_x D'_x| + |E_y D'_y| + |E_z D'_z|) \tag{10.46}$$

Similarly, eqn. 10.39 becomes, for linear isotropic material,

$$\tfrac{1}{4} H'^* \cdot B = \tfrac{1}{4}(|H'_x B_x| + |H'_y B_y| + |H'_z B_z|) \tag{10.47}$$

It is clear from eqns. 10.46 and 10.47 that $\tfrac{1}{4} E \cdot D'$ and $\tfrac{1}{4} H'^* \cdot B$ are in fact real, and this completes the proof of eqns. 10.40 for linear isotropic material.

We see therefore that, for linear isotropic material, the mean electric and magnetic energy densities may be calculated directly from the complex electromagnetic field vectors by means of eqns. 10.40. Using eqns. 10.44, we may write eqns. 10.40 in the form

$$\bar{w}_e = \tfrac{1}{4}\epsilon E \cdot E^*, \qquad \bar{w}_m = \frac{1}{4\mu} B \cdot B^* \tag{10.48}$$

Using eqns. 10.31 and 10.32, we see that eqns. 10.48 imply that

$$\bar{w}_e = \tfrac{1}{2}\epsilon\bar{E}^2, \qquad \bar{w}_m = \frac{1}{2\mu}\bar{B}^2 \tag{10.49}$$

For non-isotropic material, it still follows that the real parts of the expressions on the right-hand sides of eqns. 10.42 and 10.43 are the expressions for $\bar{w}_e$ and $\bar{w}_m$ given in eqns. 10.35 and 10.39. But to show that the imaginary parts of the expressions on the right-hand sides of eqns. 10.42 and 10.43 vanish we must now replace eqns. 10.44 by eqns. 4.55 and 6.64. We then substitute for D'_x, D'_y and D'_z into the relation

$$\tfrac{1}{4}E \cdot D'^* = \tfrac{1}{4}(E_x D'^*_x + E_y D'^*_y + E_z D'^*_z) \tag{10.50}$$

and obtain

$$\begin{aligned}
\tfrac{1}{4}E \cdot D'^* = \tfrac{1}{4}\{ & \epsilon_{xx}|E_x|^2 + \epsilon_{yy}|E_y|^2 + \epsilon_{zz}|E_z|^2 \\
& + 2\epsilon_{yz}|E_y E_z| \cos(\arg E^0_y - \arg E^0_z) \\
& + 2\epsilon_{zx}|E_z E_x| \cos(\arg E^0_z - \arg E^0_x) \\
& + 2\epsilon_{xy}|E_x E_y| \cos(\arg E^0_x - \arg E^0_y)\}
\end{aligned} \tag{10.51}$$

The fact that this is real completes the proof of the first of eqns. 10.40, and also shows that the right-hand side of this equation is given by expr. 10.51. The truth of the second of eqns. 10.40 is proved in the same way and its right-hand side is given by

$$\begin{aligned}
\tfrac{1}{4}H'^* \cdot B = \tfrac{1}{4}\{ & \mu_{xx}|H'_x|^2 + \mu_{yy}|H'_y|^2 + \mu_{zz}|H'_z|^2 \\
& + 2\mu_{yz}|H'_y H'_z| \cos(\arg H'^0_y - \arg H'^0_z) \\
& + 2\mu_{zx}|H'_z H'_x| \cos(\arg H'^0_z - \arg H'^0_x) \\
& + 2\mu_{xy}|H'_x H'_y| \cos(\arg H'^0_x - \arg H'^0_y)\}
\end{aligned} \tag{10.52}$$

If we need to discuss the time-variation of energy density, it is necessary to use formulas such as that appearing in eqn. 10.34 or vector diagrams such as that illustrated in Fig. 10.2. But if knowledge of the mean value is sufficient, eqns. 10.40 give the mean values of electric and magnetic density for linear material, isotropic or non-isotropic, and do so directly in terms of the complex electromagnetic vectors.

Chapter 11

Energy flow in oscillatory electromagnetic fields

11.1 Introduction

In the previous chapter oscillatory electromagnetic fields of angular frequency ω were described in terms of complex electromagnetic vectors. Each three-dimensional cartesian component of a vibrating electromagnetic vector was represented in the complex plane by means of a two-dimensional vector rotating in the counterclockwise direction with angular velocity ω. It is the reference components of the rotating vectors in the complex plane that constitute the actual electromagnetic vibrations.

We also found that both the electric and magnetic energies stored per unit volume at a particular location in the field could be represented as the reference components of a sum of non-rotating vector $\overrightarrow{OA}$ (Fig. 10.2) and a vector $\overrightarrow{AD}$ rotating with angular velocity 2ω. The magnitude of the rotating vector was never bigger than that of the non-rotating vector; neither the electric nor the magnetic energy density ever became negative. Moreover, it was possible to express the mean electric and magnetic energy densities at a particular location directly in terms of the complex electromagnetic fields at that location (equations. 10.40).

The same procedure can be employed for flow of energy in an oscillatory electromagnetic field. However, it is quite common for energy flow to become negative for part of each half-cycle of oscillation. Moreover, as described in Section 9.3, it is often important to be able to recognize the occurrence of such backflow of energy and to know how to eliminate it. We saw in Section 9.4 that, for circuits, this is conveniently achieved with the aid of the concept of complex power. We need to know to what extent this concept is applicable to oscillatory electromagnetic fields in general.

11.2 The time-variation of the rate of energy supply and consumption per unit volume in oscillatory electromagnetic fields

As described in connection with eqns. 4.80, 8.1 and 8.28, the work done per unit volume per unit time by an electromagnetic field on a free electric current is the scalar product of the electric vector and the current density vector. When this is negative, it is work done per unit volume per unit time on the electromagnetic field by the electric current. Let us consider what is involved in obtaining the complete information on the time-variation of the electromagnetic power dissipated per unit volume in an oscillatory electromagnetic field of angular frequency ω. The complex electric vector is given by eqn. 10.9, and we take the complex current density vector to be

$$\boldsymbol{J} = (J_x^0, J_y^0, J_z^0) \exp(j\omega t) \tag{11.1}$$

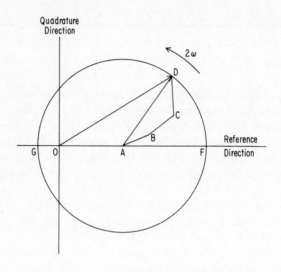

Fig. 11.1 *Vector diagram representing the time-variation of the power dissipated per unit volume*

Proceeding as in Section 10.4, we find that the time-variation of the power dissipated per unit volume may be represented as the reference component of the vector $\overrightarrow{OD}$ in Fig. 11.1, where (c.f. eqns. 10.35 and 10.36)

$$
\left.
\begin{aligned}
\overrightarrow{OA} &= \tfrac{1}{2}\{|E_x J_x| \cos(\arg E_x^0 - \arg J_x^0) \\
&\quad + |E_y J_y| \cos(\arg E_y^0 - \arg J_y^0) + |E_z J_z| \cos(\arg E_z^0 - \arg J_z^0)\} \\
\overrightarrow{AB} &= \tfrac{1}{2}|E_x J_x| \exp\{j(2\omega t + \arg E_x^0 + \arg J_x^0)\} \\
\overrightarrow{BC} &= \tfrac{1}{2}|E_y J_y| \exp\{j(2\omega t + \arg E_y^0 + \arg J_y^0)\} \\
\overrightarrow{CD} &= \tfrac{1}{2}|E_z J_z| \exp\{j(2\omega t + \arg E_z^0 + \arg J_z^0)\}
\end{aligned}
\right\} \tag{11.2}
$$

In Fig. 10.2, which refers to energy density, the point G is to the right of the origin because energy density is never negative. But for the power dissipated per unit volume, the point G in Fig. 11.1 may be to the left of the origin as shown in Fig. 11.2*a* or to the right as shown in Fig. 11.2*b*. If the free current is supplying energy to the electromagnetic field instead of withdrawing it, Figs. 11.2*a, b* take the form shown in Fig. 11.2*c, d*.

Considering the cases shown in the left half of Fig. 11.2, let us draw tangents DE and DE′ from the point D to a circle having centre A and radius OA. In this way the power per unit volume may be separated into resistive and reactive parts (c.f. Fig. 9.8). We may regard the vector $\overrightarrow{OD}$ either as the sum of the vectors $\overrightarrow{OA}$, $\overrightarrow{AE}$ and $\overrightarrow{ED}$, or as the sum of the vectors $\overrightarrow{OA}$, $\overrightarrow{AE'}$ and $\overrightarrow{E'D}$. The reactive exchange of energy per unit volume per unit time between the electromagnetic field and the current is then the reference component of either $\overrightarrow{ED}$ or $\overrightarrow{E'D}$, and it has amplitude

$$(AD^2 - OA^2)^{1/2} \tag{11.3}$$

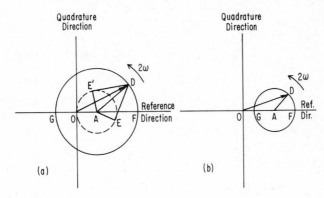

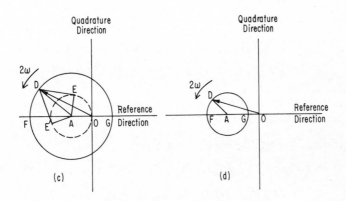

Fig. 11.2 *Vector diagrams representing the time-variation of the power density delivered (diagrams a and b) or received (diagrams c and d)*

The resistive work done per unit volume per unit time by the electromagnetic field on the current is the reference component of either $\overrightarrow{OE}$ or $\overrightarrow{OE'}$, and it has the mean value $\overrightarrow{OA}$ given by the first of eqns. 11.2.

Let us calculate the amplitude (11.3) of the reactive exchange of energy per unit volume per unit time between the field and the current for the situations illustrated in the left half of Fig. 11.2. From eqns. 11.2 we have

$$
\begin{aligned}
AD^2 = \ &\tfrac{1}{4}\{|E_x J_x|\cos(2\omega t + \arg E_x^0 + \arg J_x^0) \\
&+ |E_y J_y|\cos(2\omega t + \arg E_y^0 + \arg J_y^0) \\
&+ |E_z J_z|\cos(2\omega t + \arg E_z^0 + \arg J_z^0)\}^2 \\
&+ \tfrac{1}{4}\{|E_x J_x|\sin(2\omega t + \arg E_x^0 + \arg J_x^0) \\
&+ |E_y J_y|\sin(2\omega t + \arg E_y^0 + \arg J_y^0) \\
&+ |E_z J_z|\sin(2\omega t + \arg E_z^0 + \arg J_z^0)\}^2 \\
= \ &\tfrac{1}{4}\{|E_x J_x|^2 + |E_y J_y|^2 + |E_z J_z|^2 \\
&+ 2|E_y E_z J_y J_z|\cos(\arg E_y^0 + \arg J_y^0 - \arg E_z^0 - \arg J_z^0) \\
&+ 2|E_z E_x J_z J_x|\cos(\arg E_z^0 + \arg J_z^0 - \arg E_x^0 - \arg J_x^0) \\
&+ 2|E_x E_y J_x J_y|\cos(\arg E_x^0 + \arg J_x^0 - \arg E_y^0 - \arg J_y^0)\}
\end{aligned} \tag{11.4}
$$

From the first of eqns. 11.2 we have

$$
\begin{aligned}
OA^2 = \ &\tfrac{1}{4}\{|E_x J_x|^2 \cos^2(\arg E_x^0 - \arg J_x^0) \\
&+ |E_y J_y|^2 \cos^2(\arg E_y^0 - \arg J_y^0) \\
&+ |E_z J_z|^2 \cos^2(\arg E_z^0 - \arg J_z^0) \\
&+ 2|E_y E_z J_y J_z|\cos(\arg E_y^0 - \arg J_y^0)\cos(\arg E_z^0 - \arg J_z^0) \\
&+ 2|E_z E_x J_z J_x|\cos(\arg E_z^0 - \arg J_z^0)\cos(\arg E_x^0 - \arg J_x^0) \\
&+ 2|E_x E_y J_x J_y|\cos(\arg E_x^0 - \arg J_x^0)\cos(\arg E_y^0 - \arg J_y^0)\}
\end{aligned} \tag{11.5}
$$

We now subtract expr. 11.5 from expr. 11.4 and take the square root, making use of identities of the form

$$
\begin{aligned}
&\cos(\arg E_x^0 - \arg J_x^0 - \arg E_y^0 + \arg J_y^0) \\
&= \cos(\arg E_x^0 - \arg J_x^0)\cos(\arg E_y^0 - \arg J_y^0) \\
&\quad + \sin(\arg E_x^0 - \arg J_x^0)\sin(\arg E_y^0 - \arg J_y^0)
\end{aligned} \tag{11.6}
$$

and

$$
\begin{aligned}
&\cos(\arg E_x^0 - \arg J_x^0 - \arg E_y^0 + \arg J_y^0) \\
&\quad - \cos(\arg E_x^0 + \arg J_x^0 - \arg E_y^0 - \arg J_y^0) \\
&= 2\sin(\arg E_x^0 - \arg E_y^0)\sin(\arg J_x^0 - \arg J_y^0)
\end{aligned} \tag{11.7}
$$

We obtain

$$
(AD^2 - OA^2)^{1/2} = (p_Q^2 - q)^{1/2} \tag{11.8}
$$

where

$$
\begin{cases}
p_Q = \tfrac{1}{2}\{|E_x J_x| \sin(\arg E_x^0 - \arg J_x^0) \\
\qquad + |E_y J_y| \sin(\arg E_y^0 - \arg J_y^0) + |E_z J_z| \sin(\arg E_z^0 - \arg J_z^0)\} \quad (11.9) \\[4pt]
q = \tfrac{1}{2}\{|E_y E_z J_y J_z| \sin(\arg E_y^0 - \arg E_z^0) \sin(\arg J_y^0 - \arg J_z^0) \\
\qquad + |E_z E_x J_z J_x| \sin(\arg E_z^0 - \arg E_x^0) \sin(\arg J_z^0 - \arg J_x^0) \\
\qquad + |E_x E_y J_x J_y| \sin(\arg E_x^0 - \arg E_y^0) \sin(\arg J_x^0 - \arg J_y^0)\} \quad (11.10)
\end{cases}
$$

In accordance with expr. 11.3, the expression on the right-hand side of eqn. 11.8 is the amplitude of the reactive exchange of energy per unit volume per unit time between the field and the current. On the other hand, in accordance with Fig. 11.1, the mean resistive flow of energy per unit volume per unit time from the field to the current is, from the first of eqns. 11.2

$$
p_R = \tfrac{1}{2}\{|E_x J_x| \cos(\arg E_x^0 - \arg J_x^0) \tag{11.11}
$$
$$
+ |E_y J_y| \cos(\arg E_y^0 - \arg J_y^0) + |E_z J_z| \cos(\arg E_z^0 - \arg J_z^0)\}
$$

In Fig. 11.2, diagrams (*a*) and (*b*) correspond to $p_R > 0$ (dissipation of energy), and diagrams (*c*) and (*d*) corresponding to $p_R < 0$ (generation of energy). Diagrams (*a*) and (*c*) correspond to $q < p_Q^2$ so that the expr. 11.8 is real and so are the tangents DE and DE'. There is then backflow of energy during a portion of each half-cycle of oscillation, and the amplitude of the reactive exchange of energy per unit volume per unit time between the field and the current is $(p_Q^2 - q)^{1/2}$. Diagrams (*b*) and (*d*) correspond to $q > p_Q^2$ and no backflow of energy then takes place.

11.3 Complex power per unit volume

Let us now ask whether the key features of the resistive and reactive interaction between the field and the current can be obtained directly from the rotating vector (10.9) for the electric field and the rotating vector (11.1) for the volume current density. We form the complex power per unit volume defined by (c.f. eqn. 9.57)

$$
\tilde{p} = \tfrac{1}{2} E \cdot J^* \tag{11.12}
$$

where J^* is the complex conjugate of J. Written out in full, this is (c.f. eqns. 10.42 and 10.43)

$$
\tilde{p} = \tfrac{1}{2}[|E_x J_x| \exp\{j(\arg E_x^0 - \arg J_x^0)\} \tag{11.13}
$$
$$
+ |E_y J_y| \exp\{j(\arg E_y^0 - \arg J_y^0)\} + |E_z J_z| \exp\{j(\arg E_z^0 - \arg J_z^0)\}]
$$

The reference and quadrature components of this vector in the complex plane are p_R and p_Q given by eqns. 11.11 and 11.9, respectively. Hence the mean resistive power per unit volume from the field to the current is given by the reference component p_R of the complex power per unit volume $\tilde{p}$ defined in eqn. 11.12.

For checking whether the amplitude of the reactive exchange of energy between the field and the current is given by the quadrature component p_Q of the complex power $\tilde{p}$ we need to distinguish between the left and the right halves of Fig. 11.2. In diagrams (a) and (c) the amplitude of the reactive exchange is given by expr. 11.8, and we see that this only reduces to p_Q if

$$q = 0 \tag{11.14}$$

This condition is satisfied in a wide range of important situations, but not always. Whenever the currents flow on thin wires, the current density vector is linearly polarised. The phases of all three Cartesian components of the current vibration are then the same, and we have

$$\arg J_x^0 = \arg J_y^0 = \arg J_z^0 \tag{11.15}$$

Eqn. 11.10 shows that, in these circumstances, the condition $q = 0$ is satisfied. Even if the current flows in a surface or volume and is elliptically polarised, the condition $q = 0$ is still satisfied if the electric field is linearly polairsed:

$$\arg E_x^0 = \arg E_y^0 = \arg E_z^0 \tag{11.16}$$

If, however, both the electric vector and the current density vector are elliptically polarised and backflow of energy occurs as shown in diagrams (a) and (c) of Fig. 11.2, then it must be checked that q vanishes before it is assumed that the quadrature component of the complex power density $\tilde{p}$ in eqn. 11.12 describes the reactive exchange of energy between the field and the current.

On the other hand, for diagrams (b) and (d) in Fig. 11.2, there is no backflow of energy and it is simply a matter of making sure that the complex power density in eqns. 11.12 and 11.13 is in fact purely resistive ($p_Q = 0$). The most important situation encountered in practice is that in which the relation between the current density vector and the electric vector is that appropriate to conducting material. Such material, if isotropic, is described by means of its conductivity σ, and eqn. 11.12 then gives

$$\tilde{p} = \tfrac{1}{2}\sigma E \cdot E^* = \sigma \bar{E}^2 \tag{11.17}$$

where $\bar{E}$ is the RMS magnitude of the electric vector (eqn. 10.31). If the conducting material is non-isotropic, it is described by means of its conductivity tensor, and substitution from eqns. 4.83 into eqn. 11.12 gives

$$\begin{aligned}
\tilde{p} = \tfrac{1}{2}\{ &\sigma_{xx}|E_x|^2 + \sigma_{yy}|E_y|^2 + \sigma_{zz}|E_z|^2 \\
&+ 2\sigma_{yz}|E_y E_z|\cos(\arg E_y^0 - \arg E_z^0) \\
&+ 2\sigma_{zx}|E_z E_x|\cos(\arg E_z^0 - \arg E_x^0) \\
&+ 2\sigma_{xy}|E_x E_y|\cos(\arg E_x^0 - \arg E_y^0)\}
\end{aligned} \tag{11.18}$$

Eqns. 11.17 and 11.18 verify that $\tilde{p}$ is in fact purely resistive ($p_Q = 0$) for resistive material.

We see that the amplitude (expr. 11.8) of the reactive oscillation of energy per unit volume per unit time between the field and the current is equal to the magnitude of the quadrature component p_Q of the complex power per unit volume in eqn. 11.12 if:

either (i) $q = 0$, including all fields in which either the current density vector or the electric vector is linearly polarised,

or (ii) $p_Q = 0, q \geqslant 0$, including all linear resistive material.

On the other hand, eqn. 11.11 shows that the mean resistive rate of flow of energy per unit volume per unit time from the field to the current is equal, without exception, to the reference component p_R of the complex power per unit volume in eqn. 11.12.

11.4 The time-variation of the energy flow per unit cross-sectional area per unit time in oscillatory electromagnetic fields

At a particular point in an oscillatory electromagnetic field we are interested in calculating, for a particular direction, the flow of energy per unit cross-sectional area per unit time. The time-variation of this power density is given by the Poynting vector which, in accordance with eqn. 7.41, is obtained by forming the vector product of the electric vector whose cartesian components are expr. 10.13 with the magnetic vector whose cartesian components are (c.f. expr. 10.14)

$$[|H'_x|\cos(\omega t + \arg H'^0_x), \ |H'_y|\cos(\omega t + \arg H'^0_y), \ |H'_z|\cos(\omega t + \arg H'^0_z)]$$
(11.19)

Exprs. 11.19 are the reference components of the complex magnetic vector (c.f. eqn. 10.10)

$$H' = (H'^0_x, H'^0_y, H'^0_z)\exp(j\omega t)$$
(11.20)

Let us take the origin of cartesian co-ordinates at the point where we are studying the energy flow, and let the positive z axis be in the direction in which the energy flow is to be calculated. We are then interested only in the z component of the Poynting vector and, from exprs. 10.13 and 11.19, this is

$$|E_x H'_y|\cos(\omega t + \arg E^0_x)\cos(\omega t + \arg H'^0_y)$$

$$-|E_y H'_x|\cos(\omega t + \arg E^0_y)\cos(\omega t + \arg H'^0_x)$$
(11.21)

Notice that, in general, the components in the xy plane of both the electric vector and the magnetic vector are elliptically polarised and the two ellipses have different shapes and orientations.

Expr. 11.21 may be rewritten

$$\tfrac{1}{2}|E_x H'_y|\{\cos(\arg E^0_x - \arg H'^0_y) + \cos(2\omega t + \arg E^0_x + \arg H'^0_y)\}$$

$$-\tfrac{1}{2}|E_y H'_x|\{\cos(\arg E^0_y - \arg H'^0_x) + \cos(2\omega t + \arg E^0_y + \arg H'^0_x)\}$$
(11.22)

and the time-variation of this power density may be studied with the aid of a vector diagram of the type shown in Fig. 11.3. The vector $\overrightarrow{OA}$ corresponds to the mean value of expr. 11.22, so that

$$\overrightarrow{OA} = \tfrac{1}{2}|E_x H_y'| \cos(\arg E_x^0 - \arg H_y'^0)$$
$$-\tfrac{1}{2}|E_y H_x'| \cos(\arg E_y^0 - \arg H_x'^0) \qquad (11.23)$$

The vectors $\overrightarrow{AB}$ and $\overrightarrow{BD}$ in Fig. 11.3 are defined by

$$\left.\begin{aligned}
\overrightarrow{AB} &= \tfrac{1}{2}|E_x H_y'| \exp\{j(2\omega t + \arg E_x^0 + \arg H_y'^0)\}\\
\overrightarrow{BD} &= \tfrac{1}{2}|E_y H_x'| \exp\{j(2\omega t + \arg E_y^0 + \arg H_x'^0)\}
\end{aligned}\right\} \qquad (11.24)$$

The triangle ABC rotates counterclockwise round the point A with angular velocity 2ω and without changing its shape. The reference component of the vector $\overrightarrow{OD}$ is expr. 11.22. As D moves round the circle, therefore, the reference component of $\overrightarrow{OD}$ describes the time-variation of the energy flow per unit area per unit time. By addition of the vectors $\overrightarrow{AB}$ and $\overrightarrow{BD}$ in eqns. 11.24 we see that the square of the length of the vector $\overrightarrow{AD}$ is

$$\begin{aligned}
AD^2 &= \tfrac{1}{4}\{|E_x H_y'|\cos(2\omega t + \arg E_x^0 + \arg H_y'^0)\\
&\quad - |E_y H_x'|\cos(2\omega t + \arg E_y^0 + \arg H_x'^0)\}^2\\
&\quad + \tfrac{1}{4}\{|E_x H_y'|\sin(2\omega t + \arg E_x^0 + \arg H_y'^0)\\
&\quad - |E_y H_x'|\sin(2\omega t + \arg E_y^0 + \arg H_x'^0)\\
&= \tfrac{1}{4}\{|E_x H_y'|^2 + \tfrac{1}{4}|E_y H_x'|^2\\
&\quad - \tfrac{1}{2}|E_x E_y H_x' H_y'|\cos(\arg E_x^0 - \arg H_y'^0 + \arg E_y^0 - \arg H_x'^0)\}
\end{aligned} \qquad (11.25)$$

The square of the length of the vector $\overrightarrow{OA}$ is, from eqn. 11.23,

$$\begin{aligned}
OA^2 &= \tfrac{1}{4}\{|E_x H_y'|^2 \cos^2(\arg E_x^0 - \arg H_y'^0) + |E_y H_x'|^2 \cos^2(\arg E_y^0 - \arg H_x'^0)\\
&\quad - 2|E_x E_y H_x' H_y'| \cos(\arg E_x^0 - \arg H_y'^0)\cos(\arg E_y^0 - \arg H_x'^0)\}
\end{aligned} \qquad (11.26)$$

If $AD < OA$, there is no reactive oscillation of energy per unit area per unit time, and the vector diagram in Fig. 11.3 takes the form shown in the right-hand half of Fig. 11.2. On the other hand, if $AD > OA$, the vector $\overrightarrow{AD}$ may be resolved into a perpendicular pair of vectors $\overrightarrow{AE}$, $\overrightarrow{ED}$ or $\overrightarrow{AE'}$ $\overrightarrow{E'D}$ in the manner shown in the left-hand half of Fig. 11.2, and there is then a reactive oscillation of energy.

Using identities in the form (11.6) and (11.7), the amplitude of the reactive oscillation of energy per unit area per unit time evaluates to (c.f. eqns. 11.8, 11.9 and 11.10)

$$(AD^2 - OA^2)^{1/2} = (f_Q^2 - g)^{1/2} \qquad (11.27)$$

where

$$\begin{cases} f_Q = \tfrac{1}{2}\{|E_x H_y'| \sin (\arg E_x^0 - \arg H_y'^0) \\ \qquad - |E_y H_x'| \sin (\arg E_y^0 - \arg H_x'^0)\} \hspace{2cm} (11.28) \\ g = \tfrac{1}{2}|E_x E_y H_x' H_y'| \sin (\arg E_x^0 - \arg E_y^0) \sin (\arg H_x'^0 - \arg H_y'^0) \quad (11.29) \end{cases}$$

On the other hand, the mean resistive flow of energy per unit cross-sectional area in the positive z direction is (eqn. 11.23)

$$p_R = \tfrac{1}{2}\{|E_x H_y'| \cos (\arg E_x^0 - \arg H_y'^0) \\ - |E_y H_x'| \cos (\arg E_y^0 - \arg H_x'^0)\} \hspace{2cm} (11.30)$$

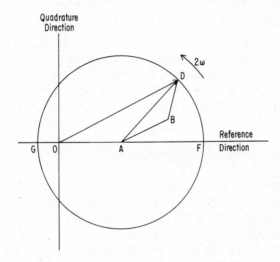

Fig. 11.3   *Vector diagram representing the time-variation of the power per unit cross-sectional area given by the Poynting vector*

In Fig. 11.2, diagrams (*a*) and (*b*) correspond to $f_R > 0$ (flow of energy in the positive z direction), and diagrams (*c*) and (*d*) correspond to $f_R < 0$ (flow of energy in the negative z direction). Diagrams (*a*) and (*c*) correspond to $g < f_Q^2$ (involving backflow of energy during part of each half-cycle), while diagrams (*b*) and (*d*) correspond to $g > f_Q^2$ (no backflow of energy).

11.5 Complex power per unit cross-sectional area

We can now see to what extent the key features of resistive and reactive flow of energy in an oscillatory electromagnetic field can be obtained directly from the complex electric and magnetic vectors. We form a complex Poynting vector defined by (c.f. eqn. 9.57)

$$\tilde{f} = \tfrac{1}{2} E \times H'^* \hspace{2cm} (11.31)$$

where H'^* is the complex conjugate of H'. Using eqns. 10.9 and 11.20, we obtain for the z component of $\tilde{f}$

$$\tilde{f}_z = \tfrac{1}{2}|E_x H'_y|\exp\{j(\arg E^0_x - \arg H'^0_y)\}$$

$$- \tfrac{1}{2}|E_y H'_x|\exp\{j(\arg E^0_y - \arg H'^0_x)\} \tag{11.32}$$

A discussion similar to that in the preceding section shows that, in an oscillatory electromagnetic field, the reference component of the complex Poynting vector always gives the mean resistive flow of energy per unit area per unit time.

Likewise, the quadrature component of the complex Poynting vector gives the amplitude of the reactive oscillation of energy per unit area per unit time if, for the z component of flow at the plane $z = 0$:

either (i) $g = 0$ (see eqn. 11.29), including all situations in which either the electric vector or the magnetic vector is linearly polarised,

or (ii) $f_Q = 0, g \geqslant 0$ (see eqns. 11.28 and 11.29), including all situations in which reduction of the tangential component of H' to zero just beyond the surface $z = 0$ can be achieved with a resistive sheet in the plane $z = 0$.

11.6 The energy theorem for complex power

At a point in an oscillatory electromagnetic field of angular frequency ω let the complex electromagnetic vectors involved in version 2 of the electromagnetic equations for linear materials be (E, D', H', B) and let the complex free electric-current-density vector be J. Then the mean electric and magnetic energies stored per unit volume are given by eqns. 10.40. Let V be a volume bounded by a surface S at rest relative to the observer. Then the total mean electric and magnetic energies stored in V (element of volume $d\tau$) are

$$\bar{W}_e = \int_V \tfrac{1}{4} E \cdot D'^* \, d\tau, \quad \bar{W}_m = \int_V \tfrac{1}{4} H'^* \cdot B \, d\tau \tag{11.33}$$

Moreover, the complex power densities out of the electromagnetic field, per unit volume of V and per unit area of S, are given by eqns. 11.12 and 11.31, respectively, so that the total complex power out of the electromagnetic field within V is

$$\tilde{P} + \tilde{F} \tag{11.34}$$

where

$$\tilde{P} = \int_V \tfrac{1}{2} E \cdot J^* \, d\tau, \quad \tilde{F} = \int_S \tfrac{1}{2}(E \times H'^*) \cdot dS \tag{11.35}$$

and dS is a vector element of area of S pointing along the outward normal. The energy theorem that we shall establish for complex power states that

$$\tilde{P} + \tilde{F} = -j2\omega(\bar{W}_m - \bar{W}_e) \tag{11.36}$$

This result should be compared and contrasted with eqn. 8.7.

From Maxwell's equations for the complex electromagnetic fields it follows that (c.f. eqns. 8.10–8.13)

$$\int_V \tfrac{1}{2} E \cdot J^* \, d\tau + \int_S \tfrac{1}{2}(E \times H'^*) \cdot dS = \int_V \tfrac{1}{2} E \cdot J^* \, d\tau + \int_V \tfrac{1}{2} \operatorname{div}(E \times H'^*) \, d\tau$$

$$= \int_V \tfrac{1}{2}\{ E \cdot (J^* - \operatorname{curl} H'^*)$$

$$+ H'^* \cdot \operatorname{curl} E \} \, d\tau$$

$$= \int_V \tfrac{1}{2}\{ E \cdot (j\omega D'^*) + H'^* \cdot (-j\omega B) \} \, d\tau$$

Hence

$$\int_V \tfrac{1}{2} E \cdot J^* \, d\tau + \int_S \tfrac{1}{2}(E \times H'^*) \cdot dS = -j2\omega \int_V (\tfrac{1}{4} H'^* \cdot B - \tfrac{1}{4} E \cdot D'^*) \, d\tau$$

$$(11.37)$$

Use of eqns. 11.33 and 11.35 then establishes eqn. 11.36.

The reference and quadrature components of eqn. 11.36 give

$$\left. \begin{aligned} P_R + F_R &= 0 \\ P_Q + F_Q &= -2\omega(\bar{W}_m - \bar{W}_e) \end{aligned} \right\} \tag{11.38}$$

or

$$\left. \begin{aligned} -F_R &= P_R \\ -F_Q &= P_Q + 2\omega(\bar{W}_m - \bar{W}_e) \end{aligned} \right\} \tag{11.39}$$

The first of eqns. 11.39 states that the net resistive flow of energy into a volume V across its surface S is equal to the net resistive flow of energy from the electromagnetic field to the electric currents in V. If written in the form $-P_R = F_R$, the first of eqns. 11.39 states that the net resistive flow of energy from the electric currents in the volume V is equal to the net resistive flow of energy out of V through the bounding surface S.

On the other hand, the second of eqns. 11.39 requires more careful interpretation. As described in Section 11.3, P_Q does not always give the reactive exchange of energy between the electromagnetic field in V and the electric currents in V. Likewise, as described in Section 11.5, F_Q does not always give the reactive exchange of energy between the electromagnetic field in V and the electromagnetic field outside V. However, if

either (i) the electric vector is linearly polarised everywhere,
or   (ii) the magnetic vector is linearly polarised everywhere and the current density vector is linearly polarised everywhere,

then the second of eqns. 11.39 states that the reactive oscillation of energy into the volume V across its bounding surface S exceeds the reactive oscillation of energy from the field to the electric currents in V by 2ω times the excess of the mean stored magnetic energy in V over the mean stored electric energy in V. This means that, if there is no reactive oscillation of energy across the bounding surface S, and it is desired to have no reactive exchange of energy between the currents in V and the electromagnetic field, then the mean stored electric and magnetic energies in V must balance (c.f. eqn. 9.44).

11.7 Complex power in a plasma

As described in Section 8.6, if the free charge carriers in a conductor possess mass, they store kinetic energy. In a plasma the stored kinetic energy is quite important.

For free electrons of charge e and mass m in an electromagnetic field (E, B) in a conductor, the drift-velocity v is determined by the relation (eqn. 8.20)

$$m\frac{dv}{dt} = e(E + v \times B) - F \tag{11.40}$$

where F is the average drag on a particle due to collisions with atoms. If in eqn. 11.40 the quantities E, B, v and F all oscillate with angular frequency ω, the term $v \times B$ involves a product of vibrations and is therefore non-linear. This non-linear term must be dropped if we are to represent the oscillations by rotating vectors in the complex plane, because the real part of the product of two complex quantities is not in general equal to the product of their real parts. There are many circumstances in which the term $v \times B$ is in fact unimportant. For electromagnetic waves, the ratio of E to B is of the order of the velocity of propagation, so that the magnitude of $v \times B$, as a ratio to E, is v divided by the velocity of propagation. The drift-velocity of the free particles in a conductor is highly non-relativistic. Consequently, the term $v \times B$ is usually negligible unless the plasma is subject to a strong imposed magnetic field (a magnetoplasma). If we drop the term $v \times B$ and at the same time let each cartesian component of v, E and F now be a vector rotating counterclockwise in the complex plane with angular velocity ω, eqn. 11.40 becomes

$$j\omega mv = eE - F \tag{11.41}$$

Let there be N free electrons per unit volume, and let us multiply eqn. 11.41 scalarly by Nv^* where v^* is the complex conjugate of v. We obtain

$$j\omega Nmv\cdot v^* = NeE\cdot v^* - NF\cdot v^* \tag{11.42}$$

This may be written (c.f. eqn. 8.28)

$$\tilde{p} = \tfrac{1}{2}E\cdot J^* - j2\omega\bar{w}_k \tag{11.43}$$

where (c.f. eqns. 8.25, 8.26 and 8.27)

$$\begin{cases} \bar{w}_k = Nmv\cdot v^* = \tfrac{1}{2}Nm\bar{v}^2 & (11.44) \\[2mm] J = Nev & (11.45) \\[2mm] \tilde{p} = \tfrac{1}{2}NF\cdot v^* & (11.46) \end{cases}$$

The quantity $\bar{w}_k$ is the mean kinetic energy stored per unit volume, the vector J is the complex current density, and $\tilde{p}$ is the complex power dissipated per unit volume as a result of collisions. This power is in fact resistive so that $\tilde{p}$ has no imaginary part. If several types of charge carriers are involved, summation signs must be included in eqns. 11.42, 11.44, 11.45 and 11.46, but eqn. 11.43 is unaffected.

Integrating eqn. 11.43 throughout a volume V of the plasma, we obtain for the total complex power dissipated in collisions (c.f. eqn. 8.29)

$$\tilde{P} = \int_V \tfrac{1}{2}E \cdot J^* \, d\tau - j2\omega \bar{W}_k \tag{11.47}$$

where

$$\bar{W}_k = \int_V \bar{w}_k \, d\tau \tag{11.48}$$

Using eqn. 11.47 in the result (11.37) derived from Maxwell's equations, we find that, for a plasma, eqn. 11.36 becomes

$$\tilde{P} + \tilde{F} = -j2\omega(\bar{W}_M - \bar{W}_e) \tag{11.49}$$

where

$$\bar{W}_M = \bar{W}_m + \bar{W}_k \tag{11.50}$$

It follows that, when relating the complex power across the surface of a volume V of a plasma to that dissipated by collisions within V, it is the total motional energy stored in V (magnetic plus kinetic) that must be used, not just the magnetic energy.

As described in Section 8.7, it is frequently convenient, for a plasma, to employ version 3 of the electromagnetic equations (Table 2.1). These involve the electric-flux-density vector D'' and the magnetic vector H'' used in eqns. 8.33, which take the form given in eqns. 8.36 when free magnetic moment per unit volume is neglected. For an oscillatory electromagnetic field of angular frequency ω, the relation between the complex electric-current-density vector J and the complex free electric moment per unit volume P_f is, from the first of eqns. 8.35,

$$J = j\omega P_f \tag{11.51}$$

Subsitution into eqn. 11.43 then gives

$$\tilde{p} = \tfrac{1}{2}E \cdot (-j\omega P_f^*) - j2\omega \bar{w}_k \tag{11.52}$$

or

$$\tfrac{1}{4}E \cdot P_f^* = -\bar{w}_k + j\tilde{p}/(2\omega) \tag{11.53}$$

Using the complex version of the first of eqns. 8.36 we obtain

$$\tfrac{1}{4}E \cdot D''^* = \tfrac{1}{4}E \cdot (D'^* + P_f^*) \tag{11.54}$$

and substitution from eqns. 10.40 and 11.53 then gives

$$\tfrac{1}{4}E \cdot D''^* = (\bar{w}_e - \bar{w}_k) + j\tilde{p}/(2\omega) \tag{11.55}$$

If use is made of version 3 of Maxwell's equations instead of version 2, eqn. 11.37 becomes

$$\int_S \tfrac{1}{2}(E \times H''^*) \cdot d\vec{S} = -j2\omega \int_V (\tfrac{1}{4}H''^* \cdot B - \tfrac{1}{4}E \cdot D''^*) \, d\tau \tag{11.56}$$

The term $\frac{1}{4}\boldsymbol{E}\cdot\boldsymbol{D}''^*$ in this equation is not simply $\bar{w}_e$, the mean electric energy of the plasma per unit volume. In accordance with eqn. 11.55, $\frac{1}{4}\boldsymbol{E}\cdot\boldsymbol{D}''^*$ involves $\tilde{p}$, the complex power dissipated per unit volume of the plasma by collisions. Moreover, eqn. 11.55 also shows that the term $\frac{1}{4}\boldsymbol{E}\cdot\boldsymbol{D}''^*$ involves $\bar{w}_k$, which is part of the mean motional energy of the plasma per unit volume.

The quantity $\frac{1}{4}\boldsymbol{E}\cdot\boldsymbol{D}''^*$ in the energy equation (11.56) is the sum of:

(i) the excess of the mean stored electric energy per unit volume (including the potential energy per unit volume due to polarisation of bound electrons) over the mean stored kinetic energy per unit volume associated with free charges, and

(ii) $j/(2\omega)$ times the mean rate of dissipation of energy per unit volume due to collisional drag on the vibrating free charges.

The quantity $\frac{1}{4}\boldsymbol{H}''^*\cdot\boldsymbol{B}$ in the energy equation (11.56) is the mean motional energy per unit volume associated only with the free-space magnetic field, modified to allow for the (normally negligible) contributions from the bound and free magnetic moments per unit volume.

The difference between $\boldsymbol{H}'$ and $\boldsymbol{H}''$ is always numerically small, but it does arise in a magnetoplasma. This is a plasma which is subject to a strong steady magnetic field round which the free charges spiral in their thermic motion, thereby creating a little magnetic moment per unit volume (see Problem 6.12). To incorporate this free magnetic moment per unit volume not only does the second of eqns. 10.40 have to be replaced by

$$\bar{w}_m = \frac{1}{4}\boldsymbol{H}''^*\cdot\boldsymbol{B} \tag{11.57}$$

but eqn. 11.31 has to be replaced by

$$\tilde{f} = \frac{1}{2}\boldsymbol{E}\times\boldsymbol{H}''^* \tag{11.58}$$

However, it should be noted that a magnetoplasma involves much more serious complications than free magnetic moment per unit volume. In a magnetoplasma there is a term $e\boldsymbol{v}\times\boldsymbol{B}$ in eqn. 11.40 that cannot be dropped as was done in eqn. 11.41; see eqn. 2.71. Moreover, in a magnetoplasma elliptical polarisation is frequently involved, so that quantities such as g in eqn. 11.29 do not vanish and the quadrature component of the complex power does not necessarily give the amplitude of the reactive oscillation of energy.

11.8 The reciprocity theorem for oscillatory electromagnetic fields

Reciprocity theorems for electrostatics and magnetostatics were developed in Sections 3.10 and 5.9. A similar reciprocity theorem may be developed for oscillatory electromagnetic systems that are such that access is available via a seris of terminal pairs, or ports. The system could be an oscillatory electric network with n ports numbered $1, 2, \ldots, n$. However, the system could be a system of n antennae. Radiation links between the ports are not excluded.

Non-linear material is excluded. Moreover, unlike the reciprocity theorem for electrostatics and magnetostatics, non-isotropic material is excluded. However, for linear isotropic material, any degree of inhomogeneity is permitted. In particular, the geomtrical configuration of antennae connected to the terminals is irrelevant so long as they do not involve non-linear or non-isotropic material.

All electromagnetic vector components in the system, and all electric currents and voltages, are supposed to be oscillating sinusoidally with a given angular frequency ω. Moreover, all oscillations are represented as the reference components of vectors rotating counterclockwise in the complex plane. In particular, a complex voltage and a complex current are associated with each port and we take the positive direction of the current to be that entering the system at the positive terminal.

Because non-linear behaviour is not involved, the complex voltages and currents at the various ports are related by linear equations involving complex coefficients. If V_r, I_r are the complex voltage and current at port number r, the linear relations between the complex voltages and the complex currents may either be written

$$V_r = \sum_{s=1}^{n} Z_{rs} I_s \tag{11.59}$$

or

$$I_s = \sum_{s=1}^{n} Y_{rs} V_s \tag{11.60}$$

The set of coefficients Z_{rs} is known as the impedance matrix of the system, and the set of coefficients Y_{rs} is known as the admittance matrix. Each set of linear equations (11.59) and (11.60) is the solution of the other.

A given electromagnetic system to which access is available through a given set of n ports may be excited at a given angular frequency in many ways. Let us compare two of these methods of excitation. In the first, let the complex voltages applied at ports numbered 1, 2, ... be $V_1, V_2, \ldots$ and let the resulting complex currents at these ports be $I_1, I_2, \ldots$. We compare this with a situation in which complex voltages $V_1', V_2', \ldots$ are applied, resulting in complex currents $I_1', I_2', \ldots$. The reciprocity theorem that we shall prove states that (c.f. eqns. 3.49)

$$\left.\begin{array}{c} V_1 I_1' + V_2 I_2' + \ldots = V_1' I_1 + V_2' I_2 + \ldots \\ \Sigma VI' = \Sigma V'I \end{array}\right\} \tag{11.61}$$

or

As with the reciprocity theorem in electrostatics, the reciprocity theorem for oscillatory electromagnetic systems may be restated in various ways. By considering the following arrangement of complex voltages and currents (c.f. expr. 3.50)

V_1	V_2	V_3	$\cdots$
I	0	0	$\cdots$
V_1'	V_2'	V_3'	$\cdots$
0	I	0	$\cdots$

it follows that (c.f. eqn. 3.53)

$$V_2 = V_2'$$ (11.62)

and therefore that

$$Z_{21} = Z_{12}$$

or more generally (c.f. eqn. 3.54)

$$Z_{rs} = Z_{sr}$$ (11.63)

The impedance matrix is therefore symmetrical about the leading diagonal. Non-diagonal elements are called mutual impedances and diagonal elements self impedances. Likewise, by considering the following arrangement of complex voltages and currents

$$
\begin{array}{cccc}
V & 0 & 0 & \ldots \\
I_1 & I_2 & I_3 & \ldots \\
\hline
0 & V & 0 & \ldots \\
I_1' & I_2' & I_3' & \ldots
\end{array}
$$

it follows that (c.f. eqn. 3.51)

$$I_1' = I_2$$ (11.64)

and therefore that

$$Y_{12} = Y_{21}$$

or more generally that (c.f. eqn. 3.52)

$$Y_{rs} = Y_{sr}$$ (11.65)

The symmetry of the impedance and admittance matrices is important for any multi-port electromagnetic network. It is particularly important, however, when applied to the two ports at opposite ends of a radio communication link. Symmetry of the impedance matrix implies that the open-circuit complex voltage at port number 2 when a complex current I is applied at port number 1 is equal to the open-circuit voltage at port number 1 when the complex current I is applied at port number 2. Other antennas are supposed to be open-circuited, but they are often too remote to matter. Likewise, symmetry of the admittance matrix implies that the short-circuit complex current at port number 2 when a complex voltage V is applied at port number 1 is equal to the short-circuit complex current at port number 1 when the complex voltage V is applied at port number 2. Other antannae are supposed to be short-circuited, but they are often too remote to matter. In short, a given radio communication link at a given frequency works equally well for transmission in either direction provided that non linear and non-isotropic materials are not significantly involved.

It should be noted that ionospheric radio communication makes use of the Earth's ionised upper atmosphere. This is a magnetoplasma on account of the effect of the Earth's magnetic field, and therefore constitutes non-isotropic material. Care is therefore necessary in applying the concept of reciprocity in ionospheric radio communication.

11.9 Proof of the reciprocity theorem for oscillatory electromagnetic fields

The reciprocity theorem for oscillatory electromagnetic systems may be established as follows. Using version 2 of the electromagnetic equations for materials, let us compare two different ways of exciting the system at the given angular frequency ω. In the first, let the electric and magnetic vectors at a particular location in the field be (E_I, H_I') and, in the second, let them be (E_{II}, H_{II}'). Using Identity 9 in Appendix B, followed by Maxwell's equations, we find that

$$
\begin{aligned}
-\operatorname{div}(E_I \times H_{II}'') &= E_I \cdot \operatorname{curl} H_{II} - H_{II}' \cdot \operatorname{curl} E_I \\
&= E_I \cdot \left(J_{II} + \frac{\partial D_{II}'}{\partial t} \right) - H_{II}' \cdot \left(-\frac{\partial B_I}{\partial t} \right) \\
&= E_I \cdot J_{II} + j\omega(E_I \cdot D_{II}' + B_I \cdot H_{II}') \\
&= \sigma E_I \cdot E_{II} + j\omega(\epsilon E_I \cdot E_{II} + \mu^{-1} B_I \cdot B_{II})
\end{aligned} \tag{11.66}
$$

provided that, at the location under study we have either free space or linear isotropic material of conductivity σ, dielectric constant ϵ/ϵ_0 and permeability μ/μ_0. By interchanging subscripts I and II in eqn. 11.66 we deduce that, at any location in the system,

$$
\operatorname{div}(E_I \times H_{II}') = \operatorname{div}(E_{II} \times H_I') \tag{11.67}
$$

At such a location let $d\tau$ be an element of volume and let us integrate through a volume V bounded by a closed surface S of which dS is a vector element of area. We obtain

$$
\int_V \operatorname{div}(E_I \times H_{II}') \, d\tau = \int_V \operatorname{div}(E_{II} \times H_I') \, d\tau \tag{11.68}
$$

or, using the divergence theorem (Identity 17, Appendix B),

$$
\int_S (E_I \times H_{II}') \cdot dS = \int_S (E_{II} \times H_I') \cdot dS \tag{11.69}
$$

Let us now consider a two-port system as indicated in Fig. 11.4. Let S_1 be a small closed surface surrounding port number 1, S_2 a small closed surface surrounding port number 2, and S_∞ a large closed surface that will be allowed to tend to infinity. It is convenient to assume that all radiation is absorbed somewhere and that in consequence a slight exponential attenuation may be supposed to exist at sufficiently large distances. Then the integrals in eqn. 11.69 over the surface S_∞ may be neglected, and we obtain

$$\int_{S_1} (E_I \times H'_{II}) \cdot dS + \int_{S_2} (E_I \times H'_{II}) \cdot dS$$

$$= \int_{S_1} (E_{II} \times H'_I) \cdot dS + \int_{S_2} (E_{II} \times H'_I) \cdot dS \qquad (11.70)$$

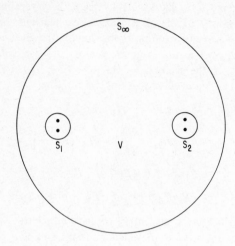

Fig. 11.4 *Volume V used for establishing the reciprocity theorem for an oscillatory electromagnetic system involved two pairs of terminals numbered 1 and 2*

We now form a transmission-line model for each port such as that appropriate to a coaxial transmission line in Fig. 7.10. Over S_1 and S_2 the electric and magnetic fields are perpendicular to each other and to dS. Using the area bounded by two adjacent lines of magnetic flux and two adjacent lines of electric force as an element of area $dx dy$, eqn. 11.70 becomes

$$\int_{S_1} E_I H'_{II} \, dx dy + \int_{S_2} E_I H'_{II} \, dx dy = \int_{S_1} E_{II} H'_I \, dx dy + \int_{S_2} E_{II} H'_I \, dx dy$$
$$(11.71)$$

and this may be rewritten as

$$\iint_{S_1} (E_I \, dx)(H'_{II} \, dy) + \iint_{S_2} (E_I \, dx)(H'_{II} \, dy)$$

$$= \iint_{S_1} (E_{II} \, dx)(H'_I \, dy) + \iint_{S_2} (E_{II} \, dx)(H'_I \, dy) \qquad (11.72)$$

Evaluating the x and y integrals separately, we derive

$$(V_1)_I(I_1)_{II} + (V_2)_I(I_1)_{II}$$
$$= (V_1)_{II}(I_2)_I + (V_2)_{II}(I_2)_I \tag{11.73}$$

For n ports we obtain the result stated in eqn. 11.61.

Chapter 12

The concept of impedance in oscillatory electromagnetic fields

12.1 Introduction

We saw in Section 9.6 that, for an oscillatory electric network, there is a close relation between the concept of complex power and that of impedance and admittance. If, at a pair of terminals of an oscillatory electric network, the voltage and current are represented as the reference components of counterclockwise rotating vectors V and I in the complex plane, then the impedance and admittance looking in the direction of the positive current on the positive conductor are (eqns. 9.56 and 9.61)

$$Z = V/I, \qquad Y = I/V \tag{12.1}$$

The complex power in this direction is (eqns. 9.58 and 9.62)

$$\tilde{P} = \tfrac{1}{2}VI^* = \tfrac{1}{2}Z|I|^2 = \tfrac{1}{2}Y^*|V|^2 \tag{12.2}$$

or

$$\tilde{P} = Z\bar{I}^2 = Y^*\bar{V}^2 \tag{12.3}$$

where $\bar{V}$ and $\bar{I}$ are the RMS voltage and current at the terminals. Hence the impedance looking in the direction of the positive current on the positive conductor is the complex power in this direction measured per unit mean square current. The admittance is the conjugate of the complex power measured per unit mean square voltage.

In the previous chapter we have also seen that, for material that behaves in a linear manner, the concept of complex power is useful not only in connection with electric networks but also (i) for describing the exchange of energy between an electromagnetic field and an electric current, and (ii) for describing the flow of energy in the field per unit cross-sectional area per unit time. At a particular location this complex power per unit cross-sectional area is given in magnitude and direction by (eqn. 11.31)

$$\tilde{f} = \tfrac{1}{2}E \times H'^* \tag{12.4}$$

where E and H' are the complex electric and magnetic vectors as used in version 2 of the electromagnetic equations for the material (Chapter 2). The question arises as to whether a concept of impedance may be associated with the complex power density $\frac{1}{2}E \times H'^*$ in a manner similar to that by which the concept of circuit impedance is associated with the complex power $\frac{1}{2}VI^*$.

12.2 Definition of field impedance

The field impedance in an electromagnetic field in linear material at a location where the complex electric and magnetic vectors are E and H' may be defined in terms of normalised complex power density as follows.

At this location in the material let dS be a vector element of area lying in a surface L and pointing in the direction of a unit vector l. Then the complex power across dS in the direction of l is

$$\tfrac{1}{2}(E \times H'^*) \cdot dS$$

or, measured per unit area,

$$\tfrac{1}{2}(E \times H'^*) \cdot l \tag{12.5}$$

The components of E and H' parallel to the surface L are

$$E_{\parallel} = l \times E$$
$$H'_{\parallel} = l \times H' \tag{12.6}$$

and their RMS values are (c.f. eqns. 10.31 and 10.32)

$$\left.\begin{array}{l} \bar{E}_{\parallel} = \dfrac{1}{\sqrt{2}}\{(l \times E) \cdot (l \times E^*)\}^{1/2} \\[2mm] \bar{H}'_{\parallel} = \dfrac{1}{\sqrt{2}}\{(l \times H') \cdot (l \times H'^*)\}^{1/2} \end{array}\right\} \tag{12.7}$$

We define the field impedance Z looking across L in the direction l as expr. 12.5 measured per unit mean square $H'_{\parallel}$, so that

$$Z = \frac{l \cdot E \times H'^*}{(l \times H') \cdot (l \times H'^*)} \tag{12.8}$$

Likewise, we define the complex conjugate of the field admittance Y looking across L in the direction l as expr. 12.5 measured per unit mean square $E_{\parallel}$, so that

$$Y = \frac{l \cdot E^* \times H'}{(l \times E) \cdot (l \times E^*)} \tag{12.9}$$

With these definitions of field impedance and field admittance, the complex power per unit area across L in the direction l is

$$\tilde{f} = Z\bar{H}'^2_{\parallel} = Y^*\bar{E}^2_{\parallel} \tag{12.10}$$

These expressions may be compared with the expressions $Z\bar{I}^2$ and $Y^*\bar{V}^2$ in eqns. 12.3 for the complex power into a port of an electric network. The first of exprs. 12.10 shows that the complex power per unit cross-sectional area in any direction in any oscillatory electromagnetic field, measured per unit mean square magnetic field strength at right angles to the direction, is equal to the field impedance looking in that direction. The second of exprs. 12.10 shows that the complex power per unit cross-sectional area in any direction in any oscillatory electromagnetic field, measured per unit mean square electric field strength at right angles to the direction, is equal to the complex conjugate of the field admittance looking in that direction.

It should be noticed that, if it is desired to take account of the phenomenon of free magnetic moment per unit volume, then H' is replaced by H'' (eqn. 11.58) and the definitions of the field impedance and field admittance looking in the direction l become

$$Z = \frac{l \cdot E \times H''^*}{(l \times H'') \cdot (l \times H''^*)}, \qquad Y = \frac{l \cdot E^* \times H''}{(l \times E) \cdot (l \times E^*)}. \tag{12.11}$$

On the other hand, in non-magnetic material and in free space, the definitions are

$$Z = \frac{l \cdot E \times H^*}{(l \times H) \cdot (l \times H^*)}, \qquad Y = \frac{l \cdot E^* \times H}{(l \times E) \cdot (l \times E^*)}. \tag{12.12}$$

In the remainder of this book the primes will be dropped from the magnetic vector. This makes no difference if magnetic material is not present. However, if bound magnetic moment per unit volume is involved it must be remembered that H will now mean H', while if free magnetic moment is also involved, then H will mean H'' (Chapter 2). On the other hand, the distinctions between D, D' and D'' corresponding to free space, non-conducting materials and conducting materials will not be dropped.

A particularly simple situation exists if we are considering flow of energy across a surface L in circumstances when the components of E and H lying in L are mutually perpendicular. In Fig. 12.1, let the origin 0 be the point in the oscillatory electromagnetic field at which we are examining the flow of energy, and let the surface L across which we are studying the flow coincide at 0 with the xy plane. Let the positive directions of the components of E and H at 0 parallel to L be along the x and y axes. The positive directions of $E_{\parallel}$, $H_{\parallel}$ and l then form a right-handed system of axes (x, y, z). We have

$$l = (0, 0, 1), \quad E = (E_{\parallel}, 0, E_{\perp}), \quad H = (0, H_{\parallel}, H_{\perp})$$

where $E_{\perp}$ and $H_{\perp}$ are the components of the complex electric and magnetic vectors perpendicular to L. It follows that

$$E \times H^* = (-E_{\perp}H_{\parallel}^*, -E_{\parallel}H_{\perp}^*, E_{\parallel}H_{\parallel}^*), E^* \times H = (-E_{\perp}^*H_{\parallel}, -E_{\parallel}^*H_{\perp}, E_{\parallel}^*H_{\parallel})$$

so that

$$l \cdot (E \times H^*) = E_{\parallel}H_{\parallel}^*, \ l \cdot (E^* \times H) = E_{\parallel}^*H_{\parallel} \tag{12.13}$$

Moreover,

$$l \times H = (-H_{\parallel}, 0, 0), \; l \times E = (0, E_{\parallel}, 0)$$

so that

$$(l \times H) \cdot (l \times H^*) = H_{\parallel} H_{\parallel}^*, (l \times E) \cdot (l \times E^*) = E_{\parallel} E_{\parallel}^* \qquad (12.14)$$

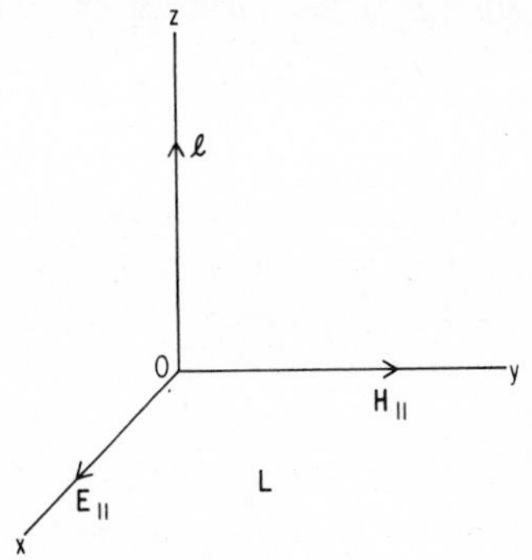

Fig. 12.1. *Illustrating a situation in which $E_{\parallel}$, $H_{\parallel}$ and l form a mutually perpendicular right-handed system of directions*

Substitution into eqns. 12.12 from eqns. 12.13 and 12.14 then gives

$$Z = \frac{E_{\parallel}}{H_{\parallel}}, \qquad Y = \frac{H_{\parallel}}{E_{\parallel}} \qquad (12.15)$$

In this important case the field impedance and the field admittance are mutually reciprocal whereas, in general, exprs. 12.12 are not mutually reciprocal.

The situation illustrated in Fig. 12.1 is one in which we are looking in a particular direction (the positive z axis) across a plane L (the xy plane) which in practice might be an interface between two oscillatory electromagnetic fields, possibly fields existing in different materials. The electric and magnetic fields do not necessarily vibrate in the plane L, but the components $E_{\parallel}$ and $H_{\parallel}$ of these vectors parallel to the plane are assumed to be mutually perpendicular. The positive directions of $E_{\parallel}$, $H_{\parallel}$ and l are arranged to form a right-handed system of directions, so that rotation from $E_{\parallel}$ to $H_{\parallel}$ through a right angle is right-handed about the direction of look. In these circumstances the field impedance looking across L in the direction l is $E_{\parallel}/H_{\parallel}$ and the field admittance is $H_{\parallel}/E_{\parallel}$.

There are many practical situations in which the general definitions of field impedance and field admittance given in eqns. 12.12 reduce to the simpler version given in eqns. 12.15. If in Fig. 12.1 the parallel component of magnetic field just above the plane L were reduced to zero without modifying the electromagnetic field in the region below L ($z < 0$), a sheet of oscillatory electric current would be needed in L with complex amplitude $H_{\parallel}$ and with positive direction of flow parallel to the positive x axis. This electric current would be flowing in the presence of an oscillatory electric field of complex amplitude $E_{\parallel}$ also vibrating parallel to the x axis, positive in the direction of the positive x axis. The complex power per unit area into the terminating current sheet on L from the undisturbed electromagnetic field below L would be

$$\tilde{f} = \tfrac{1}{2}E_\| H_\|^* \tag{12.16}$$

Using the field impedance and field admittance appearing in eqns. 12.15, this complex power per unit area may be written

$$\tilde{f} = \tfrac{1}{2}Z|H_\||^2 = \tfrac{1}{2}Y^*|E_\||^2 \tag{12.17}$$

In terms of the mean square tangential electric and magnetic fields, $\bar{E}_\|$ and $\bar{H}_\|$, the complex power per unit area into the terminating current sheet on L from the undisturbed electromagnetic field below L becomes (c.f. eqns. 12.10)

$$\tilde{f} = Z\bar{H}_\|^2 = Y^*\bar{E}_\|^2 \tag{12.18}$$

If $E_\|$ and $H_\|$ in Fig. 12.1 were not mutually perpendicular, the surface current in L required to terminate the magnetic field existing below L would not have the same direction as the electric field in L. A resolving factor would be needed in calculating the complex power per unit area of L, and this is what is allowed for by the cross product in the definition of the complex Poynting vector. Consequently, the simpler definitions of field impedance and field admittance given in eqns. 12.15 must, in general, be replaced by the definitions given in eqns. 12.12.

As in the case of circuit impedance, the unit of field impedance is the ohm, being the ratio of volt metre^{-1} to ampere metre^{-1} as shown in eqns. 12.15.

12.3   The relation between circuit impedance and boundary conditions at an interface between oscillatory electric networks

Field impedance as defined in the previous section may be used to describe the resistive and reactive flow of energy at any point in any direction in a sinusoidally oscillating electromagnetic field. However, field impedance frequently has a particularly important role to play at an interface between two different materials. This role is similar to that played by circuit impedance at a junction between two sinusoidally oscillating electric networks.

Fig. 12.2 illustrates two linear electric networks each with a pair of terminals, and each oscillating sinusoidally at the same frequency. Network number 1 has a pair of terminals at which the complex voltage is V_1 and the corresponding complex current is I_1. The positive direction of I_1 is out of network number 1 at the positive terminal, so that the port is formed by what is usually called a pair of output terminals. From knowledge of the way in which network number 1 is constructed the relation between V_1 and I_1 may be calculated. The relation is

$$V_1 = Z_1 I_1 \tag{12.19}$$

where Z_1 is the impedance of network number 1 looking out of the port. Network number 2 in Fig. 12.2 has a port at which the complex voltage is V_2 and the corresponding complex current is I_2. The positive direction of I_2 is into network number 2 at the positive terminal, so that the port is formed by a pair of input terminals. From knowledge of the way in which network number 2 is constructed the relation between V_2 and I_2 may be calculated. The relation is

$$V_2 = Z_2 I_2 \qquad (12.20)$$

where Z_2 is the impedance of network number 2 looking into the port.

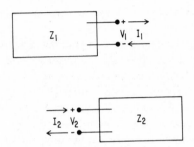

Fig. 12.2.   *Illustrating a network with an output impedance Z_1, and a network with an input impedance Z_2*

We now ask whether it is possible to connect the output terminals of network number 1 to the input terminals of network number 2 without upsetting any of the voltages and currents in either network. The answer is in the affirmative if

$$V_1 = V_2, \qquad I_1 = I_2 \qquad (12.21)$$

Since these are relations between complex voltages and currents, they state that both the amplitudes and phases of the voltage and current oscillations at the output terminals of network number 1 must be the same as those at the input terminals of network number 2.

From eqns. 12.19, 12.20 and 12.21, it follows that $Z_1 = Z_2$. This condition may be used to replace either of conditions (12.21). Hence the conditions that permit interconnection of the networks without upsetting their behaviour may either be stated in the form given in eqns. 12.21, or in the form

$$Z_1 = Z_2, \qquad I_1 = I_2 \qquad (12.22)$$

or in the form

$$Z_1 = Z_2, \qquad V_1 = V_2 \qquad (12.23)$$

The condition

$$Z_1 = Z_2 \qquad (12.24)$$

appearing in eqns. 12.22 and 12.23 is a condition on how the networks are constructed. This is the key condition to be satisfied if the two networks are to be interconnected without upsetting the behaviour of either. It is, of course, also true that the degree of excitation of the two networks must be appropriately related in both amplitude and phase, and this is what the second condition in either eqns. 12.22 or 12.23 assures. However, this second condition is trivial and may be largely taken for granted, whereas the condition (12.24) involving the way in which the networks are constructed is non-trivial. Hence the two boundary conditions concerning equality of the complex voltages and currents at the interface between the two networks reduce to a trivial condition concerning the relative degree of excitation of the two networks and a non-trivial condition concerning the equality of the output impedance of network number 1 and the input impedance of network number 2.

The equality of these two impedances implies the equality of both their resistive and reactive components, and these are interpretable in terms of energy flow as described in Section 9.6. Equality of the resistive components of the two impedances means that the energy of which network number 1 must rid itself permanently will be accepted by network number 2 and will not be repaid by it to network number 1. Equality of the reactive components of the two impedances means that the energy that network number 1 must temporarily deposit on loan for repayment a quarter of a cycle later will be accepted on deposit by network number 2 and will be repaid by it to network number 1 at the appropriate time. In short, the boundary conditions at an interface between two electric networks describe the match that must exist between the powers on the two sides of the interface. This match must exist both for resistive power and for reactive power, that is, for complex power. The key feature in this match is the equality of the output impedance (or admittance) of the first network and the input impedance (or admittance) of the second.

12.4   The relation between field impedance and boundary conditions at an interface between oscillatory electromagnetic fields

Let us now suppose that the ports to the oscillatory networks in Fig. 12.2 involve coaxial connectors of the same size. The connection from each port to the interior of its network constitutes a short piece of coaxial transmission line. In a plane perpendicular to the length of such a line the electromagnetic field takes the form shown in Fig. 7.10. The pattern of the electromagnetic field in the plane of the port is the same for both networks. Moreover, the amplitude and phase of the tangential magnetic field in the plane of the port for network number 1 may easily be adjusted to be the same as that for network number 2. But for interconnection without upsetting the behaviour of either network, the tangential electric field in the plane of the port for network number 1 must also be the same as that for network number 2 in amplitude and phase. This requires that, at corresponding points in the planes of the two ports, the field impedance looking out of network number 1 must be equal to the field impedance looking into network number 2. Interpreted in terms of resistive and reactive power per unit area per unit mean square tangential magnetic field strength as described in Section 12.2, this means that both the resistive and reactive flow of energy at corresponding points in the planes of the two ports must match. After interconnection, the energy that the electromagnetic field on one side of the interface needs to deliver (or receive) must be accepted (or provided) by the electromagnetic field on the other side of the interface, and this must be true both for energy transferred permanently and for energy transferred temporarily to be repaid a quarter of cycle later.

For more general oscillatory electromagnetic fields than those associated with a coaxial connector, the matching of tangential electric and magnetic fields at an interface between the electromagnetic fields is of particular interest at an interface between two different materials. At each point of such an interface the normal component of the complex power out of material number 1 must be the same as the normal component of the complex power into material number 2. Measuring these complex powers per unit mean square tangential magnetic field strength at the interface, it follows that the field impedances looking normally across the interface from material number 1 into material number 2 must be the same on the two sides of the interface, that is, they must match. If, alternatively, we measure the normal components of the complex powers per unit mean square tangential electric field strength at the interface, we may say that there must be a match between the two field admittances looking normally across the interface from material number 1 into material number 2.

While equality of the field impedance or field admittance looking normally across an interface between two materials is a key condition to be satisfied at the interface, it is not the only condition. The pattern of the electromagnetic field over the interface must be the same on the two sides. If, for example, the electromagnetic field in material number 2 involves a wave travelling over the interface with a certain velocity, having a certain field-pattern in the interface, then this must be true also for the electromagnetic field in material number 1.

As we shall see in detail in the following chapter, the laws of optical refraction at a plane interface between two dielectric materials are designed to make the pattern of electromagnetic field at the interface move over the interface at the same speed on both sides of the interface. This does not, however, assure that the impedance looking across the interface from material number 1 into material number 2 for the incident wave is equal to that for the transmitted wave. Even with appropriate adjustment of the relative amplitude and phase of the incident and transmitted waves, the field impedances looking across the interface do not match in general. This is why there is, in general, a reflected wave. The amplitude and phase of the reflected wave have to be chosen in such a way as to make the field impedance looking across the interface for the combined incident and reflected waves in material number 1 identical with the field impedance looking in the same direction for the transmitted wave in material number 2. Alternatively, if it is desired to have no reflected wave, then the direction of incidence and the state of elliptical polarisation of the incident wave must be adjusted so that the field impedance for this wave alone looking normally across the interface matches that for the transmitted wave looking in the same direction.

At any interface L between (i) an oscillatory electromagnetic field in a volume V_1, and (ii) an oscillatory electromagnetic field of the same frequency in an adjacent volume V_2, the boundary conditions to be satisfied are the continuity of the components of the complex electric and magnetic vectors parallel to the interface:

$$E_{\|1} = E_{\|2}, \qquad H_{\|1} = H_{\|2} \tag{12.25}$$

In many practical problems, however, a number of features necessary to ensure these continuity conditions are easily taken into account in the process of setting up the problem. But the one condition that always requires substantial attention is the continuity of either the field impedance or the field admittance looking normally across the interface:

$$Z_1 = Z_2, \qquad Y_1 = Y_2 \tag{12.26}$$

where Z and Y are defined by eqns. 12.12, taking l to be a unit vector directed normal to the interface L.

12.5   Field impedance for a plane wave

Let us suppose that an infinite plane electromagnetic wave of angular frequency ω is propagating through linear isotropic homogeneous material of permeability μ/μ_0, dielectric constant ϵ/ϵ_0 and conductivity σ. We combine the dielectric constant and the conductivity into a complex dielectric constant defined at angular frequency ω as $\tilde{\epsilon}/\epsilon_0$ where (eqn. 2.44)

$$\tilde{\epsilon} = \epsilon - j\sigma/\omega \tag{12.27}$$

The refractive index of the material is defined as (eqn. 2.30)

$$n = \left(\frac{\mu}{\mu_0} \frac{\epsilon}{\epsilon_0} \right)^{1/2} \tag{12.28}$$

and the complex refractive index as (eqn. 2.47)

$$\tilde{n} = \left(\frac{\mu}{\mu_0} \frac{\tilde{\epsilon}}{\epsilon_0} \right)^{1/2} \tag{12.29}$$

Let the plane wave be travelling in the direction of the positive z axis, and let it be linearly polarised with the electric vector vibrating along the x axis and the magnetic vector vibrating along the y axis. If E and H are the complex electric and magnetic field strengths at the origin, the complex electromagnetic field of the wave at all points may be written

$$\begin{aligned} E &= (E, 0, 0) \exp \{j(\omega t - kz)\} \\ H &= (0, H, 0) \exp \{j(\omega t - kz)\} \end{aligned} \tag{12.30}$$

The factor

$$\exp\{j(\omega t - kz)\} \tag{12.31}$$

is known as the wave function, and k as the propagation constant. In conducting material the propagation constant is complex and its imaginary part is negative. Hence we may write

$$k = \beta - j\alpha \tag{12.32}$$

where β and α are real, so that the wave function (12.31) becomes

$$\exp(-\alpha z)\exp\{j(\omega t - \beta z)\} \tag{12.33}$$

We see that α is the rate of attenuation of the wave in the direction of propagation in nepers per unit distance, and that β is the rate of change of phase in the direction of propagation in radians per unit distance. We also see that, if an observer moves along the z axis so as to keep the phase constant, then

$$\omega t - \beta z = \text{constant}$$

and his velocity is

$$v = \frac{dz}{dt} = \frac{\omega}{\beta} \tag{12.34}$$

The planes $z = $ constant are planes of constant phase, and they move in the direction of the positive z axis with the velocity (12.34), which is known as the phase velocity. Hence, eqns. 12.30 describe a wave of angular frequency ω propagating in the direction of the positive z axis with a phase velocity given by the ratio of ω to the real part of k, and a rate of attenuation given by the negative of the imaginary part of k.

Let us substitute from eqns. 12.30 into version 3 of the electromagnetic equations in Table 2.3, remembering that we are now dropping primes on the magnetic vector. All the field equations are satisfied identically except the x component of the electric curl equation and the y component of the magnetic curl equation, which give

$$kE = \omega\mu H, \qquad kH = \omega\tilde{\epsilon}E$$

or

$$k = \omega(\mu\tilde{\epsilon})^{1/2}, \; E/H = (\mu/\tilde{\epsilon})^{1/2} \tag{12.35}$$

The first of these equations gives the propagation constant k for the wave as a function of its angular frequency ω. In connection with the second of eqns. 12.35, it is convenient to define quantities ζ and η such that

$$\zeta = (\mu/\tilde{\epsilon})^{1/2}, \qquad \eta = (\tilde{\epsilon}/\mu)^{1/2} \tag{12.36}$$

These quantities are determined by the permeability, the dielectric constant and the conductivity of the material, and are known, respectively, as the impedance and admittance of the material. They are mutually reciprocal. In terms of them, eqns. 12.30 may be written either as

$$E = \zeta H(1, 0, 0) \exp \{j(\omega t - kz)\}$$
$$H = H(0, 1, 0) \exp \{j(\omega t - kz)\} \tag{12.37}$$

or as

$$E = E(1, 0, 0) \exp \{j(\omega t - kz)\}$$
$$H = \eta E(0, 1, 0) \exp \{j(\omega t - kz)\} \tag{12.38}$$

The field of impedance and field admittance at any point in the wave looking across a plane of constant phase in the direction of propagation may be obtained by substituting from eqns. 12.37 or 12.38 into eqns. 12.12 using

$$l = (0, 0, 1) \tag{12.39}$$

However, the electromagnetic field in a plane of constant phase has an electric vector in the x direction and a magnetic vector in the y direction in accordance with eqns. 12.37 and 12.38. Since these directions are mutually perpendicular we may use the simpler definitions of field impedance and field admittance given in eqns. 12.15. From eqns. 12.37 and 12.38, the field impedance and field admittance at any point in the wave looking across a plane of constant phase in the direction of propagation are

$$Z = \zeta, \qquad Y = \eta \tag{12.40}$$

These equations state that, for a plane wave travelling through homogeneous linear isotropic material, the field impedance and field admittance looking across any plane of constant phase in the direction of propagation are equal, respectively, to the impedance and admittance of the material.

In free space the impedance defined in eqn. 12.36 becomes

$$\zeta_0 = (\mu_0/\epsilon_0)^{1/2} = 377 \, \Omega \tag{12.41}$$

approximately and is known as the impedance of free space. For a plane harmonic electromagnetic wave travelling through free space, the ratio of the complex amplitude of the electric vector at any point, measured in volt metre^{-1}, to the complex amplitude of the magnetic vector at the same point, measured in ampere metre^{-1}, is 377 Ω.

Let us now turn the direction l of look through an angle θ from the direction of phase propagation towards the positive direction of the electric vector as shown in Fig. 12.3. We now have

$$l = (\sin \theta, 0, \cos \theta) \tag{12.42}$$

and a plane L perpendicular to this direction makes an angle θ with the planes of constant phase. The complex electric and magnetic field strengths parallel to L are

$$E_{\parallel} = E \cos \theta, \qquad H_{\parallel} = H \tag{12.43}$$

and the corresponding vectors are mutually perpendicular. From eqns. 12.15 and 12.43, it follows that the field impedance and field admittance looking in the direction l in Fig. 12.3 are

$$Z = \zeta \cos \theta, \qquad Y = \eta \sec \theta \tag{12.44}$$

Instead of considering the field impedance in a direction perpendicular to the magnetic field as shown in Fig. 12.3 let us now use a direction perpendicular to the electric field as shown in Fig. 12.4. We then have

$$l = (0, \sin \theta, \cos \theta) \tag{12.45}$$

and

$$E_{\parallel} = E, \qquad H_{\parallel} = H \cos \theta \tag{12.46}$$

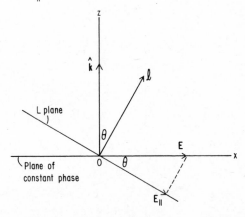

Fig. 12.3.   *Illustrating, for a linearly polarised plane wave travelling in the direction of a unit vector $\hat{k}$, calculation of the field impedance and field admittance looking in the direction of a unit vector l perpendicular to the magnetic field*

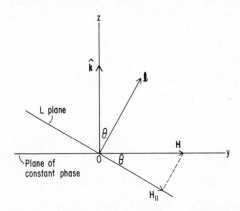

Fig. 12.4.   *Illustrating, for a linearly polarised plane wave travelling in the direction of a unit vector $\hat{k}$, calculation of the field impedance and field admittance looking in the direction of a unit vector l perpendicular to the electric field*

The components of the electric and magnetic fields in a plane L perpendicular to the direction l given by eqn. 12.45 are mutually perpendicular. It therefore follows from eqns. 12.15 and 12.46 that the field impedance and the field admittance looking in the direction l in Fig. 12.4 are

$$Z = \zeta \sec \theta, \qquad Y = \eta \cos \theta \tag{12.47}$$

For the plane wave given by eqns. 12.37 or 12.38 let us now calculate the field impedance looking in an arbitrary direction l specified by spherical polar angles (θ, ψ) as shown in Fig. 12.5. For a plane L perpendicular to this direction, the tangential components of the electric and magnetic fields of the wave are no longer mutually perpendicular in general. We cannot therefore use the simpler definitions of field impedance and field admittance in eqns. 12.15, and must use instead the definitions in eqns. 12.12. We obtain

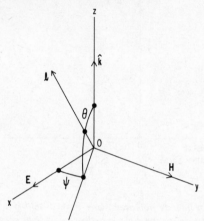

Fig. 12.5. *Illustrating calculation of the field impedance and field admittance of a plane wave looking in an arbitrary direction specified by spherical polar angles θ, ψ ($\theta = 0$ is the direction of propagation; $\theta = \frac{1}{2}\pi$, $\psi = 0$ is the positive direction of the electric vector; $\theta = \frac{1}{2}\pi$, $\psi = \frac{1}{2}\pi$ is the positive direction of the magnetic vector)*

$$\begin{cases} l = (\sin\theta \cos\psi, \sin\theta \sin\psi, \cos\theta) \\ l \times E = \zeta H (0, \cos\theta, -\sin\theta \sin\psi) \exp\{j(\omega t - kz)\} \\ l \times H = H(-\cos\theta, 0, \sin\theta \cos\psi) \exp\{j(\omega t - kz)\} \end{cases}$$

so that

$$\begin{cases} l \cdot (E \times H^*) = \zeta H H^* \cos\theta, \qquad l \cdot (E^* \times H) = \zeta^* H H^* \cos\theta \\ (l \times E) \cdot (l \times E^*) = \zeta \zeta^* H H^* (1 - \sin^2\theta \cos^2\psi) \\ (l \times H) \cdot (l \times H^*) = H H^* (1 - \sin^2\theta \sin^2\psi) \end{cases}$$

leading to

$$Z = \zeta \frac{\cos\theta}{1 - \sin^2\theta \sin^2\psi}, \qquad Y = \eta \frac{\cos\theta}{1 - \sin^2\theta \cos^2\psi} \tag{12.48}$$

These equations reduce to eqns. 12.44 when $\psi = 0$ and to eqns. 12.47 when $\psi = \frac{1}{2}\pi$, as they should.

12.6   The relation between field impedance and field admittance

Circuit impedance and circuit admittance are always mutually reciprocal, but the same is not true for field impedance and field admittance. The field impedances and field admittances in eqns. 12.40, 12.44 and 12.47 are mutually reciprocal, and the components of E and H in a plane L at right angles to the direction l of look are mutually perpendicular. But these statements are not true for the field impedance and field admittance in eqns. 12.48. The product of the field impedance and the field admittance looking in the direction (θ, ϕ) in Fig. 12.5 is

$$ ZY = \frac{\cos^2\theta}{(1 - \sin^2\theta \, \sin^2\psi)(1 - \sin^2\theta \, \cos^2\psi)} $$

which may be rewritten

$$ ZY = \frac{\cos^2\theta}{\cos^2\theta + \sin^4\theta \, \sin^2\psi \, \cos^2\psi} \tag{12.49} $$

Eqn. 12.49 verifies that the field impedance Z and the field admittance Y looking in a particular direction are not in general mutually reciprocal. However, we also see from eqn. 12.49 that Z and Y are reciprocal if either (i) $\theta = 0$, or (ii) $\psi = 0$, or (iii) $\psi = \frac{1}{2}\pi$. These are the cases that correspond to eqns. 12.40, 12.44 and 12.47, respectively. They are also the situations in which the components of the electric and magnetic field in a plane L at right angles to the direction l of look are mutually perpendicular.

Problems involving interfaces between oscillatory electromagnetic fields (for example, a boundary between two different materials) are not always easy to solve. The tractable problems are usually ones for which the parallel components of the electric and magnetic vectors at the interface are mutually perpendicular, so that the field impedance and field admittance looking normally across the interface are mutually reciprocal (eqns. 12.15).

It should be noted, however, that problems involving an interface at which it is not true that (i) the parallel components of the electric and magnetic vectors are mutually perpendicular, and (ii) the field impedance and field admittance looking normally across the interface are mutually reciprocal, can sometimes be analysed into two fields for each of which separately these conditions are satisfied.

Let us reconsider the situation illustrated in Fig. 12.5, and let us introduce new axis (X, Y, Z) for which the Z axis coincides with the z axis but the (X, Y) axes are rotated about the common Z and z axes through an angle ψ as shown in Fig. 12.6. The complex electromagnetic field (E, H) whose cartesian components referred to the (x, y, z) axes are given in eqns. 12.37 may then be expressed as the sum of the complex electromagnetic fields (E_1, H_1) and (E_2, H_2) whose cartesian components referred to axes (X, Y, Z) are

$$ \left. \begin{aligned} E_1 &= \zeta H \cos\psi (1, 0, 0) \exp\{j(\omega t - kz)\} \\ H_1 &= H \cos\psi (0, 1, 0) \exp\{j(\omega t - kz)\} \end{aligned} \right\} \tag{12.50} $$

and

$$E_2 = \zeta H \sin \psi (0, -1, 0) \exp \{j(\omega t - kz)\}$$
$$H_2 = H \sin \psi (1, 0, 0) \exp \{j(\omega t - kz)\}$$

(12.51)

For (E_1, H_1) the field impedance and field admittance looking in the direction l in Fig. 12.6 evaluate to (c.f. eqns. 12.44)

$$Z_1 = \zeta \cos \theta, \qquad Y_1 = \eta \sec \theta$$

(12.52)

and for (E_2, H_2) they evaluate to (c.f. eqns. 12.47)

$$Z_2 = \zeta \sec \theta, \qquad Y_2 = \eta \cos \theta$$

(12.53)

The field impedance and the field admittance in eqns. 12.52 are mutually reciprocal, and so are the field impedance and the field admittance in eqns. 12.53, but the two field impedances are different from each other.

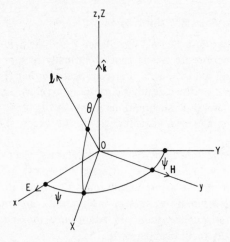

Fig. 12.6.   *Illustrating how a wave travelling in the direction of the common z and Z axes with its magnetic vector linearly polarised in the y direction can be resolved into a pair of linearly polarised waves with their magnetic vectors polarised in the X and Y directions*

What eqns. 12.50–12.53 show is the following. In Fig. 12.6, we suppose that (i) we have a plane wave travelling in the direction of the common z and Z axes with its magnetic vector linearly polarised in the y direction, (ii) an interface L exists perpendicular to the direction l, and (iii) we need to make use of the field impedance or field admittance looking across L in the direction l. This field impedance and field admittance are given by eqns. 12.48 and are not mutually reciprocal. Instead of using this impedance and admittance, we can analyse the wave into two waves. Let p be the plane defined by the direction l of look and the direction $\hat{k}$ of phase propagation. This plane is the XZ plane in Fig. 12.6. We analyse the wave into two waves, one of which is linearly polarised with its magnetic vector vibrating perpendicular to the plane p (the E_1, H_1 field in eqns. 12.50) and the other of which is linearly polarised with its magnetic vector vibrating parallel to the plane p (the E_2, H_2 field in eqns. 12.51). For each of these waves the field impedance and the field admittance looking in the direction l are mutually reciprocal. For each of these waves the components of the electric and magnetic vectors parallel to a plane L at right angles to l are mutually perpendicular. For each of these waves the field impedance looking across L is the ratio of the parallel component of the complex electric vector to the parallel component of the complex magnetic vector, and the field admittance is the reciprocal of this. But for wave 1 it is the electric vector that has to be multiplied by $\cos \theta$ to obtain the component parallel to L (c.f. Fig. 12.3), whereas for wave 2 it is the magnetic vector that is multiplied by $\cos \theta$ (c.f. Fig. 12.4). Consequently the field impedances and field admittances for the two waves looking in the direction l differ as shown in eqns. 12.52 and 12.53.

12.7 Impedances and admittances in a strip transmission line

A good approximation to a plane wave in a homogeneous dielectric of impedance ζ and admittance η may be constructed by means of a strip transmission line using conducting strips of width b and separation a $(a \ll b)$ with the dielectric between the strips. Some fringing occurs at the edges of the strips; in principle, it could be reduced as shown in Fig. 7.8. Neglecting fringing, let us take the length of the line to be in the direction of the z axis and the strips to be perfectly conducting plates perpendicular to the x axis. Then a harmonic wave of angular frequency ω travelling along the line may be taken to have a complex electromagnetic field between the strips given by eqns. 12.37 or 12.38. It follows that eqns. 12.40 give the field impedance and field admittance at any point between the strips looking along the line in the direction of propagation, while eqns. 12.48 give the field impedance and field admittance looking in any other direction.

Particular interest attaches to eqns. 12.44 and 12.47. Eqns. 12.44 show how, at any point between the strips, the field impedance and field admittance change as the direction of look swings through an angle θ from the direction of propagation towards the direction perpendicular to the conducting strips. When $\theta = \frac{1}{2}\pi$ we are looking perpendicular to the perfectly conducting strips, and eqns. 12.44 give

$$Z = 0, \qquad Y = \infty \qquad\qquad (12.54)$$

The field impedance looking into a perfect conductor vanishes as for a short circuit in circuit theory.

Likewise, eqns. 12.47 show how, at any point between the strips, the field impedance and field admittance change as the direction of look swings from the direction of propagation towards the direction perpendicular to the open edge of the transmission line. When $\theta = \frac{1}{2}\pi$ one is looking perpendicular to the open edge, and eqns. 12.47 give

$$Z = \infty, \qquad Y = 0 \qquad\qquad (12.55)$$

corresponding to an open circuit.

When looking in the direction of the length of a transmission line, it is often more convenient to use circuit impedance rather than field impedance. The relation between the two may be calculated in terms of the geometry of the line. At a particular normal cross-section of the line we have (i) a complex voltage V between the conductors, (ii) a complex electric current I in the conductors, (iii) a complex electric field strength E at a point between the conductors, and (iv) a complex magnetic field strength H at the same point. The directions of the electric and magnetic vectors at each point are mutually perpendicular for all perfectly conducting transmission lines (c.f. Figs. 7.10 and 7.11). For a strip transmission line, the electric and magnetic fields between the conductors are uniform if fringing is neglected. The complex voltage per unit distance between the conductors is E and the complex current per unit width on strips is H. For a strip transmission line with strips of width b at a distance a apart ($a \ll b$), we have

$$V = Ea, \qquad I = Hb \qquad (12.56)$$

Hence

$$\frac{V}{I} = \frac{E}{H}\frac{a}{b} \qquad (12.57)$$

Circuit impedance and field impedance looking along the line are proportional to each other, the factor of proportionality depending on the geometry of a normal cross-section of the line.

Consider a wave propagating along the line in the direction of the z axis. Let this direction also be the direction of positive current on the positive conductor. Circuit impedance looking in this direction is then what is known as the characteristic impedance Z_0 of the line (Problem 8.2), and the field impedance is the impedance ζ of the material between the conductors. Hence, for a strip transmission having perfectly conducting strips of width b at a distance a apart ($a \ll b$), eqn. 12.57 gives

$$Z_0 = \zeta a/b \qquad (12.58)$$

Using eqns. 12.56 and 12.58, eqns. 12.37 may be expressed in terms of complex voltage and current as

$$\left.\begin{array}{l} V = Z_0 I_0 \exp\{j(\omega t - kz)\} \\ I = I_0 \exp\{j(\omega t - kz)\} \end{array}\right\} \qquad (12.59)$$

where the propagation constant k is still given by the first of eqns. 12.35. The complex power along the line in the z direction is (eqns. 12.2 and 12.3)

$$\widetilde{P} = \tfrac{1}{2}VI^* = \tfrac{1}{2}Z_0|I|^2 = Z_0\overline{I}^2 \qquad (12.60)$$

where $\overline{I}$ is the RMS current. From eqns. 12.37, the complex power in the z direction per unit cross-sectional area at any point between the strips is (eqn. 12.4)

$$\widetilde{f} = \tfrac{1}{2}EH^* = \tfrac{1}{2}\zeta|H|^2 = \zeta\overline{H}^2 \qquad (12.61)$$

where $\bar{H}$ is the RMS magnetic field strength. Because of the neglect of fringing in the strip transmission line, the relation between $\tilde{P}$ and $\tilde{f}$ should be

$$\tilde{P} = \tilde{f} \, ab \tag{12.62}$$

Use of eqns. 12.56 and 12.58 in eqns. 12.60 and 12.61 verifies that this is true.

12.8   Dielectric loss

Dielectric loss causes attenuation of a plane wave in a dielectric, or of a wave in a transmission line that possesses dielectric insulation. Dielectric loss can arise from dielectric hysteresis associated with non-linear behaviour of the material (Section 4.11), and also from conductivity of the material. The latter is easier to treat, and is taken into account by the appearance of σ in the imaginary part of $\tilde{\epsilon}$ in eqn. 12.27.

At an angular frequency ω, isotropic material is classified as predominantly a dielectric or predominantly a conductor according as the real or imaginary parts of $\tilde{\epsilon}$ are predominant in eqn. 12.27. For a wave to propagate through material without undue loss, the imaginary part of $\tilde{\epsilon}$ must be small compared with the real part. Expanding by the binomial theorem, we deduce from eqns. 12.35 and 12.36 the approximate relations

$$
\begin{cases}
k = \omega(\mu\epsilon)^{1/2} \left(1 - j \dfrac{\sigma}{2\omega\epsilon}\right) & (12.63) \\[2mm]
\varsigma = \left(\dfrac{\mu}{\epsilon}\right)^{1/2} \left(1 + j \dfrac{\sigma}{2\omega\epsilon}\right) & (12.64)
\end{cases}
$$

It is usually satisfactory to drop the small term $\sigma/(2\omega\epsilon)$ in eqn. 12.64 and to write the impedance of the dielectric as

$$\varsigma = (\mu/\epsilon)^{1/2} \tag{12.65}$$

However, the term $\sigma/(2\omega\epsilon)$ in eqn. 12.63 cannot be dropped because it is the imaginary part of k that describes the attenuation of the wave.

Comparing eqn. 12.63 with eqn. 12.32 we obtain

$$\beta = \omega(\mu\epsilon)^{1/2}, \qquad \alpha = \tfrac{1}{2}\sigma\varsigma \tag{12.66}$$

and ς is given by eqn. 12.65. The first of eqns. 12.66 shows that propagation of phase is unaffected by the presence of a small amount of conductivity in the dielectric. The second of eqns. 12.66 yields an attenuation factor (c.f. expr. 12.33)

$$\exp(-\alpha z) = \exp(-\tfrac{1}{2}\sigma\varsigma z) \tag{12.67}$$

This applies to field strength, electric or magnetic. For energy density or energy flow the attenuation factor is

$$\exp(-2\alpha z) = \exp(-\sigma\varsigma z) \tag{12.68}$$

Eqn. 12.68 implies that, in a dielectric that possesses a small amount of conductivity, a fraction

$$\sigma \zeta \qquad (12.69)$$

of the energy flowing in the direction of propagation is abstracted for each unit of distance that the wave propagates. This may be verified by considering a unit cube of the lossy dielectric with one pair of faces perpendicular to the electric field and another pair perpendicular to the direction of propagation. Let $\bar{E}$ be the RMS electric field strength at this location. In terms of the admittance η of the dielectric (the reciprocal of ζ in eqn. 12.65), the mean rate of flow of energy through the unit cube is (eqns. 12.18 and 12.36)

$$\eta \bar{E}^2 \qquad (12.70)$$

But the mean rate of absorption of energy in the unit cube is (eqn. 11.17)

$$\sigma \bar{E}^2 \qquad (12.71)$$

Dividing expr. 12.71 by expr. 12.70, we see that the fraction of energy abstracted per unit distance is σ/η, and this is identical with expr. 12.69.

If we are discussing, not propagation of an infinite plane wave through an infinite homogeneous dielectric, but propagation along a transmission line possessing dielectric uniformly distributed between the conductors, it is often more convenient to calculate loss due to slight conductivity of the dielectric by using circuit impedance rather than field impedance. The complex voltage and current for a wave travelling along the line are then given by eqns. 12.59, where Z_0 is the characteristic impedance of line (reciprocal Y_0). At a location along the line where the RMS oscillatory voltage is $\bar{V}$, the mean rate of flow of energy along the line is (c.f. expr. 12.70)

$$Y_0 \bar{V}^2 \qquad (12.72)$$

If G is the conductance per unit length of the line, the mean rate of absorption of energy in the dielectric per unit length of the line is (c.f. expr. 12.71)

$$G\bar{V}^2 \qquad (12.73)$$

Dividing expr. 12.73 by expr. 12.72, we see that the fraction of energy abstracted per unit distance along the line is G/Y_0 or (c.f. expr. 12.69)

$$GZ_0 \qquad (12.74)$$

The attentuation along the line in nepers per unit distance is therefore (c.f. eqn. 12.67)

$$\alpha = \tfrac{1}{2}GZ_0 \qquad (12.75)$$

For a strip transmission with strips of width b at a small distance a apart, Z_0 is given by eqn. 12.58 and the conductance per unit length is (Problem 2.2)

$$G = \sigma a/b \qquad\qquad (12.76)$$

so that expr. 12.75 agrees with the second of eqns. 12.66.

12.9 Metal loss

For waves of angular frequency ω guided by metal surfaces, as in a strip transmission line, loss also arises from the imperfect conductivity of the metal. A fraction, normally small, of the energy flowing along the transmission line is abstracted sideways, and is absorbed in the metallic guides. The wave produced in the metal is a highly attenuated wave travelling almost normally into the interior of the metal. As indicated in Fig. 12.7, there is a wave between the strips travelling along the line. But when the strips are imperfectly conducting, there are also waves travelling in each metal strip almost at right angles to the length of the line. The closeness of the angle to a right angle is considerably more precise in practice than can be shown in the diagram. At radio frequencies the waves in the metal are attenuated rapidly. Usually they are completely absorbed before they have an opportunity to emerge from the opposite surface of the metal, and we shall assume that this condition is satisfied. The waves in the metal cause energy to be sapped from the wave in the dielectric. By calculating the fraction of energy removed per unit length of the line we can estimate in nepers per unit distance the rate of attenuation along the line due to metal loss.

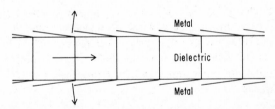

Metal

Dielectric

Metal

Fig. 12.7.   *Illustrating, for a strip transmission line having imperfectly conducting metal strips, propagation of phase along the line in the insulating dielectric, combined with propagation of phase in the metal almost normal to the metal-dielectric interface*
Arrows indicate direction of phase propagation. It is assumed that a wave entering the metal from the dielectric is absorbed before it reaches the opposite face of the strip

For the wave travelling in the metal, eqns. 12.37 may be applied if the direction of the z axis now points normally into the metal. Let the conductivity of the metal be σ_m, let its permeability be μ_m/μ_0 (usually unity), and let its dielectric constant be ϵ_m/ϵ_0 (usually negligible). It follows from eqns. 12.35 and 12.36 that the propagation constant k_m and the impedance ζ_m of the metal at angular frequency ω are

$$\begin{cases} k_m = \omega\{\mu_m(\epsilon_m - j\sigma_m/\omega)\}^{1/2} & (12.77) \\ \zeta_m = \{\mu_m/(\epsilon_m - j\sigma_m/\omega)\}^{1/2} & (12.78) \end{cases}$$

For a metal, the conductivity is normally so high that ϵ_m is negligible compared with σ_m/ω, and eqns. 12.77 and 12.78 become

$$\begin{cases} k_m = (\omega\mu_m\sigma_m)^{1/2} \exp(-j\pi/4) = (\tfrac{1}{2}\omega\mu_m\sigma_m)^{1/2}(1-j) & (12.79) \\ \zeta_m = (\omega\mu_m/\sigma_m)^{1/2} \exp(j\pi/4) = (\tfrac{1}{2}\omega\mu_m/\sigma_m)^{1/2}(1+j) & (12.80) \end{cases}$$

Eqns. 12.79 may be written

where
$$k_m = \beta_m - j\alpha_m \tag{12.81}$$
$$\beta_m = \alpha_m = (\tfrac{1}{2}\omega\mu_m\sigma_m)^{1/2} \tag{12.82}$$

Hence the wavelength in the metal is $2\pi\delta$, and the distance for one neper attenuation is δ, where

$$\delta = (\tfrac{1}{2}\omega\mu_m\sigma_m)^{-1/2} \tag{12.83}$$

This quantity is known as the skin depth in the metal and, at radio frequencies, it is normally small compared with the thickness of the metal. For a perfect conductor the current flows on the surface, but for an imperfect conductor it is spread over a depth of the order of δ given by eqn. 12.83.

Eqns. 12.80 may be written

where
$$\zeta_m = R_m + jX_m \tag{12.84}$$
$$R_m = X_m = (\tfrac{1}{2}\omega\mu_m/\sigma_m)^{1/2} \tag{12.85}$$

These equations give the field impedance ζ_m at any point of the metal looking normally away from the interface with the dielectric, and this is also the field impedance in the dielectric at the interface looking normally into the metal. If $\bar{H}$ is the RMS magnetic field strength in the dielectric, the complex power into the metal from the dielectric, per unit area of the interface, is $\zeta_m\bar{H}^2$ and the resistive component of this is

$$R_m\bar{H}^2 \tag{12.86}$$

where R_m is given by eqns. 12.85.

Expr. 12.86 gives the mean rate per unit time at which energy is being abstracted sideways from the transmission line per unit area of the metal surface. From this we can calculate the rate at which energy is being abstracted sideways per unit length of the line. It is by expressing this as a fraction the total mean rate of flow of energy along the line that we calculate the rate of attenuation along the line due to metal loss.

Let us consider a strip transmission line with strips of width b at a small distance a apart. For a unit length of the line there is an area b of metal on each strip, making $2b$ in all. Expr. 12.86 therefore shows that the rate of sideways abstraction of energy per unit length of the line is

$$2bR_m\bar{H}^2 \tag{12.87}$$

or
$$2(R_m/b)\bar{I}^2 \tag{12.88}$$

if expressed in terms of the RMS electric current $\bar{I}$ along the line with the aid of the second of eqns. 12.56. But, from eqns. 12.60, the mean rate of flow of energy along the line is

$$Z_0 \bar{I}^2 \tag{12.89}$$

where Z_0 is the characteristic impedance of the line given, in accordance with eqns. 12.58 and 12.65, by

$$Z_0 = (\mu/\epsilon)^{1/2}(a/b) \tag{12.90}$$

Dividing expr. 12.88 by expr. 12.89, we see that the fraction of energy abstracted from the line per unit distance because of metal loss is

$$\frac{2R_m}{bZ_0} \tag{12.91}$$

Hence the rate of attenuation along the line in nepers per unit distance due to metal loss is

$$\frac{R_m}{bZ_0} \tag{12.92}$$

where R_m is given by eqns. 12.85 and Z_0 by eqn. 12.90. The coefficient of $\bar{I}^2$ in expr. 12.88 is known as the high frequency resistance of the transmission line per unit length.

When a wave travelling along a transmission line suffers attenuation due both to metal loss and to dielectric loss, both types of loss must be taken into account. Provided that the two rates of attenuation are small, the combined rate of attenuation is simply the sum of the two.

Problems

12.1   A twin-wire transmission line consists of a pair of thin metal wires, each of radius a, at a distance d apart. The metal is highly conducting, and it may be assumed that current is concentrated near the surfaces. The impedance of the metal is ζ_m, given by eqns. 12.84 and 12.85. The rest of space is filled with homogeneous loss-free material of impedance ζ ($\zeta \gg |\zeta_m|$). Show that the characteristic impedance of the transmission line is approximately

$$Z_0 = \frac{\zeta}{\pi} \ln \frac{d}{a}$$

and that the high-frequency resistance of each wire per unit length is approximately

$$R_m/(2\pi a)$$

Deduce that the attention along the line in nepers per unit distance due to metal loss is approximately

$$\frac{1}{\pi a} \frac{R_m}{Z_0}$$

12.2   A coaxial transmission line has an inner conductor of external radius a and a coaxial outer conductor of internal radius b. The metal is highly conducting, and it may be assumed that current is concentrated near the surfaces. The impedance of the metal is ζ_m, given by eqns. 12.84 and 12.85. The space between the conductors is filled with homogeneous loss-free material of impedance ζ ($\zeta \gg |\zeta_m|$). Show that the characteristic impedance of the transmission line is approximately

$$Z_0 = \frac{\zeta}{2\pi} \ln \frac{b}{a}$$

and that the high-frequency resistances per unit length of the inner and outer conductors respectively are approximately

$$R_m/(2\pi a), \qquad R_m/(2\pi b)$$

Deduce that the attenuation along the line in nepers per unit distance due to metal loss is approximately

$$\frac{1}{2\pi} \left(\frac{1}{a} + \frac{1}{b} \right) \frac{R_m}{Z_0}.$$

Chapter 13

Reflection and refraction of electromagnetic waves at a plane interface

13.1 Introduction

If a plane harmonic electromagnetic wave is incident upon a plane interface between two homogeneous non-conducting materials at an angle θ_1 with the normal to the interface, then in general a reflected wave occurs in the first material at an angle θ_1 with the normal, and a transmitted wave occurs in the second material at a different angle θ_2 with the normal. This is the process of reflection and refraction that is familiar in optics. In this chapter we shall study the process of reflection and refraction of electromagnetic waves at a plane interface, not only for insulating materials but also for materials possessing conductivity.

13.2 The exponential wave-function

Let us suppose that the incident, reflected and transmitted waves each involve complex electromagnetic fields for which the variations with the time t and with the cartesian space co-ordinates (x, y, z) can each be described with the aid of the exponential wave-function

$$\exp\{j(\omega t - k_x x - k_y y - k_z z)\} \tag{13.1}$$

In this wave-function the vector

$$k = (k_x, k_y, k_z) \tag{13.2}$$

is known as the propagation vector. If the position vector (x, y, z) is denoted by r, the wave-function may be written

$$\exp\{j(\omega t - k \cdot r)\} \tag{13.3}$$

The wave-function (13.1) is the product of a function of t, a function of x, a function of y and a function of z, each of which is a rotating vector in the complex plane (Section 2.9). At a given point (x, y, z) in space, the vector (13.1) rotates in the complex plane in the counterclockwise direction at the rate ω radians per unit time, assuming ω to be real. However, as we move in the direction of the x co-ordinate at a fixed time, expr. 13.1 describes a vector that rotates in the complex plane in the clockwise direction at the rate k_x radians per unit distance, assuming k_x to be real. Similar statements apply for the y and z co-ordinates. If (ω, k_x, k_y, k_z) are complex, the vectors spiral in the complex plane instead of simply rotating, and the reference components represent sinusoidal oscillations in time and space with amplitudes that vary exponentially.

Let ω be real, but let (k_x, k_y, k_z) have real parts $(\beta_x, \beta_y, \beta_z)$ and imaginary parts $-(\alpha_x, \alpha_y, \alpha_z)$, so that

$$k = \beta - j\alpha = (\beta_x - j\alpha_x, \beta_y - j\alpha_y, \beta_z - j\alpha_z) \tag{13.4}$$

The wave-function (13.1) then becomes

$$\exp\{-(\alpha_x x + \alpha_y y + \alpha_z z)\} \exp\{j(\omega t - \beta_x x - \beta_y y - \beta_z z)\} \tag{13.5}$$

An equivalent statement is that the wave-function (13.3) becomes

$$\exp(-\alpha \cdot r) \exp\{j(\omega t - \beta \cdot r)\} \tag{13.6}$$

In the complex plane this is a vector for which the natural logarithm of the magnitude is

$$N = -\alpha \cdot r \tag{13.7}$$

and for which the counterclockwise angle with the reference direction is

$$\phi = \omega t - \beta \cdot r \tag{13.8}$$

For the wave described by expr. 13.6, N is its amplitude in nepers at position r, while ϕ is its phase in radians at position r and time t.

From eqn. 13.7 we see that the surfaces of constant amplitude are the planes

$$\alpha \cdot r = \text{constant} \tag{13.9}$$

which are perpendicular to the vector α as shown in Fig. 13.1. The perpendicular distance from the origin onto one of the planes of constant amplitude is

$$p = \hat{\alpha} \cdot r \tag{13.10}$$

where $\hat{\alpha}$ is a unit vector in the direction of α, and r is the position vector of a point on the plane. Hence eqn. 13.7 may be written

$$N = -\alpha p \tag{13.11}$$

and we see that the amplitude decreases in the direction α at the rate α nepers per unit distance.

From eqn. 13.8 we see that the surfaces of constant phase are the planes

$$\beta \cdot r = \text{constant} \tag{13.12}$$

which are perpendicular to the direction of the vector β as shown in Fig. 13.1. The perpendicular distance from the origin onto one of the planes of constant phase is

$$q = \hat{\beta} \cdot r \tag{13.13}$$

where $\hat{\boldsymbol{\beta}}$ is a unit vector in the direction of $\boldsymbol{\beta}$, and r is the position vector of a point on the plane. Hence eqn. 13.8 may be written

$$\phi = \omega t - \beta q \qquad (13.14)$$

From this equation we see that, at a given time t, the phase-lag increases in the direction $\boldsymbol{\beta}$ at the rate β radians per unit distance.

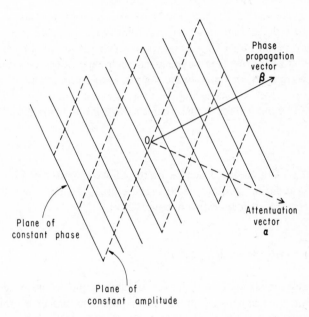

Fig. 13.1   *Illustrating the planes of constant phase and the planes of constant amplitude associated with an exponential wave-function*

From eqn. 13.14, we also see that a plane on which the phase has a given value ϕ is moving in the direction of the vector $\boldsymbol{\beta}$ at the speed

$$v = \frac{dq}{dt} = \frac{\omega}{\beta} \qquad (13.15)$$

This is known as the phase velocity of the wave. The wavelength λ of the wave is the distance that a plane of constant phase advances during a single period of oscillation $2\pi/\omega$. Since it advances at the speed given by eqn. 13.15, we have

$$\lambda = 2\pi/\beta \qquad (13.16)$$

Eqn. 13.15 states that the product of the frequency f of the wave in hertz $[\omega/(2\pi)]$ and the wavelength λ of the wave $[2\pi/\beta]$ is equal to the phase velocity v.

Fig. 13.1 illustrates the fact that, if $\boldsymbol{\alpha}$ and $\boldsymbol{\beta}$ are real vectors and k is given by eqn. 13.4, then the wave-function (13.3) represents a wave of angular frequency ω and wavelength $2\pi/\beta$ that is travelling with phase velocity ω/β in the direction $\boldsymbol{\beta}$

and that is being attenuated at the rate α nepers per unit distance in the direction $\boldsymbol{\alpha}$. The planes of constant phase and the planes of constant amplitude intersect at an angle as shown in Fig. 13.1. At a fixed point in space, the phase increases with time at the rate ω radians per unit time. At a fixed time, the phase-lag increases in the direction $\boldsymbol{\beta}$ at the rate β radians per unit distance; β is known as the phase constant. The amplitude decreases exponentially in the direction $\boldsymbol{\alpha}$ at the rate α nepers per unit distance; α is known as the attenuation constant.

Although the wave illustrated in Fig. 13.1 is composed of planes of constant phase and planes of constant amplitude, it is not what is known as a plane wave. The wave illustrated in Fig. 13.1 has an exponential variation of amplitude across a plane of constant phase. A plane wave has no such exponential variation of amplitude across a wave front. In a plane wave the planes of constant amplitude coincide with the plane of constant phase, and the vectors $\boldsymbol{\alpha}$ and $\boldsymbol{\beta}$ in Fig. 13.1 have a common direction. If a unit vector in this common direction is denoted by $\hat{k}$, eqn. 13.4 then becomes

$$k = (\beta - j\alpha)\hat{k} \tag{13.17}$$

If the unit vector $\hat{k}$ is in the direction of the positive z axis, then $\beta - j\alpha$ in eqn. 13.17 is the propagation constant appearing in eqn. 12.32, and the wave-function (13.3) is the wave-function appearing in eqns. 12.30.

13.3 The concept of a dispersion relation

For linear homogeneous material, solutions of the complex electromagnetic equations (Table 2.3) exist in which each cartesian component of each complex electromagnetic vector varies in time and space proportionally to the exponential wave-function (13.1). However, arbitrary values for the parameters ω, k_x, k_y, and k_z are not possible. The electromagnetic equations for a particular material, and even for free space, require a relation between these parameters. This is known as the dispersion relation. For complicated material the functional dependence involved in the dispersion relation is complicated. However, it is not difficult to derive the dispersion relation for linear isotropic material whose properties are described by a permeability μ/μ_0, a dielectric constant ϵ/ϵ_0 and a conductivity σ, even if these quantities are permitted to depend on the angular frequency ω (c.f. Section 2.12).

Consider a homogeneous material for which the complex dielectric constant at angular frequency ω is $\tilde{\epsilon}/\epsilon_0$ and the permeability is μ/μ_0. Using version 3 of the electromagnetic equations in Table 2.3, let us see whether these equations have a solution in which each cartesian component of the complex electromagnetic field varies in time and space in accordance with the wave-function (13.1). In these circumstances, differentiation with respect to t is equivalent to multiplication by $j\omega$ as described in Section 2.9. Moreover, differentiation with respect to x, y, or z is equivalent to multiplication by $-jk_x$, $-jk_y$, or $-jk_z$, respectively. We may summarise these statements by using the wave-function (13.3) and then writing

$$\frac{\partial}{\partial t} \equiv j\omega, \quad \nabla \equiv -jk \tag{13.18}$$

The second of eqns. 13.18 means that, when all electromagnetic field quantities vary in time and space proportionally to the wave-function (13.3), all three of the following statements are true:

(i) $-jk$ times any scalar function of position $\phi(x, y, z)$ is equal to $\nabla\phi$;
(ii) the scalar product of $-jk$ with any vector field A is equal to $\nabla \cdot A$; and
(iii) the vector product of $-jk$ with any vector field A is equal to $\nabla \times A$.

It follows that, for the wave-function (13.3), the Maxwell differential curl relations simply become vector algebraic relations. Making use of the connection relations in Table 2.3, the differential field equations in the last column of Table 2.3 yield the vector algebraic equations

$$H = \frac{1}{\omega\mu}k \times E, \quad E = -\frac{1}{\omega\tilde{\epsilon}}k \times H \tag{13.19}$$

where the primes on the magnetic vector have been dropped. These equations imply that

$$k \cdot E = 0, \quad k \cdot H = 0, \quad E \cdot H = 0 \tag{13.20}$$

The dispersion relation that we are seeking is obtained by eliminating the six quantities (E_x, E_y, E_z), (H_x, H_y, H_z) between the six algebraic equations corresponding to the cartesian components of eqns. 13.19. If we substitute for H from the first of the vector eqns. 13.19 into the second, we then have to eliminate the cartesian components of E from the three algebraic equations implied by the vector equation

$$E + \frac{1}{\omega^2\mu\tilde{\epsilon}}\{k \times (k \times E)\} = 0$$

that is (Identity 10, Appendix B),

$$E + \frac{1}{\omega^2\mu\tilde{\epsilon}}\{(k \cdot E)k - k^2 E\} = 0$$

Using the first of eqns. 13.20, we obtain for the eliminant

$$k^2 = \omega^2\mu\tilde{\epsilon} \tag{13.21}$$

That this is a relation between the quantities (ω, k_x, k_y, k_z) appearing in the wave-function (13.1) is seen by writing eqn. 13.21 in the form

$$k_x^2 + k_y^2 + k_z^2 = \omega^2\mu\tilde{\epsilon} \tag{13.22}$$

This is the dispersion relation for linear isotropic material of complex dielectric constant $\tilde{\epsilon}/\epsilon_0$ and permeability μ/μ_0.

If the wave under consideration is a linearly polarised plane wave, the above discussion reduces to that given in Section 12.5. The first of eqns. 12.35 is the dispersion relation (13.21). For a plane wave the vectors $\boldsymbol{\alpha}$ and $\boldsymbol{\beta}$ in Fig. 13.1 point in the same direction, and in Section 12.5 this common direction is taken as that of the positive z axis. Moreover, in Section 12.5 the electric vector is taken as vibrating along the x axis and the magnetic vector along the y axis (eqns. 12.30), so that the electric vector, the magnetic vector and the direction of propagation form a right-handed system of directions. The same information is conveyed by eqns. 13.19 and 13.20 in a form independent of the orientation of the axes of co-ordinates. In terms of the impedance ζ or admittance η of the material as defined in eqns. 12.36, eqns. 13.19 may be written in either of the following forms:

$$\left.\begin{array}{ll} H = \dfrac{\eta}{k}k \times E, & E = -\dfrac{1}{\eta k}k \times H \\[4mm] H = \dfrac{1}{\zeta k}k \times E, & E = -\dfrac{\zeta}{k}k \times H \end{array}\right\} \qquad (13.23)$$

Substitution into the last of these equations from eqns. 12.30 then gives the second of eqns. 12.35.

It should be noticed that, if we are not dealing with a linearly polarised plane wave, care is necessary in the interpretation of eqns. 13.19 and 13.20. In general, the vector k in these equations does not connote a single direction in three-dimensional space. It connotes two directions, those of the vectors α and β in Fig. 13.1. This is because the propagation vector k is a vector in two senses; it has cartesian components in three-dimensional space, and each cartesian component is a vector in the complex number sense; see eqns. 13.4. Again, if the wave is elliptically polarised (see Section 10.3), then the complex electric vector connotes a pair of different directions in three-dimensional space as shown in eqns. 10.21, and the same is true for the magnetic vector. In general, therefore, each of the vectors k, E and H in eqns. 13.19 and 13.20 is associated with twin directions in three-dimensional space.

In spite of this complication, an elliptically polarised wave in which the planes of constant amplitude do not coincide with the planes of constant phase has the dispersion relation (13.21), the same as that for a linearly polarised plane wave (first of eqns. 12.35). It should be noted however that, in the general case, k in eqn. 13.21 means the three-dimensional magnitude of the propagation vector k in eqn. 13.4 as defined in eqn. 10.4. Hence the dispersion relation (13.21) or (13.22), if written out in full, reads

$$(\beta_x - j\alpha_x)^2 + (\beta_y - j\alpha_y)^2 + (\beta_z - j\alpha_z)^2 = \omega\mu(\epsilon - j\sigma/\omega) \qquad (13.24)$$

It is often convenient to express the dispersion relation with the aid of the complex refractive index of the material defined in eqns. 2.47 and 12.29. If this is done, the dispersion relation (13.21) becomes

$$k = \tilde{n}\omega/c \qquad (13.25)$$

Since ω/c is the propagation constant at angular frequency ω in free space $(\tilde{n} = 1)$, eqn. 13.25 states that the complex refractive index of a material at any frequency is the ratio of the three-dimensional magnitude of the propagation vector k to the propagation constant in free space.

The dispersion relation (13.21) gives the propagation constant k at angular frequency ω for a material whose permeability is μ/μ_0 and whose complex dielectric constant is $\tilde{\epsilon}/\epsilon_0$. However, it is often convenient to invert this statement by using the dispersion relation in conjunction with the definition of the impedance and admittance of the material given in eqns. 12.36. Eqns. 12.36 and 13.21 may be solved for $\tilde{\epsilon}$ and μ in either of the following forms:

$$\left.\begin{aligned} \tilde{\epsilon} &= k/(\omega\zeta), & \mu &= k\zeta/\omega \\ \tilde{\epsilon} &= k\eta/\omega, & \mu &= k/(\omega\eta) \end{aligned}\right\} \tag{13.26}$$

We therefore see that, at angular frequency ω, the propagation constant k and the impedance ζ (or its reciprocal η) completely characterise linear isotropic material for electromagnetic purposes. Consequently, when studying wave propagation in a linear isotropic material at a stated frequency, it is often convenient to describe the properties of the material by quoting its propagation constant and its impedance (or admittance).

13.4 Evanescent waves

Let us consider the significance of the exponential wave-function (13.3) in linear isotropic material that is non-conducting. This includes free space as a special case. The dispersion relation is obtained by putting $\sigma = 0$ in eqn. 13.24. The imaginary part of this equation then gives

$$\alpha_x\beta_x + \alpha_y\beta_y + \alpha_z\beta_z = 0$$

or

$$\boldsymbol{\alpha} \cdot \boldsymbol{\beta} = 0 \tag{13.27}$$

This states that either (i) $\boldsymbol{\alpha} = 0$, or (ii) $\boldsymbol{\beta} = 0$, or (iii) the vectors $\boldsymbol{\alpha}$ and $\boldsymbol{\beta}$ in Fig. 13.1 are at right angles.

When $\boldsymbol{\alpha} = 0$ there is no exponential variation of amplitude in Fig. 13.1. There are simply planes of constant phase propagating in the direction $\boldsymbol{\beta}$ with velocity ω/β and wavelength $2\pi/\beta$. This is a plane wave travelling in the direction $\boldsymbol{\beta}$. Travelling plane waves are of great importance for transmission of energy from one location to another.

But when $\boldsymbol{\alpha} \neq 0$, any propagation of phase that may take place in non-conducting material must, in accordance with eqn. 13.27, be at right angles to the direction of exponential attenuation of amplitude. In practice such a wave is frequently associated with a plane interface. The direction of exponential attenuation is away from the interface in the perpendicular direction, and the direction of phase propagation (if any) is parallel to the interface. In consequence, there is then no exponential variation of amplitude in the direction of propagation. Such a wave is called an evanescent wave. It creates an electromagnetic field that is localised near the interface. The function of evanescent waves is to incorporate fine structure into an electromagnetic field.

Any electromagnetic field in homogeneous material, or in free space, can be Fourier analyzed into waves represented by exponential wave-functions of the type appearing in expr. 13.1. Some of these waves can be plane waves that are concerned with the transmission of energy. Others can be evanescent waves that are concerned with local storage of energy (c.f. an inductor or a capacitor). It is by adjusting and controlling the evanescent waves that many practical objectives are achieved (tuning, matching, filtering, etc.)

13.5 The incident, reflected and transmitted waves associated with a plane interface

Let us consider a fixed plane interface between two materials whose propagation constants and impedances are k_1, ζ_1 and k_2, ζ_2, and let the corresponding admittances be η_1 and η_2. Let the interface be the plane $z = 0$ in cartesian co-ordinates (x, y, z), and let the propagation vectors for the incident, reflected and transmitted waves be, respectively,

$$\left. \begin{aligned} k_1 &= k_1(S_1, 0, C_1) \\ k_{-1} &= k_1(S_1, 0, -C_1) \\ k_2 &= k_2(S_2, 0, C_2) \end{aligned} \right\} \tag{13.28}$$

where
$$S_1^2 + C_1^2 = 1, \quad S_2^2 + C_2^2 = 1 \tag{13.29}$$

It will be noticed that the propagation vectors for the incident and reflected waves in eqns. 13.28 have identical components parallel to the interface $z = 0$, but reversed components perpendicular to the interface as indicated in Fig. 13.2.

If the waves are plane waves (so that the planes of constant amplitude coincide with the planes of constant phase) then, in accordance with eqn. 13.17, the quantities S_1 and C_1 in eqns. 13.28 and 13.29 are the sine and cosine of the angle of incidence θ_1 shown in Fig. 13.2, while S_2 and C_2 are the sine and cosine of the angle of refraction θ_2. However, we shall see that, even if the incident and reflected waves are plane waves, the transmitted wave may not be a plane wave. In these circumstances the quantities S_2 and C_2 in eqns. 13.28 and 13.29 are not real, and no useful purpose is served by regarding them as the sine and cosine of a complex angle of refraction. Likewise, if it happens that the planes of constant amplitude for the incident wave do not coincide with the planes of constant phase, the quantities S_1 and C_1 in eqns. 13.28 and 13.29 are not real, and no useful purpose is then served by regarding them as the sine and cosine of a complex angle of incidence.

It is convenient to study separately the situations when (i) the magnetic vector H vibrates parallel to the interface and (ii) the electric vector E vibrates parallel to the interface. If H is vibrating parallel to the interface, let the complex magnetic vectors at the origin for the incident, reflected and transmitted waves be, respectively,

$$H_1 = H_1(0, 1, 0) \exp(j\omega t)$$
$$H_{-1} = H_{-1}(0, 1, 0) \exp(j\omega t) \qquad (13.30)$$
$$H_2 = H_2(0, 1, 0) \exp(j\omega t)$$

The quantities H_1, H_{-1} and H_2 are the complex amplitudes for the incident, reflected and transmitted waves at the origin. Using eqns. 13.28 and 13.30 in eqns. 13.23 we deduce that the complex electric vectors at the origin for the incident, reflected and transmitted waves are, respectively,

$$E_1 = \zeta_1 H_1(C_1, 0, S_1) \exp(j\omega t)$$
$$E_{-1} = \zeta_1 H_{-1}(-C_1, 0, -S_1) \exp(j\omega t) \qquad (13.31)$$
$$E_2 = \zeta_2 H_2(C_2, 0, -S_2) \exp(j\omega t)$$

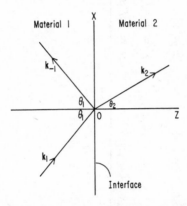

Fig. 13.2 *Illustrating reflection and refraction at an interface between different materials*

The complete expressions for the complex electromagnetic fields as functions of time and position are obtained by replacing the exponential time-functions in eqns. 13.30 and 13.31 by exponential wave-functions, using the appropriate propagation vectors appearing in eqns. 13.28. The results are listed in the left half of Table 13.1.

If, on the other hand, E is vibrating parallel to the interface, the complex electric vectors at the origin for the incident, reflected and transmitted waves may be taken as (c.f. eqns. 13.30)

$$E_1 = E_1(0, 1, 0) \exp(j\omega t)$$
$$E_{-1} = E_{-1}(0, 1, 0) \exp(j\omega t) \qquad (13.32)$$
$$E_2 = E_2(0, 1, 0) \exp(j\omega t)$$

and the complex magnetic vectors at the origin are then (eqns. 13.23)

Table 13.1   *The incident, reflected and transmitted waves at a fixed interface (oblique incidence)*

	H parallel to interface	E parallel to interface
Incident wave, $z \leqslant 0$	$E = \zeta_1 H_1(C_1, 0, -S_1) \exp\{j(\omega t - k_1 S_1 x - k_1 C_1 z)\}$ $H = H_1(0, 1, 0) \exp\{j(\omega t - k_1 S_1 x - k_1 C_1 z)\}$	$E = E_1(0, 1, 0) \exp\{j(\omega t - k_1 S_1 x - k_1 C_1 z)\}$ $H = \eta_1 E_1(-C_1, 0, S_1) \exp\{j(\omega t - k_1 S_1 x - k_1 C_1 z)\}$
Reflected wave, $z \leqslant 0$	$E = \zeta_1 H_{-1}(-C_1, 0, -S_1) \exp\{j(\omega t - k_1 S_1 x + k_1 C_1 z)\}$ $H = H_{-1}(0, 1, 0) \exp\{j(\omega t - k_1 S_1 x + k_1 C_1 z)\}$	$E = E_{-1}(0, 1, 0) \exp\{j(\omega t - k_1 S_1 x + k_1 C_1 z)\}$ $H = \eta_1 E_{-1}(C_1, 0, S_1) \exp\{j(\omega t - k_1 S_1 x + k_1 C_1 z)\}$
Transmitted wave, $z \geqslant 0$	$E = \zeta_2 H_2(C_2, 0, -S_2) \exp\{j(\omega t - k_2 S_2 x - k_2 C_2 z)\}$ $H = H_2(0, 1, 0) \exp\{j(\omega t - k_2 S_2 x - k_2 C_2 z)\}$	$E = E_2(0, 1, 0) \exp\{j(\omega t - k_2 S_2 x - k_2 C_2 z)\}$ $H = \eta_2 E_2(-C_2, 0, S_2) \exp\{j(\omega t - k_2 S_2 x - k_2 C_2 z)\}$

$$
\left.
\begin{aligned}
H_1 &= \eta_1 E_1(-C_1, 0, S_1) \exp(j\omega t) \\
H_{-1} &= \eta_1 E_{-1}(C_1, 0, S_1) \exp(j\omega t) \\
H_2 &= \eta_2 E_2(-C_2, 0, S_2) \exp(j\omega t)
\end{aligned}
\right\}
\qquad (13.33)
$$

The complete expressions for the complex electromagnetic fields of the waves are therefore as listed in the right half of Table 13.1.

We need to understand the relations between the amplitudes and phases of the incidence, reflected and transmitted waves. These amplitudes and phases are conveniently measured at the origin, using whichever electromagnetic vector is vibrating parallel to the interface. Hence we need to calculate the ratios $E_1 : E_{-1} : E_2$ when E is vibrating parallel to the interface, and the ratios $H_1 : H_{-1} : H_2$ when H is vibrating parallel to the interface.

13.6 The relation between the propagation vectors for the incident, reflected and transmitted waves

The boundary conditions at $z = 0$ cannot be satisfied unless we have, over the interface, the same spatial variations with x for the incident, reflected and transmitted waves. This requires that the x components of the propagation vectors in Table 13.1 are the same in both materials, that is, that

$$
k_1 S_1 = k_2 S_2 \qquad (13.34)
$$

If expressed in terms of the complex refractive indices $\tilde{n}_1$ and $\tilde{n}_2$ of the two materials, eqn. 13.34 becomes (eqn. 13.25)

$$
\tilde{n}_1 S_1 = \tilde{n}_2 S_2 \qquad (13.35)
$$

If both materials are non-conducting, the refractive indices are real, and may be written as n_1 and n_2. Moreover, if the waves in both materials are plane waves, S_1 and S_2 may be written as $\sin \theta_1$ and $\sin \theta_2$ (Fig. 13.2). Eqn. 13.35 then becomes

$$
n_1 \sin \theta_1 = n_2 \sin \theta_2 \qquad (13.36)
$$

which is Snell's law as encountered in optics. However, in the absence of these special conditions, Snell's law is better written in the form given either in eqn. 13.35 or in eqn. 13.34.

Eqn. 13.34 states that the components of the propagation vectors parallel to the interface must be the same for the incident, reflected and transmitted waves. However, the components of these propagation vectors normal to the interface are different. Using the first two of eqns. 13.28, we obtain for the z components of the propagation vectors for the incident and reflected waves

$$
\pm k_1 C_1 = \pm k_1 (1 - S_1^2)^{1/2} \qquad (13.37)
$$

Using the third of eqns. 13.28, together with eqn. 13.34, we obtain for the z component of the propagation vector for the transmitted wave

$$k_2 C_2 = k_1 \{(k_2/k_1)^2 - S_1^2\}^{1/2} \tag{13.38}$$

It will be convenient to write this equation as

$$k_2 C_2 = k_1 q \tag{13.39}$$

where

$$q = \{(k_2/k_1)^2 - S_1^2\}^{1/2} \tag{13.40}$$

We can then write the propagation vectors for the incident, reflected and transmitted waves respectively as (eqns. 13.28)

$$\left.\begin{aligned}
k_1 &= \beta_1 - j\alpha_1 = k_1(S_1, 0, C_2) \\
k_{-1} &= \beta_{-1} - j\alpha_{-1} = k_1(S_1, 0, -C_1) \\
k_2 &= \beta_2 - j\alpha_2 = k_1(S_1, 0, q)
\end{aligned}\right\} \tag{13.41}$$

Because there is no source of waves at $z = +\infty$, the sign of the radical in eqn. 13.40 must be interpreted so that the z components of β_2 and α_2 are non-negative.

13.7 The field impedances of the waves looking normally across the interface

When the magnetic field in Fig. 13.2 is vibrating parallel to the interface $z = 0$, the boundary conditions at the interface are

$$E_x \text{ is continuous,} \quad H_y \text{ is continuous} \tag{13.42}$$

In satisfying these conditions it is convenient to make use of field impedances looking normally across the interface from the first material to the second. Using the first of eqns. 12.44, these impedances may be written down for the incident, reflected and transmitted waves. They are given by E_x/H_y, using the electromagnetic fields for the incident, reflected and transmitted waves listed in the left half of Table 13.1. We obtain, respectively,

$$Z_1 = \zeta_1 C_1, \quad Z_{-1} = -\zeta_1 C_1, \quad Z_2 = \zeta_2 C_2 \tag{13.43}$$

Using eqn. 13.39, these impedances may be written

$$Z_1 = -Z_{-1} = \zeta_1 C_1, \quad Z_2 = \zeta_2(k_1/k_2)q \tag{13.44}$$

where q is given by eqn. 13.40.

In terms of the impedances Z_1 and Z_2 for the incident and transmitted waves looking across the interface towards the second material, the boundary conditions in eqns. 13.42 become

$$\left.\begin{aligned}
Z_1 H_1 - Z_1 H_{-1} &= Z_2 H_2 \\
H_1 + H_{-1} &= H_2
\end{aligned}\right\} \tag{13.45}$$

These are linear algebraic equations for the ratios $H_1 : H_{-1} : H_2$, and the solution is

$$\frac{H_1}{Z_1 + Z_2} = \frac{H_{-1}}{Z_1 - Z_2} = \frac{H_2}{2Z_1} \tag{13.46}$$

We have here derived the ratios of the complex amplitudes H_1, H_{-1} and H_2 for the incident, reflected and transmitted waves, as measured at a point of the interface by their magnetic vectors, which are assumed parallel to the interface. The ratio H_2/H_1 is known as the complex transmission coefficient; its magnitude is the ratio of the amplitude of the transmitted wave close to a point of the interface to the amplitude of the incident wave close to the same point of the interface, and its argument is the phase-advance that occurs on transmission through the interface. The ratio H_{-1}/H_1 is known as the complex reflection coefficient; its magnitude is the ratio of the amplitude of the reflected wave close to a point of the interface to the amplitude of the incident wave close to the same point of the interface, and its argument is the phase-advance on reflection at the interface.

When the electric field is vibrating parallel to the interface, we make use of the field admittances Y_1, Y_{-1} and Y_2 for the incident, reflected and transmitted waves looking across the interface from the first material towards the second. These are (c.f. eqns. 13.43)

$$Y_1 = -Y_{-1} = \eta_1 C_1, \quad Y_2 = \eta_2 C_2 \qquad (13.47)$$

and, using eqn. 13.39, they may be written (c.f. eqns. 13.44)

$$Y_1 = -Y_{-1} = \eta_1 C_1, \quad Y_2 = \eta_2(k_1/k_2)q \qquad (13.48)$$

The boundary conditions require the continuity of E_y and H_x at the interface. This gives (c.f. eqns. 13.45)

$$\left.\begin{array}{c} Y_1 E_1 - Y_1 E_{-1} = Y_2 E_2 \\ E_1 + E_{-1} = E_2 \end{array}\right\} \qquad (13.49)$$

of which the solution is (c.f. eqns. 13.46)

$$\frac{E_1}{Y_1 + Y_2} = \frac{E_{-1}}{Y_1 - Y_2} = \frac{E_2}{2Y_1} \qquad (13.50)$$

The ratio E_{-1}/E_1 is the complex reflection coefficient and the ratio E_2/E_1 is the complex transmission coefficient.

13.8 Reflection and transmission of a plane wave incident normally on a plane interface

If the incident wave is a plane wave and it is incident normally on the interface $z = 0$, then the incident and transmitted waves propagate parallel to the z axis in the positive direction, and the reflected wave propagates parallel to the z axis in the negative direction. We have

$$\left.\begin{array}{cc} S_1 = 0, & C_1 = 1 \\ S_2 = 0, & C_2 = 1 \end{array}\right\} \qquad (13.51)$$

Table 13.2   *The incident, reflected and transmitted waves at a fixed interface (normal incidence)*

	Positive H in the same direction for all waves	Positive E in the same direction for all waves
Incident wave, $z \leqslant 0$	$E = \zeta_1 H_1(1, 0, 0) \exp\{j(\omega t - k_1 z)\}$ $H = H_1(0, 1, 0) \exp\{j(\omega t - k_1 z)\}$	$E = E_1(0, 1, 0) \exp\{j(\omega t - k_1 z)\}$ $H = \eta_1 E_1(-1, 0, 0) \exp\{j(\omega t - k_1 z)\}$
Reflected wave, $z \leqslant 0$	$E = \zeta_1 H_{-1}(-1, 0, 0) \exp\{j(\omega t + k_1 z)\}$ $H = H_{-1}(0, 1, 0) \exp\{j(\omega t + k_1 z)\}$	$E = E_{-1}(0, 1, 0) \exp\{j(\omega t + k_1 z)\}$ $H = \eta_1 E_{-1}(1, 0, 0) \exp\{j(\omega t + k_1 z)\}$
Transmitted wave, $z \geqslant 0$	$E = \zeta_2 H_2(1, 0, 0) \exp\{j(\omega t - k_2 z)\}$ $H = H_2(0, 1, 0) \exp\{j(\omega t - k_2 z)\}$	$E = E_2(0, 1, 0) \exp\{j(\omega t - k_2 z)\}$ $H = \eta_2 E_2(-1, 0, 0) \exp\{j(\omega t - k_2 z)\}$

and Table 13.1 becomes Table 13.2. The waves are linearly polarised. The two halves of Table 13.2 in fact represent the same electromagnetic fields; the axes of co-ordinates used on the right merely have to be rotated in the right-handed sense about the positive z axis to make them coincide with the axes used on the left. The left half is appropriate if one wishes to have the positive directions of the magnetic vectors the same for all waves, and to use the magnetic field for specifying the amplitudes and phases of the waves. The right half is appropriate if one wishes to have the positive directions of the electric vectors the same for all waves, and to use the electric fields for specifiying the amplitudes and phases of the waves. The field impedances and admittances for the incident and transmitted waves looking normally across the interface into the second material are now (eqns. 13.43 and 13.47)

$$\left.\begin{array}{ll} Z_1 = \zeta_1, & Z_2 = \zeta_2 \\ Y_1 = \eta_1, & Y_2 = \eta_2 \end{array}\right\} \tag{13.52}$$

These equations also follow from eqns. 12.40. Substitution from eqns. 13.52 into eqns. 13.46 and 13.50 gives, for the complex reflection and transmission co-efficients at normal incidence,

$$\left.\begin{array}{l} \dfrac{H_1}{\zeta_1 + \zeta_2} = \dfrac{H_{-1}}{\zeta_1 - \zeta_2} = \dfrac{H_2}{2\zeta_1} \\[2mm] \dfrac{E_1}{\eta_1 + \eta_2} = \dfrac{E_{-1}}{\eta_1 - \eta_2} = \dfrac{E_2}{2\eta_1} \end{array}\right\} \tag{13.53}$$

These two sets of equations describe the same reflection and transmission phenomenon, but in the first set the positive direction of the magnetic vector is the same for all waves, whereas in the second set it is the positive direction of the electric vector that is the same for all waves.

From eqns. 13.53 we notice that there is no reflected wave if

$$\zeta_1 = \zeta_2, \quad \eta_1 = \eta_2 \tag{13.54}$$

A plane wave incident normally on a plane interface between two different homogeneous materials suffers no reflection if the impedances (and therefore the admittances) of the two materials are the same. To prevent reflection there is no need for the propagation constants to be the same for the two materials, or for the complex refractive indices to be the same for the two materials. It is only the impedances of the two materials that have to match in order to prevent reflection at normal incidence.

Let us dissect the electromagnetic field described in Table 13.2 by a pair of perfectly conducting planes perpendicular to the electric vectors at a distance a apart, and let us retain only the electromagnetic field between the perfectly conducting sheets. We then have a perfectly conducting strip transmission line that is filled with homogeneous material of propagation constant k_1 and impedance ζ_1 for $z < 0$, and with homogeneous material of propagation constant k_2 and impedance ζ_2 for $z > 0$. The properties of such a transmission line change discontinuously at the position

$z = 0$. Table 13.2 is still applicable, and we may derive the complex reflection and transmission coefficients from eqns. 13.53.

When dealing with transmission lines it is usually more convenient to work in terms of voltage and current rather than electric and magnetic field, and to use the characteristic impedance of the line rather than the impedance of the material between the conductors. Let the strips be of breadth b ($\gg a$), and let fringing at the edges of the strips be neglected. Then the characteristic impedances of the transmission line for $z < 0$ and $z > 0$ are, respectively, (eqn. 12.58)

$$Z_1 = \zeta_1 a/b, \quad Z_2 = \zeta_2 a/b \tag{13.55}$$

Table 13.3   *The incident, reflected and transmitted waves associated with a junction between two transmission lines*

	Positive I in same direction for all waves	Positive V in same direction for all waves
Incident wave, $z \leqslant 0$	$V = Z_1 I_1 \exp\{j(\omega t - k_1 z)\}$ $I = I_1 \exp\{j(\omega t - k_1 z)\}$	$V = V_1 \exp\{j(\omega t - k_1 z)\}$ $I = Y_1 V_1 \exp\{j(\omega t - k_1 z)\}$
Reflected wave, $z \leqslant 0$	$V = -Z_1 I_{-1} \exp\{j(\omega t + k_1 z)\}$ $I = I_{-1} \exp\{j(\omega t + k_1 z)\}$	$V = V_{-1} \exp\{j(\omega t + k_1 z)\}$ $I = -Y_1 V_{-1} \exp\{j(\omega t + k_1 z)\}$
Transmitted wave, $z \geqslant 0$	$V = Z_2 I_2 \exp\{j(\omega t - k_2 z)\}$ $I = I_2 \exp\{j(\omega t - k_2 z)\}$	$V = V_2 \exp\{j(\omega t - k_2 z)\}$ $I = Y_2 V_2 \exp\{j(\omega t - k_2 z)\}$

Moreover the complex voltage and current for each wave are deduced from the corresponding complex electric and magnetic field strength in accordance with eqns. 12.56. In this way we may rewrite Table 13.2 in terms of voltages, currents and characteristic impedances are shown in Table 13.3 ($Y_1 = Z_1^{-1}$, $Y_2 = Z_2^{-1}$). In the left half of Table 13.3 the positive directions of the currents on a conductor are the same for all waves, whereas in the right half the positive directions of the voltages between the conductors are the same for all waves. In terms of voltages, currents and characteristic impedances, eqns. 13.53 become

$$\left.\begin{array}{c} \dfrac{I_1}{Z_1 + Z_2} = \dfrac{I_{-1}}{Z_1 - Z_2} = \dfrac{I_2}{2Z_1} \\[2mm] \dfrac{V_1}{Y_1 + Y_2} = \dfrac{V_{-1}}{Y_1 - Y_2} = \dfrac{V_2}{2Y_1} \end{array}\right\} \tag{13.56}$$

These equations ensure that the boundary conditions at the junction $z = 0$ are satisfied. If expressed in terms of complex voltage and current, these boundary conditions are:

$$V \text{ is continuous}, \quad I \text{ is continuous} \tag{13.57}$$

Table 13.3 and eqns. 13.56 apply to a junction between any pair of homogeneous transmission lines. It is merely a matter of replacing the characteristic impedances in eqns. 13.55 by those appropriate to the geometry of the transmission lines in use. From eqns. 13.56 we see that there is no reflection at a junction between homogeneous transmission lines if the characteristic impedances (and therefore the characteristic admittances) of the lines match:

$$Z_1 = Z_2, \quad Y_1 = Y_2 \tag{13.58}$$

It may be noticed that the expressions for the incident and reflected waves in Table 13.3 still apply if a transmission line of characteristic impedance Z_1 is terminated by a lumped impedance Z_2. In these circumstances the complex reflection coefficient is still given by eqns. 13.56, the values of V_2 and I_2 then being the complex voltage and current for the terminating impedance. In particular, if a loss-free transmission line of characteristic resistance R is terminated by a loss-free lumped reactance X as shown in Fig. 13.3, then the complex reflection coefficient of the termination is (eqns. 13.56)

$$\frac{I_{-1}}{I_1} = \frac{R - jX}{R + jX} \tag{13.59}$$

so that

$$|I_{-1}/I_1| = 1, \quad \arg(I_{-1}/I_1) = -2\tan^{-1}(X/R) \tag{13.60}$$

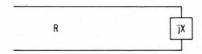

Fig. 13.3 *Illustrating a transmission line of characteristic resistance R terminated by a lumped impedance jX*

The first of eqns. 13.60 states that the amplitude of the reflected wave is equal to the amplitude of the incidence wave. The second of eqns. 13.60 states that the phase-lag on reflection is equal to $2\tan^{-1}(X/R)$. The sign of the phase-change depends on whether the reactance is positive (inductive) or negative (capacitive), that is, on whether it stores more magnetic energy than electric energy, or more electric energy than magnetic energy (Section 9.3).

13.9 Calculation of the Fresnel reflection coefficients

Let us now apply eqns. 13.46 and 13.50 to reflection of a plane wave from a plane interface at oblique incidence. The complex electromagnetic fields for the incident, reflected and transmitted waves are then as shown in Table 13.1. For simplicity, let us assume that both materials are non-magnetic insulators. This is the situation that is of great importance in optics.

In the absence of conductivity in both materials, all the quantities $k_1, k_2, \zeta_1, \zeta_2,$

η_1, η_2 are real. The refractive indices of the two materials are also real, and it is convenient to denote by n the ratio of the refractive index of the second material to that of the first. Because the permeabilities of the two materials are now the same, we have

$$n = \frac{n_2}{n_1} = \left(\frac{\epsilon_2}{\epsilon_1}\right)^{1/2} = \frac{k_2}{k_1} = \frac{\zeta_1}{\zeta_2} = \frac{\eta_2}{\eta_1} \tag{13.61}$$

Moreover, if θ_1 is the angle of incidence of the plane wave at the interface as shown in Fig. 13.2, eqns. 13.41 for the propagation vectors of the incident, reflected and transmitted waves become

$$\left.\begin{array}{l} k_1 = \beta_1 - j\alpha_1 = k_1(\sin\theta_1, 0, \cos\theta_1) \\[4pt] k_{-1} = \beta_{-1} - j\alpha_{-1} = k_1(\sin\theta_1, 0, -\cos\theta_1) \\[4pt] k_2 = \beta_2 - j\alpha_2 = k_1(\sin\theta_1, 0, q) \end{array}\right\} \tag{13.62}$$

where, from eqn. 13.40,

$$q = (n^2 - \sin^2\theta_1)^{1/2} \tag{13.63}$$

Furthermore, the field impedances Z_1 and Z_2 for the incident and transmitted waves looking normally across the interface when the magnetic field is vibrating parallel to the interface become (eqns. 13.44)

$$Z_1 = \zeta_1 \cos\theta_1, \quad Z_2 = \zeta_2 q/n \tag{13.64}$$

Likewise, the field admittances Y_1 and Y_2 for the incident and transmitted waves looking normally across the interface when the electric field is vibrating parallel to the interface become (eqns. 13.48)

$$Y_1 = \eta_1 \cos\theta_1, \quad Y_2 = \eta_2 q/n \tag{13.65}$$

For the incident and reflected waves we see from eqns. 13.62 that

$$\alpha_1 = \alpha_{-1} = 0 \tag{13.66}$$

But for the transmitted wave it is not necessarily true that α_2 vanishes even though the second material is non-conducting. This is because, in accordance with eqn. 13.63,

$$q = \begin{cases} (n^2 - \sin^2\theta_1)^{1/2} & \text{if } n \geq \sin\theta_1 \tag{13.67} \\[6pt] -j(\sin^2\theta_1 - n^2)^{1/2} & \text{if } n \leq \sin\theta_1 \tag{13.68} \end{cases}$$

Using eqn. 13.68, the propagation vector for the transmitted wave in eqns. 13.62 gives

$$\left.\begin{array}{l} \beta_2 = k_1(\sin\theta_1, 0, 0) \\[4pt] \alpha_2 = k_1\{0, 0, (\sin^2\theta_1 - n^2)^{1/2}\} \end{array}\right\} \tag{13.69}$$

The sign of the radical in eqn. 13.68 was chosen so that the z component of α_2 in the second of eqns. 13.69 is positive. This equation then implies exponential decay of amplitude as we move away from the interface into the second material. On the other hand, the first of eqns. 13.69 implies propagation of phase parallel to the interface. The planes of constant phase are therefore perpendicular to the planes of constant amplitude, and we have an evanescent wave (Section 13.4) that creates in the second material only a local electromagnetic field near the interface. There is no propagation away from the interface to large distances in the second material.

We need to consider separately the two situations described by eqns. 13.67 and 13.68:

(i) *Total reflection* $(n \leqslant \sin\theta_1)$: This is the situation in which we use eqn. 13.68. It only occurs if $n < 1$, that is, if the refractive index of the second non-magnetic insulator is less than that of the first. Even then, it is further necessary for the angle of incident θ_1 to exceed the critical angle of incidence θ_C defined by

$$\theta_C = \sin^{-1}n \tag{13.70}$$

Substitution for q from eqn. 13.68 into eqns. 13.64 and 13.65 gives:

For H parallel to the interface:
$$Z_1 = \zeta_1\cos\theta_1, \quad Z_2 = -j\zeta_2(\sin^2\theta_1 - n^2)^{1/2}/n \tag{13.71}$$

For E parallel to the interface:
$$Y_1 = \eta_1\cos\theta_1, \quad Y_2 = -j\eta_2(\sin^2\theta_1 - n^2)^{1/2}/n \tag{13.72}$$

From the second of eqns. 13.71 and the second of eqns. 13.72 we see that the field impedance looking across the interface into the second material is purely reactive. From the first of eqns. 13.71 and the first of eqns. 13.72 we see that the field impedance for the incident wave looking across the interface into the second material is purely resistive. The same is also true for the reflected wave looking in the reverse direction (c.f. eqns. 13.44). We thus have a transmission system of the type illustrated in Fig. 13.3, and therefore one for which reflection is total (c.f. eqns. 13.59 and 13.60). The evanescent wave in the second material constitutes local storage of energy just beyond the interface at $z = 0$, and this is a reactive termination to the transmission system in the region $z < 0$. Flow of energy across the interface is purely reactive. Energy that crosses the interface into the second material during one quarter-cycle of an oscillation is returned during the succeeding quarter-cycle. There is no resistive flow of energy into the second material, and this is why reflection is total.

The sign of Z_2 in the second of eqns. 13.71 shows that, when the magnetic field is vibrating parallel to the interface, the field impedance looking across the interface into the evanescent wave is not only purely reactive but is capacitively reactive (see Fig. 9.10*a*). This means that the evanescent wave stores more electric energy than magnetic energy when the magnetic field is vibrating parallel to the interface. On the other hand, the sign of Y_2 in the second of eqns. 13.72 shows that, when the electric field vibrates parallel to the interface, the field admittance looking across the interface into the evanescent wave is inductively reactive (see Fig. 9.10*b*). This means that the evanescent wave stores more magnetic energy than electric energy when the electric field is vibrating parallel to the interface.

We obtain the complex reflection coefficients by substitution from eqns. 13.71 and 13.72 into eqns. 13.46 and 13.50. If we replace ζ_2 by ζ_1/n and η_2 by $\eta_1 n$ in accordance with eqns. 13.61, we obtain:

For H parallel to the interface:

$$\frac{H_{-1}}{H_1} = \frac{n^2 \cos\theta_1 + j(\sin^2\theta_1 - n^2)^{1/2}}{n^2 \cos\theta_1 - j(\sin^2\theta_1 - n^2)^{1/2}} \tag{13.73}$$

For E parallel to the interface:

$$\frac{E_{-1}}{E_1} = \frac{\cos\theta_1 + j(\sin^2\theta_1 - n^2)^{1/2}}{\cos\theta_1 - j(\sin^2\theta_1 - n^2)^{1/2}} \tag{13.74}$$

For each of the expressions on the right-hand sides of eqns. 13.73 and 13.74 the numerator is the complex conjugate of the denominator. This verifies the fact that, if $n < \sin\theta_1$, the magnitude of the reflection coefficient is unity both when the magnetic field is vibrating parallel to the interface and when the electric field is vibrating parallel to the interface.

(ii) *Partial reflection* $(n > \sin\theta_1)$: This is the situation in which we use eqn. 13.67. If $n < 1$ we encounter the condition $n > \sin\theta_1$ when the angle of incidence θ_1 is less than the critical angle θ_C defined in eqn. 13.70. But if $n > 1$, we encounter the condition $n > \sin\theta_1$ for all angles of incidence. Substitution for q from eqn. 13.67 into eqns. 13.64 and 13.65 gives:

For H parallel to the interface:

$$Z_1 = \zeta_1 \cos\theta_1, \quad Z_2 = \zeta_2(n^2 - \sin^2\theta_1)^{1/2}/n \tag{13.75}$$

For E parallel to the interface:

$$Y_1 = \eta_1 \cos\theta_1, \quad Y_2 = \eta_2(n^2 - \sin^2\theta_1)^{1/2}/n \tag{13.76}$$

From the second of eqns. 13.75 and the second of eqns. 13.76 we see that the field impedance in the second material looking normally away from the interface is resistive. In particular, the field impedance looking across the interface into the second material is resistive. There is now purely resistive flow of energy across the interface and through the second material to large values of z. Reflection of the incident wave from the interface is now partial.

By substituting from eqns. 13.75 and 13.76 into eqns. 13.46 and 13.50, and replacing ζ_2 by ζ_1/n and η_2 by $\eta_1 n$ in accordance with eqns. 13.61, we see that the complex reflection coefficients are:

For H parallel to the interface:

$$\frac{H_{-1}}{H_1} = \frac{n^2 \cos\theta_1 - (n^2 - \sin^2\theta_1)^{1/2}}{n^2 \cos\theta_1 + (n^2 - \sin^2\theta_1)^{1/2}} \tag{13.77}$$

For E parallel to the interface:

$$\frac{E_{-1}}{E_1} = \frac{\cos\theta_1 - (n^2 - \sin^2\theta_1)^{1/2}}{\cos\theta_1 + (n^2 - \sin^2\theta_1)^{1/2}} \tag{13.78}$$

With $n > \sin \theta_1$, both of these reflection coefficients are real. Their absolute values are less than unity, although they tend to unity as θ_1 tends to the critical angle θ_C (eqn. 13.70) and to the glancing angle $\frac{1}{2}\pi$.

Equations. 13.77 and 13.78 are known as the Fresnel reflection coefficients. They apply to reflection of plane waves incident upon a fixed plane interface between homogeneous non-magnetic insulators. They take the form given in eqns. 13.73 and 13.74 under conditions of total reflection ($n \leqslant \sin \theta_1$).

13.10 Behaviour of the Fresnel reflection coefficients

In accordance with eqns. 13.46 and 13.50, the complex reflection coefficients depend on the field impedances or field admittances for the incident and transmitted waves looking across the interface into the second material. For non-magnetic insulators, the impedances when the magnetic field vibrates parallel to the interface are given by eqns. 13.71 if $n \leqslant \sin \theta_1$ and by eqns. 13.75 if $n \geqslant \sin \theta_1$. The variation of these impedances with angle of incidence is illustrated in the left half of Fig. 13.4 both for $n > 1$ and $n < 1$ (actually for $n = 5/3$ and $n = 3/5$). On the other hand, the admittances when the electric field vibrates parallel to the interface are given by eqns. 13.72 if $n \leqslant \sin \theta_1$ and by eqns. 13.76 if $n \geqslant \sin \theta_1$. The variation of these admittances with angle of incidence is illustrated in the right half of Fig. 13.4.

In the right half of Fig. 13.4 we see that there is no angle of incidence for which $Y_1 = Y_2$. In accordance with the first of eqns. 13.50, it follows that there is no real angle of incidence for which the Fresnel reflection coefficient vanishes when the electric field vibrates parallel to the interface. On the other hand, when the magnetic field vibrates parallel to the interface, we see from the left half of Fig.13.4 that there is a real angle of incidence θ_B, known as the Brewster angle, for which $Z_1 = Z_2$. At the Brewster angle the impedance for the incident plane wave looking across the interface matches the impedance of the transmitted plane wave looking across the interface. Consequently, in accordance with the first of eqns. 13.46, there is no reflected wave. By equating the numerator of the expression on the right-hand side of eqn. 13.77 to zero, it follows that the Brewster angle is

$$\theta_B = \tan^{-1} n \qquad (13.79)$$

Values of the Fresnel reflection coefficients in eqns. 13.73, 13.74, 13.77 and 13.78 are listed in Table 13.4 for four values of the angle of incidence. For $\theta_1 = \frac{1}{2}\pi$ (glancing incidence), reflection is perfect and there is a phase-change of $\pm\pi$ in the component of the electromagnetic field vibrating parallel to the interface; this statement assumes that the positive direction for this component is the same for both the incident wave and the reflected wave. For the critical angle of incidence ($\theta_1 = \theta_C$), of interest when $n < 1$, reflection is perfect and there is no change of phase, using the same sign convention. When the angle of incidence is equal to θ_B defined in eqn. 13.79, the reflection coefficient vanishes when the magnetic field vibrates parallel to the interface, but not when the electric field vibrates parallel to the interface. For normal incidence ($\theta_1 = 0$), the two reflection coefficients listed in Table 13.4 connote the same physical situation. In the left-hand column the reflection coefficient refers to the magnetic vector, with the positive direction for this vector taken to be the same for the incident and reflected waves; in the right-hand column the reflection coefficient refers to the electric vector, with the positive direction for this vector taken to be the same for the incident and reflected waves.

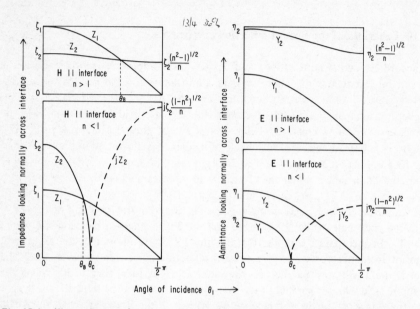

Fig. 13.4 *Illustrating, as function of the angle of incidence, the field impedances or field admittances for the incident and transmitted waves looking across the interface into the second material*

A complex reflection coefficient may be represented in the complex plane by means of a vector drawn from the origin. The length of the vector is the fraction by which the amplitude of the incident wave at the interface must be multiplied in order to obtain the amplitude of the reflected wave at the interface. The counter-clockwise angle of the vector is the angle by which the phase of the incident wave at the interface must be advanced in order to obtain the phase of the reflected wave at the interface. This phase is to be measured for the component of the electro-magnetic field vibrating parallel to the interface (the magnetic vector with one linear polarization and the electric vector with the other). Moreover, we take the positive directions for this component to be the same for both the incident and the reflected waves.

With these conventions, vector diagrams drawn in the complex plane for the reflection coefficients in eqns. 13.73, 13.74, 13.77 and 13.78 take the form shown in Fig. 13.5. The tip of the vector from the origin O starts at normal incidence from the point on the real axis marked $\theta_1 = 0$, corresponding to the first entry in Table 13.4. The tip of the vector finishes at glancing incidence at the point -1 on the real axis marked $\theta_1 = \frac{1}{2}\pi$, corresponding to the last entry in Table 13.4. If $n > 1$, the tip of the vector moves negatively along the real axis from the point marked $\theta_1 = 0$ to the point marked $\theta_1 = \frac{1}{2}\pi$ as the angle of incidence increases; the reflection coefficient varies with angle of incidence as shown in the upper half of Fig. 13.6.

Table 13.4   *Particular values of the Fresnel reflection coefficients*

Angle of incidence θ_1	Reflection coefficient	
	H_{-1}/H_1	E_{-1}/E_1
0	$\dfrac{n-1}{n+1}$	$\dfrac{1-n}{1+n}$
$\theta_B = \tan^{-1}n$	0	$\dfrac{1-n^2}{1+n^2}$
$\theta_C = \sin^{-1}n$	1	1
$\frac{1}{2}\pi$	-1	-1

But if $n < 1$, the tip of the vector in Fig. 13.5 moves along the real axis in the opposite direction to the point $+1$ as θ_1 increases from zero to the critical angle θ_C, corresponding to the third entry in Table 13.4; between $\theta_1 = 0$ and $\theta_1 = \theta_C$ the reflection coefficient varies with angle of incidence as shown in the lower half of Fig. 13.6. When $n < 1$, the tip of the vector in Fig. 13.5 passes from the point $+1$ on the positive real axis to the point -1 on the negative real axis as θ_1 increases from θ_C to $\frac{1}{2}\pi$, not by moving along the real axis, but by moving round a semi-circle of unit radius. In these circumstances the magnitude of the reflection coefficient is unity, and the phase-advance on reflection increases from 0 at $\theta = \theta_C$ to π at $\theta_1 = \frac{1}{2}\pi$.

If the magnitude of the complex reflection coefficient is plotted as a function of angle of incidence, we obtain the curves shown in Fig. 13.7A. If the counterclockwise angle of the complex reflection coefficient is plotted as a function of the angle of incidence, we obtain either the curves shown in Fig. 13.7B or those shown in Fig. 13.7C. To draw these curves we have to decide whether, when the tip of the vector in Fig. 13.5 passes the origin, it does so on the top side or the bottom side. This requires that we take into account the fact that the materials in reality possess at least slight conductivity (see Problem 13.1). Fig. 13.7B is drawn on the assumption that the conductivity of the second material is more important. In Fig. 13.7C it is the conductivity of the first material that is assumed to be more important. However, both diagrams assume that the effect of conductivity is slight in both materials.

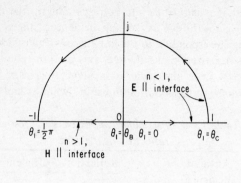

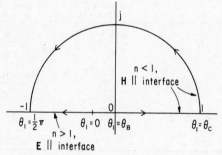

Fig. 13.5   *Illustrating in the complex plane the variation of the Fresnel reflection coefficients with angle of incidence θ_1*

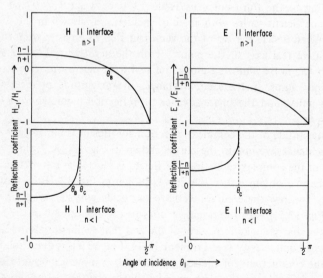

Fig. 13.6   *Illustrating the variation of the Fresnel reflection coefficients with angle of incidence*
When $n < 1$ and $\theta > \theta_C$, see Fig. 13.7

13.11 Matching at oblique incidence

For normal incidence we have seen in eqns. 13.54 that there is no reflection of a plane wave from a plane interface between homogeneous materials if the impedances of the materials are equal. In the absence of conductivity in both materials, this means that (eqns. 12.36)

$$\mu_1/\epsilon_1 \;=\; \mu_2/\epsilon_2 \qquad\qquad\qquad\qquad (13.80)$$

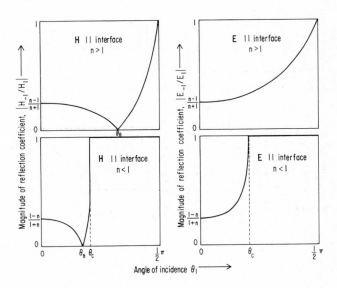

Fig. 13.7A   *Illustrating the variation of the magnitude of the Fresnel reflection coefficients with angle of incidence*

In other words the ratio of the permeabilities of the two materials must be equal to the ratio of their dielectric constants. For this to be true at least one of the materials must possess magnetic properties.

For oblique incidence we see in Table 13.4 and in Figs. 13.5, 13.6 and 13.7 that there is no reflection of a plane wave from a plane interface between homogeneous non-magnetic insulators when the angle of incidence is equal to the Brewster angle θ_B given by eqn. 13.79, provided that the magnetic field vibrates parallel to the interface. When the electric field vibrates parallel to the interface, Figs. 13.5, 13.6 and 13.7 show that there is no Brewster angle. However, this is only true when the permeabilities of the two materials are the same, including the case, commonly encountered in optics, when the permeabilities of both materials are unity. If one is dealing with a pair of non-conducting materials that not only have different dielectric properties but also have different magnetic properties, then it is quite possible for an incident plane wave to suffer no reflection even when the electric field vibrates parallel to the interface. However, we shall see that a Brewster angle never occurs simultaneoustly for both linear polarisations.

If one or both of the materials possess conductivity, then it is not in general possible to adjust the angle of incidence for a plane wave so as to eliminate reflection from the interface. However, this only means that absence of reflection is impossible for an incident wave in which the planes of constant amplitude coincide with the planes of constant phase. It is quite possible to have zero reflection, but only for an incident wave in which the planes of constant amplitude do not coincide with the planes of constant phase.

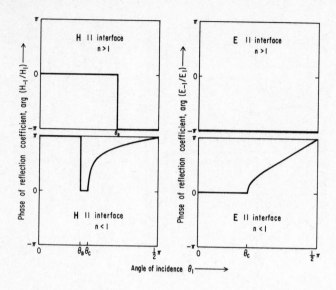

Fig. 13.7B *Illustrating the variation of the phase of the Fresnel reflection coefficients with angle of incidence if the second material possesses slight conductivity and the first material does not*

Let us consider the situation in which both of the linear isotropic materials possess magnetic, dielectric and conduction properties, and in which the incident wave is of the type illustrated in Fig. 13.1. From eqns. 13.46 and 13.50, the condition for no reflection is

$$Z_1 = Z_2(H \| \text{ interface}) \tag{13.81}$$

$$Y_1 = Y_2(E \| \text{ interface}) \tag{13.82}$$

This is the condition that the impedance (or admittance) looking across the interface for the incident wave matches that for the transmitted wave, so that no reflected wave is involved. Using eqns. 13.43 and 13.47, eqns. 13.81 and 13.82 become

$$\zeta_1 C_1 = \zeta_2 C_2(H \| \text{ interface}) \tag{13.83}$$

$$\eta_1 C_1 = \eta_2 C_2(E \| \text{ interface}) \tag{13.84}$$

We also have from Snell's law (eqn. 13.34)

$$k_1 S_1 = k_2 S_2 \tag{13.85}$$

and this is true for both linear polarisations. In eqns. 13.83, 13.84 and 13.85 the quantities ζ_1, ζ_2, η_1, η_2, k_1, k_2 are, in general, complex. So are the quantities S_1, C_1, S_2, C_2 if we are handling waves for which the planes of constant amplitude do not coincide with the planes of constant phase. However, S_1, C_1, S_2 and C_2 are such that (eqns. 13.29)

$$S_1^2 + C_1^2 = 1, \quad S_2^2 + C_2^2 = 1 \tag{13.86}$$

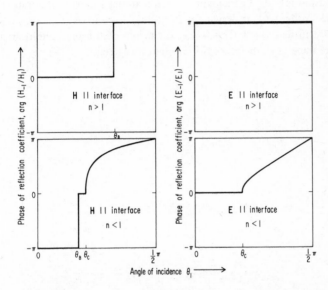

Fig. 13.7C *Illustrating the variation of the phase of the Fresnel reflection coefficients with angle of incidence if the first material possesses slight conductivity and the second material does not*

To ascertain the conditions for no reflection we solve eqns. 13.83, 13.85 and 13.86 for S_1, C_1, S_2, C_2 for one linear polarisation, and eqns. 13.84, 13.85 and 13.86 for the other linear polarisation. We obtain:

For *H* parallel to the interface:

$$\left.\begin{array}{ll} S_1 = \left\{ \dfrac{(\zeta_1/\zeta_2)^2 - 1}{(\zeta_1/\zeta_2)^2 - (k_1/k_2)^2} \right\}^{1/2} & C_1 = \left\{ \dfrac{1 - (k_1/k_2)^2}{(\zeta_1/\zeta_2)^2 - (k_1/k_2)^2} \right\}^{1/2} \\[4mm] S_2 = \left\{ \dfrac{(\zeta_2/\zeta_1)^2 - 1}{(\zeta_2/\zeta_1)^2 - (k_2/k_1)^2} \right\}^{1/2} & C_2 = \left\{ \dfrac{1 - (k_2/k_1)^2}{(\zeta_2/\zeta_1)^2 - (k_2/k_1)^2} \right\}^{1/2} \end{array}\right\} \tag{13.87}$$

For E parallel to the interface:

$$S_1 = \left\{ \frac{(\eta_1/\eta_2)^2 - 1}{(\eta_1/\eta_2)^2 - (k_1/k_2)^2} \right\}^{1/2} \quad C_1 = \left\{ \frac{1 - (k_1/k_2)^2}{(\eta_1/\eta_2)^2 - (k_1/k_2)^2} \right\}^{1/2}$$

$$S_2 = \left\{ \frac{(\eta_2/\eta_1)^2 - 1}{(\eta_2/\eta_1)^2 - (k_2/k_1)^2} \right\}^{1/2} \quad C_2 = \left\{ \frac{1 - (k_2/k_1)^2}{(\eta_2/\eta_1)^2 - (k_2/k_1)^2} \right\}^{1/2} \qquad (13.88)$$

Eqns. 13.87 give the values of S_1, C_1, S_2, C_2 that must be substituted into the left half of Table 13.1 in order to make H_{-1} vanish. Eqns. 13.88 give the values of S_1, C_1, S_2, C_2 that must be substituted into the right half of Table 13.1 in order to make E_{-1} vanish. Fig. 13.8 illustrates the resulting planes of constant phase and planes of constant amplitude for the incident wave in material 1 and for the transmitted wave in material 2. Fig. 13.8 is drawn for substantial conductivity in material 2 but no conductivity in material 1. This is why the planes of constant amplitude in material 1 are at right angles to the planes of constant phase (eqn. 13.27).

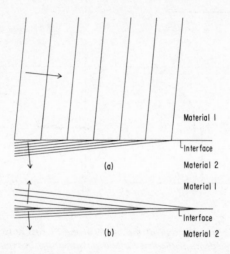

Fig. 13.8   *Illustrating (a) planes of constant phase, and (b) planes of constant amplitude, when an incident wave in a non-conducting material is matched to a transmitted wave in a conducting material*
   Arrows indicate the direction of propagation of phase in diagram (a), and the direction of exponential decay in diagram (b)

If the conductivity of material 2 in Fig. 13.8 becomes very high, propagation of phase in material 1 is nearly parallel to the interface, and propagation of phase in material 2 is nearly perpendicular to the interface. This is the situation encountered in Fig. 12.7 in connection with metal loss in a transmission line along which a harmonic wave is travelling. If material 1 in Fig. 13.8 is the atmosphere and material 2 is the Earth, the diagram illustrates how a radio wave travelling nearly horizontally over the surface of the Earth has energy slowly abstracted from it at the surface, to be absorbed in the ground or ocean. Such a wave is known as a Zenneck wave.

If neither of the materials possesses conductivity, the quantities k_1, k_2, ζ_1 and ζ_2 in eqns. 13.87 and 13.88 are all real. If it is desired that the incident and transmitted waves be plane waves, we must also have S_1, C_1, S_2 and C_2 in eqns. 13.87 and 13.88 real. This requires:

For **H** parallel to the interface:

either $\quad \zeta_1 \leqslant \zeta_2, \quad k_1 \leqslant k_2$

or $\qquad \zeta_1 \geqslant \zeta_2, \quad k_1 \geqslant k_2$ $\qquad\qquad\qquad\qquad$ (13.89)

For **E** parallel to the interface:

either $\quad \eta_1 \leqslant \eta_2, \quad k_1 \leqslant k_2$

or $\qquad \eta_1 \geqslant \eta_2, \quad k_1 \geqslant k_2$ $\qquad\qquad\qquad\qquad$ (13.90)

Expressed in terms of the ratio of the dielectric constants and the ratio of the permeabilities of the materials, these conditions become

For **H** parallel to the interface:

either $\quad \epsilon_1/\epsilon_2 \leqslant \mu_2/\mu_1 \leqslant \epsilon_2/\epsilon_1$

or $\qquad \epsilon_1/\epsilon_2 \geqslant \mu_2/\mu_1 \geqslant \epsilon_2/\epsilon_1$ $\qquad\qquad\qquad$ (13.91)

For **E** parallel to the interface:

either $\quad \mu_1/\mu_2 \leqslant \epsilon_2/\epsilon_1 \leqslant \mu_2/\mu_1$

or $\qquad \mu_1/\mu_2 \geqslant \epsilon_2/\epsilon_1 \geqslant \mu_2/\mu_1$ $\qquad\qquad\qquad$ (13.92)

The inequalities (13.91) and (13.92) are mutually exclusive, so that the condition for no reflection cannot be satisfied simultaneously for both linear polarizations. The conditions (13.92) imply that the ratio of the dielectric constants for the materials must lie between the permeability ratio and its reciprocal. This is a condition that can never be satisfied if the permeability ratio is unity, including the case when both materials are non-magnetic; see the right half of Fig. 13.7A. The conditions (13.91) imply that the ratio of the permeabilities for the materials must lie between the dielectric constant ratio and its reciprocal. This is a condition that is always satisfied if the permeability ratio is unity, including the case when both materials are non-magnetic; see the left half of Fig. 13.7A.

If conductivity is not involved and conditions (13.91) are satisified, we can replace S_1, C_1, S_2 and C_2 in eqns. 13.87 by $\sin\theta_1, \cos\theta_1, \sin\theta_2$ and $\cos\theta_2$, and deduce that

$$\frac{\tan\theta_1}{(\epsilon_2/\epsilon_1)^{1/2}} = \frac{\tan\theta_2}{(\epsilon_1/\epsilon_2)^{1/2}} = \left\{\frac{(\epsilon_2/\epsilon_1) - (\mu_2/\mu_1)}{(\epsilon_2/\epsilon_1)(\mu_2/\mu_1) - 1}\right\}^{1/2} \qquad (13.93)$$

If $\mu_1 = \mu_2$, eqns. 13.93 become

$$\tan\theta_1 = (\epsilon_2/\epsilon_1)^{1/2}, \quad \tan\theta_2 = (\epsilon_1/\epsilon_2)^{1/2} \qquad (13.94)$$

The first of these equations, if expressed in terms of the ratio n of the refractive indices in accordance with eqns. 13.61, verifies eqn. 13.79 for the Brewster angle.

13.12 The concept of coherent scattering

We have discussed in some detail in this chapter what is involved when a wave is reflected and refracted at a fixed plane interface between homogenous materials. It is clear that the phenomena of reflection and refraction may be said to take place 'at' the interface. Nevertheless, caution is required in pushing this notion too far.

Let us suppose for simplicity that the first 'material' is simply free space. A plane harmonic wave is then incident in free space upon homogeneous material with a plane interface, and it causes the free and bound charges in the material to become vibrating dipoles that reradiate electromagnetic waves. The amplitude and phase of the wave reradiated by each charged particle is controlled by the amplitude and phase of the vibrating electromagnetic field in which it is situated. The resultant electromagnetic field at a particular charged particle is composed of the undisturbed incident plane wave in free space, combined with the waves reradiated to it coherently in free space by all the other charged particles.

In the material beyond the interface, let us ask what is the sum of the undisturbed incident plane wave in free space and the waves coherently reradiated by all the other charged particles in free space. The answer is: the wave that we have called the transmitted wave. Moreover, in the free space in front of the interface, let us ask what is the direct physical cause of the reflected wave. The answer is that the reflected wave is the sum of the waves reradiated in free space by all the vibrating charged particles beyond the interface. Consequently, the reflected wave is not produced 'at the interface', but is the sum of the waves reradiated by all the charged particles in the material beyond the interface in circumstances when each of these particles is vibrating under the influence of the undisturbed incident plane wave and of the waves that it receives in free space from every other charged particle. If we wish to use this method of describing what is happening, we use version 1 of the electromagnetic equations in Tables 2.1, 2.2 and 2.3. If, however, we wish to arrive directly at a relatively simple statement of the resultant electromagnetic field, we use version 3 of the electromagnetic equations in Tables 2.1, 2.2 and 2.3. It is the latter technique that we have employed in this chapter, but this should not blind us to the fact that the fundamental physical process involved in reflection and refraction is coherent scattering by the free and bound charges in the materials involved.

As an example, consider a plane wave incident in free space on a plane interface with a homogeneous non-magnetic insulator, and let the magnetic field be vibrating parallel to the interface. The reflected wave in free space is simply what is reradiated from the vibrating electric moment per unit volume in the material. The bound electrons in the material are vibrating perpendicular to the propagation vector for the transmitted wave, as illustrated by the double arrow in Fig. 13.9.

Since an electric dipole does not radiate along its axis (Problem 1.8), it follows that there is no reflected wave in free space if the propagation vector for the reflected wave is parallel to the direction of vibration of the bound electrons, that is, if the propagation vector for the reflected wave is perpendicular to that for the transmitted wave. From Fig. 13.9 we can see that this occurs if

$$\theta_1 + \theta_2 = \tfrac{1}{2}\pi \tag{13.95}$$

This is just what is implied by the fact that $\tan\theta_1$ and $\tan\theta_2$ in eqns. 13.94 are mutually reciprocal.

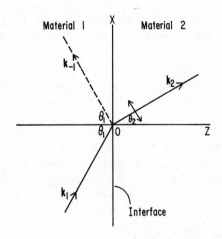

Fig. 13.9 *Illustrating the Brewster angle phenomenon in terms of lack of coherent reradiation by atoms in the direction of the reflected wave*

Problems

13.1 A plane interface separates two homogeneous non-magnetic materials. The first material has a dielectric constant ϵ_1/ϵ_0 and a conductivity σ_1. The second material has a dielectric constant ϵ_2/ϵ_0 and a conductivity σ_2. A plane harmonic electromagnetic wave is incident upon the interface at an angle θ_1 to the normal, with the magnetic field vibrating parallel to the interface. The angular frequency ω of the wave is large compared both with σ_1/ϵ_1 and σ_2/ϵ_2. The angle of incidence θ_1 is equal to $\tan^{-1}(\epsilon_2/\epsilon_1)^{1/2}$, that is, to the Brewster angle applicable when $\sigma_1 = \sigma_2 = 0$. Show that the reflection coefficient may be written

$$\frac{H_{-1}}{H_1} = j\frac{1}{4\omega}\left(\frac{\sigma_1}{\epsilon_1} - \frac{\sigma_2}{\epsilon_2}\right)\frac{\epsilon_2 - \epsilon_1}{\epsilon_2}$$

Deduce that, in a plot of the complex reflection coefficient of type shown in Fig. 13.5, the curve running close to the origin passes on the top side if $\epsilon_1 < \epsilon_2$, $\sigma_1/\epsilon_1 > \sigma_2/\epsilon_2$ or if $\epsilon_1 > \epsilon_2$, $\sigma_1/\epsilon_1 < \sigma_2/\epsilon_2$, and on the under side if $\epsilon_1 > \epsilon_2$, $\sigma_1/\epsilon_1 > \sigma_2/\epsilon_2$ or if $\epsilon_1 < \epsilon_2$, $\sigma_1/\epsilon_1 < \sigma_2/\epsilon_2$, as shown in Figs. 13.7B and C.

13.2   A plane harmonic electromagnetic wave of angular frequency ω is incident obliquely in homogeneous dielectric material upon a plane homogeneous non-magnetic metallic sheet whose thickness is large compared with its skin-depth. The metal has conductivity σ and unit permeability, and the dielectric has dielectric constant ϵ/ϵ_0 and zero conductivity. The dielectric properties of the metal may be neglected in comparison with its conduction properties. Show that, if $\omega \ll \sigma/\epsilon$, the magnitude of the impedance ζ_m of the metal is small compared with the impedance ζ of the dielectric, and that the wave transmitted into the metal travels almost normally away from the interface regardless of the angle of incidence θ of the wave in the dielectric. Deduce that the impedances of the incident and transmitted waves looking into the metal perpendicular to the interface are approximately $\zeta \sec \theta$ and ζ_m when the electric vector vibrates parallel to interface, and that they are approximately $\zeta \cos \theta$ and ζ_m when the magnetic vector vibrates parallel to the interface.

13.3   In the previous problem deduce that the complex reflection coefficient of the metallic sheet is approximately -1 for all angles of incidence when the electric vector vibrates parallel to the sheet, and is approximately

$$\frac{\phi - \phi_0}{\phi + \phi_0}$$

when the magnetic vector vibrates parallel to the sheet, where $\phi (= \frac{1}{2}\pi - \theta_1)$ is the angle in radians between the sheet and the direction of incidence, and ϕ_0 is the small ratio of the impedance ζ_m of the metal to the impedance ζ of the dielectric. Show that

$$\phi_0 = (\omega\epsilon/\sigma)^{1/2} \exp(j\tfrac{1}{4}\pi)$$

approximately. Deduce that the magnitude of the reflection coefficient for a plane harmonic electromagnetic wave incident upon the metal-dielectric interface with the magnetic vector vibrating parallel to the interface has, as a function of the direction of incidence, a minimum value when $\phi = |\phi_0|$, and that the minimum value is $\sqrt{2} - 1$.

13.4   Cartesian co-ordinates in homogeneous linear isotropic material of propagation constant k, impedance ζ and admittance η are denoted by (x, y, z). There are two plane harmonic electromagnetic waves travelling in the material, one in the direction z increasing and the other in the direction z decreasing. Both waves are linearly polarised with the electric field vibrating parallel to the x axis and the magnetic field vibrating parallel to the y axis. At two planes $z = z_1$ and $z = z_2$ ($z_1 < z_2$), the combined complex electric and magnetic vectors in the x and y directions are calculated; they are denoted by (E_1, H_1) at $z = z_1$ and by (E_2, H_2) at $z = z_2$. Show that (E_1, H_1) are related to (E_2, H_2) by the matrix

$$\begin{pmatrix} \cos\{k(z_2 - z_1)\} & j\zeta \sin\{k(z_2 - z_1)\} \\ j\eta \sin\{k(z_2 - z_1)\} & \cos\{k(z_2 - z_1)\} \end{pmatrix}$$

Deduce that the field impedances Z_1 and Z_2 at $z = z_1$ and $z = z_2$, respectively, looking in the direction z increasing, are related by the equation

$$Z_1 = \zeta \frac{Z_2 \cos\{k(z_2 - z_1)\} + j\zeta \sin\{k(z_2 - z_1)\}}{jZ_2 \sin\{k(z_2 - z_1)\} + \zeta \cos\{k(z_2 - z_1)\}}$$

13.5 If, in Problem 13.4, the material is non-conducting, and if $z_2 - z_1$ is a quarter of a wavelength in the material, show that $Z_1 Z_2 = \zeta^2$. Hence show that any two homogeneous non-conducting materials may be matched at normal incidence by means of an intervening window whose thickness is a quarter of a wavelength and whose impedance is the geometric mean of the impedances of the two materials (the quarter-wave transformer).

13.6 Show that the result of the previous problem is valid at oblique incidence if either the electric or magnetic vector vibrates parallel to the interfaces and if we use in the results of Problem 13.4:

(i) components of the complex electric and magnetic vectors parallel to the interfaces,
(ii) field impedances and admittances looking normal to the interfaces, and
(iii) propagation constants equal to the components of the propagation vectors normal to the interfaces.

13.7 Show that the relation between the impedance Z_1 and the impedance Z_2 in Problem 13.4 may be written

$$\left(\frac{Z_1 + \zeta}{Z_1 - \zeta}\right) \bigg/ \left(\frac{Z_2 + \zeta}{Z_2 - \zeta}\right) = \exp\{2jk(z_2 - z_1)\}$$

For a plane harmonic electromagnetic wave travelling through the material in the direction z increasing, let ϕ be the total phase-lag in radians between $z = z_1$ and $z = z_2$, and let N be the total attenuation in nepers between these planes. Show that

$$\left|\frac{Z_1 + \zeta}{Z_1 - \zeta}\right| \bigg/ \left|\frac{Z_2 + \zeta}{Z_2 - \zeta}\right| = \exp 2N$$

and

$$\arg\left(\frac{Z_1 + \zeta}{Z_1 - \zeta}\right) - \arg\left(\frac{Z_2 + \zeta}{Z_2 - \zeta}\right) = 2\phi$$

In the complex w plane draw the curves

$$\left|\frac{w + 1}{w - 1}\right| = \exp 2u, \quad \arg\left(\frac{w + 1}{w - 1}\right) = 2v$$

for a convenient series of constant values of u and v, and show that the curves constitute two families of circles intersecting at right angles, with the family $v = $ constant passing through the points $w = \pm 1$. If $w = Z_1/\zeta$ is the point (u_1, v_1) and $w = Z_2/\zeta$ is the point (u_2, v_2), show that

$$u_1 - u_2 = N, \quad v_1 - v_2 = \phi$$

For given values of k, ζ, z_1 and z_2, devise a method for using the circle diagram to deduce Z_1 when Z_2 is specified.

13.8   Explain why, in the previous problem, use is made in practice of the half of the diagram for which the real part of w is positive. Show that this half of the w plane may be transformed into the interior of a circle in a w' plane defined by $w' = 1/(w + 1)$. In the w' plane show that the curves $v = $ constant are radial lines spaced at equal angular intervals, while the curves $u = $ constant are circles that all touch each other at the same point. Use this circle diagram (known as a Smith chart) to verify the result derived in Problem 13.5.

Chapter 14

Storage and flow of energy in electromagnetic waves

14.1 Introduction

In Section 12.3 we saw that, if an oscillatory electric network with an output impedance Z_1 is to be connected to another such network with an input impedance Z_2 without upsetting the behaviour of either, then it is necessary for Z_1 to be equal to Z_2 at the frequency involved. In particular, we saw in Section 13.8 that, if a transmission line having characteristic impedance Z_1, and carrying a wave travelling toward its output terminals, is to be connected to another transmission line having characteristic impedance Z_2, and carrying a wave travelling away from its input terminals, then Z_1 must be equal to Z_2 if there is to be no reflected wave at the junction.

Equality of the two impedances requires equality of both the resistive and reactive components. Equality of the resistive components implies that the second transmission line is able to accept and retain the energy of which the first transmission line must rid itself permanently if no reflection is to occur at the junction. Equality of the reactive components implies that the second transmission line is able to accept on loan any energy that the first transmission line needs to deposit for repayment a quarter of a cycle later. Similar statements apply for a plane wave incident normally upon a plane interface between homogeneous materials (Section 13.8). Such statements also apply at oblique incidence (Section 13.11), although matching may then require that the planes of constant phase differ from the planes of constant amplitude in the manner illustrated in Fig. 13.8. To appreciate these phenomena more completely, further study is required of the storage and flow of energy in oscillatory electromagnetic fields.

As described in Section 2.9, it is frequently desirable to represent a quantity that varies sinusoidally with time at angular frequency ω as the reference component of a vector that rotates in the complex plane with angular velocity ω. The same applies to a quantity that varies sinusoidally in space, and this is what led to use of the exponential wave-function in the two preceding chapters. However, use of rotating vectors in the complex plane does have a disadvantage when non-linear operations are involved. Even in free space, calculations of energy per unit volume

$(\frac{1}{2}\epsilon_0 E^2 + \frac{1}{2}\mu_0 H^2)$ and rate of flow of energy per unit cross-sectional area $(E \times H)$ involve products of electromagnetic field components, and these are non-linear operations. Calculation of the full details of energy storage and flow therefore involves departing from the use of complex electromagnetic vectors; the product of the real parts of two complex numbers is not equal to the real part of their product. It is true that information about energy flow and storage can be obtained without departing from the use of complex electromagnetic vectors; it was to achieve this that we introduced and applied in Chapters 9–13 the concepts of complex power, the complex Poynting vector, and impedance. But this information, although valuable, is limited. For a complete treatment of energy storage and flow in oscillatory electromagnetic fields, use must be made of the actual vibrating electromagnetic vectors, not simply the complex electromagnetic vectors.

In this chapter we shall first calculate the complex electromagnetic vectors, and then switch to the actual vibrating electromagnetic vectors in order to examine the full details of storage and flow of energy in oscillatory electromagnetic fields.

14.2   Travelling plane wave in homogeneous linear isotropic loss-free material

Let us suppose that a linearly polarised plane electromagnetic wave of angular frequency ω is travelling through homogeneous material at rest relative to the observer. We assume that the material satisfies the connection relations given in Table 2.2, so that the material behaves in a linear isotropic manner. We further assume that the material is not only free of hysteresis loss but is also free of loss associated with conductivity. We also assume that the mass of the vibrating electrons in the material may be neglected and that no kinetic energy is generated in the material by passage of the wave. Only electric and magnetic energy is then involved, and the material may be characterised by a real dielectric constant ϵ/ϵ_0, a permeability μ/μ_0 and a zero conductivity.

In these circumstances the impedance and admittance of the material are (eqns. 12.36)

$$\zeta = (\mu/\epsilon)^{1/2}, \qquad \eta = (\epsilon/\mu)^{1/2} \tag{14.1}$$

and, in accordance with the first of eqns. 12.35, the propagation constant is

$$k = \omega(\mu\epsilon)^{1/2} \tag{14.2}$$

Because of the absence of conductivity, the quantities ζ, η and k are all real, and the velocity of propagation of phase is (eqn. 12.34)

$$v = \omega/k = (\mu\epsilon)^{-1/2} \tag{14.3}$$

The wavelength λ and the periodic time T are given by

$$\lambda = 2\pi/k, \qquad T = 2\pi/\omega \tag{14.4}$$

Let the direction of propagation of phase be along the positive z axis, so that the wave-function is (expr. 12.31)

$$\exp \{j(\omega t - kz)\} \tag{14.5}$$

Let the electric vector vibrate parallel to the x axis; let its amplitude be E_0 and its phase at the origin be zero. Then the complex electric and magnetic vectors at the point (x, y, z) at time t are (eqns. 12.38)

$$\begin{aligned}
\tilde{E} &= E_0(1, 0, 0) \exp \{j(\omega t - kz)\} \\
\tilde{H} &= \eta E_0(0, 1, 0) \exp \{j(\omega t - kz)\}
\end{aligned} \tag{14.6}$$

Hence the actual vibrating electric and magnetic vectors at the point (x, y, z) at time t are

$$\begin{aligned}
E &\doteq E_0(1, 0, 0) \cos (\omega t - kz) \\
H &= \eta E_0(0, 1, 0) \cos (\omega t - kz)
\end{aligned} \tag{14.7}$$

From eqns. 14.7 it follows that the electric and magnetic energies per unit volume at position (x, y, z) at time t are, respectively,

$$\begin{aligned}
w_e &= \tfrac{1}{2}\epsilon E^2 = \tfrac{1}{2}\epsilon E_0^2 \cos^2 (\omega t - kz) \\
w_m &= \tfrac{1}{2}\mu H^2 = \tfrac{1}{2}\epsilon E_0^2 \cos^2 (\omega t - kz)
\end{aligned} \tag{14.8}$$

where, in the latter case, use has been made of the second of eqns. 14.1. We see that the energy density is equally divided between electric and magnetic, and that the total energy density is

$$w = w_e + w_m = \epsilon E_0^2 \cos^2(\omega t - kz) \tag{14.9}$$

The Poynting vector $E \times H$ is directed parallel to the positive z axis, and the magnitude of the rate of flow of energy per unit cross-sectional area at position (x, y, z) at time t is

$$f = EH = \eta E_0^2 \cos^2(\omega t - kz) \tag{14.10}$$

In Fig. 14.1, the curves to the left of the vertical broken line illustrate energy storage and flow in a plane wave travelling to the right. The curves to the right of the vertical broken line illustrate a way of terminating the travelling wave at the broken line without causing a reflected wave; this terminating arrangement will be discussed in Section 14.4. The curves to the left of the vertical broken line depict the electric energy density w_e and the magnetic energy density w_m in eqns. 14.8, together with the total energy density w in eqns. 14.9. The curves give the variations of these quantities with distance z, marked in wavelengths. The four curves show the behaviour at intervals of one-eight of a period over half a period, after which time they repeat. The four curves show how the pattern of energy density in space moves progressively to the right with the velocity $(\mu\epsilon)^{-1/2}$ appearing in eqns. 14.3.

Comparison of eqns. 14.10 with eqns. 14.9 shows that curves for energy flow and for energy density behave in the same way both in space and time. The greatest energy flow takes place where the greatest energy density exists, as shown by the arrows in the portion of Fig. 14.1 to the left of the vertical broken line. Moreover, division of eqns. 14.10 by eqns. 14.9, followed by use of the second of eqns. 14.1, shows that

$$\frac{f}{w} = \frac{\eta}{\epsilon} = (\mu\epsilon)^{-1/2} \tag{14.11}$$

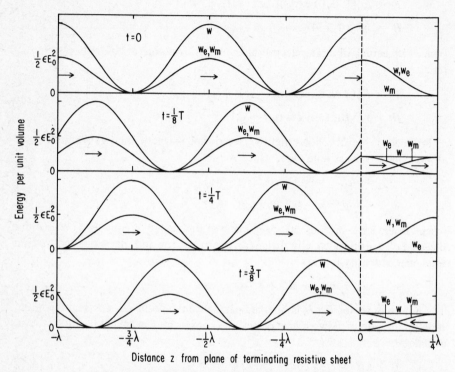

Fig. 14.1.   *Illustrating storage and flow of energy for a plane electromagnetic wave travelling in a homogeneous loss-free material of impedance ζ, terminated by a resistive sheet at z = 0 of surface resistance ζ and by a perfectly conducting sheet at z = ¼λ*

This means that the ratio of the rate of flow of energy per unit cross-sectional area to the energy per unit volume is equal to the velocity of propagation. In other words, the progressive displacement illustrated in Fig. 14.1 for the spatial pattern of stored energy precisely accounts for the flow of energy.

If eqns. 14.9 and 14.10 are averaged, either with respect to t over half a period or with respect to z over half a wavelength, we obtain

$$\bar{w} = \tfrac{1}{2}\epsilon E_0^2, \qquad \bar{f} = \tfrac{1}{2}\eta E_0^2 \tag{14.12}$$

Division of these two equations and application of the second of eqns. 14.11 verify that the ratio of the mean rate of flow of energy per unit cross-sectional area to the mean energy stored per unit volume is equal to the velocity of propagation.

14.3   Standing plane wave in homogeneous linear isotropic loss-free material

Let us now assume that the travelling wave in Fig. 14.1 is not terminated at the vertical broken line but can continue to the right-hand edge of the diagram $(z = \tfrac{1}{4}\lambda)$ where it is incident normally on an infinite perfectly conducting metal sheet. Perfect reflection occurs at the sheet with a phase-change of π in the electric field and of zero in the magnetic field. The complex electromagnetic field for the combined incident and reflected waves is

$$\left. \begin{aligned}
\tilde{E} &= E_0(1, 0, 0) \exp \{j(\omega t - kz)\} + E_0(1, 0, 0) \exp \{j(\omega t + kz)\} \\
\tilde{H} &= \eta E_0(0, 1, 0) \exp \{j(\omega t - kz)\} + \eta E_0(0, -1, 0) \exp \{j(\omega t + kz)\}
\end{aligned} \right\}$$

$$(14.13)$$

By putting $z = \tfrac{1}{4}\lambda$ $(kz = \tfrac{1}{2}\pi)$ in these equations we verify that the electric field vanishes at the reflecting sheet.

Eqns. 14.13 may be rewritten

$$\left. \begin{aligned}
\tilde{E} &= 2E_0(1, 0, 0) \cos kz \exp (j\omega t) \\
\tilde{H} &= 2\eta E_0(0, 1, 0) \sin kz \exp \{j(\omega t - \tfrac{1}{2}\pi)\}
\end{aligned} \right\}$$

$$(14.14)$$

Although the electromagnetic field oscillates in time at each location, the pattern of the combined field in space does not travel either in the direction z increasing or in the direction z decreasing. The spatial pattern stands still and constitutes what is known as a standing wave. The standing wave pattern is different for the electric field $[\propto \cos kz]$ from what it is for the magnetic field $[\propto \sin kz]$. Moreover, the phase of the magnetic vibration at each location differs from the phase of the electric vibration by $\tfrac{1}{2}\pi$.

From eqns. 14.14 we see that the field impedance and field admittance at location (x, y, z) looking in the direction z increasing are respectively

$$\frac{\tilde{E}_x}{\tilde{H}_y} = j\zeta \cot kz, \qquad \frac{\tilde{H}_y}{\tilde{E}_x} = -j\eta \tan kz \qquad (14.15)$$

The field impedance looking into the perfectly conducting sheet at $z = \tfrac{1}{4}\lambda$ $(kz = \tfrac{1}{2}\lambda)$ is zero. However, the field impedance also vanishes at $z = -\tfrac{1}{4}\lambda$ $(kz = -\tfrac{1}{2}\pi)$. Hence, an infinite perfectly conducting sheet may also be inserted at the location $z = -\tfrac{1}{4}\lambda$. We shall suppose that both conducting sheets exist, and that eqns. 14.14 describe a complex electromagnetic field existing between the two. This electromagnetic field consists of a pair of oppositely travelling waves, each of which is reflected into the other at each sheet, forming a standing wave between the sheets.

From eqns. 14.14 it follows that the actual vibrating electric and magnetic vectors are

$$E = 2E_0(1, 0, 0) \cos kz \cos \omega t$$
$$H = 2\eta E_0(0, 1, 0) \sin kz \sin \omega t$$
(14.16)

Consequently the electric and magnetic energies per unit volume at the location (x, y, z) at time t are

$$w_e = \tfrac{1}{2}\epsilon E_0^2 = 2\epsilon E_0^2 \cos^2 kz \cos^2 \omega t$$
$$w_m = \tfrac{1}{2}\mu H^2 = 2\epsilon E_0^2 \sin^2 kz \sin^2 \omega t$$
(14.17)

and the total energy density w is the sum of these two expressions. The rate of flow of energy in the positive z direction per unit cross-sectional area is

$$f = EH = \eta E_0^2 \sin 2kz \sin 2\omega t$$
(14.18)

In the left half of Fig. 14.2 is shown the variations of w, w_e and w_m with z over the range $-\tfrac{1}{4}\lambda \leqslant z \leqslant \tfrac{1}{4}\lambda$ between the two perfectly conducting reflecting sheets. Curves are shown at intervals of one-eight of a period over half a period, after which time they repeat. At $t = 0$ there is no stored magnetic energy. At this epoch, all the stored energy is electric; the density is a maximum at $z = 0$, half way between the reflecting sheets, and vanishes adjacent to the sheets. A quarter of a period later $(t = \tfrac{1}{4}T)$, all the stored energy is magnetic; the density is maximum adjacent to the sheets, and vanishes half way between them. Another quarter of a cycle later $(t = \tfrac{1}{2}T)$, the energy pattern has returned to that for $t = 0$, but the directions of the electric and magnetic vectors are reversed. Every half cycle, energy oscillates back and forth between locations roughly half way between the conducting sheets and locations near the sheets. This oscillatory flow is illustrated in the right half of Fig. 14.2 with the aid of curves for the rate of flow of energy based on eqns. 14.18. When the energy is piled up about half way between the conducting sheets, it is entirely electric; when the energy is piled up near the conducting sheets it is entirely magnetic.

14.4 Travelling and standing waves on a loss-free transmission line

The discussion in the preceding two sections concerning travelling and standing plane waves in homogeneous linear isotropic loss-free material may be adapted to describe travelling and standing waves on a loss-free transmission line. It is merely a matter of replacing (i) ϵ and μ by the capacitance of the line per unit length C and the inductance of the line per unit length L, and (ii) E and H by the voltage and current at position z along the line at time t. The impedance ζ and admittance η of the material are then replaced, respectively, by the characteristic impedance Z_0 and the characteristic admittance Y_0 of the line given by (c.f. eqns. 14.1; see Problem 8.2)

$$Z_0 = (L/C)^{1/2}, \qquad Y_0 = (C/L)^{1/2}$$
(14.19)

Both Z_0 and Y_0 are real becauses losses (dielectric and metal) are being neglected.

For a transmission line, Fig. 14.2 describes the behaviour of a section of line extending from $z = -\frac{1}{4}\lambda$ to $z = \frac{1}{4}\lambda$, short-circuited at the two ends. The central part of the half-wavelength of line near $z = 0$ functions as a capacitor that stores electric energy, and the end-sections of the half-wavelength of line near $z = -\frac{1}{4}\lambda$ and $z = \frac{1}{4}\lambda$ function as inductors that store magnetic energy. Fig. 14.2 describes the oscillation of energy that takes place in this resonant system.

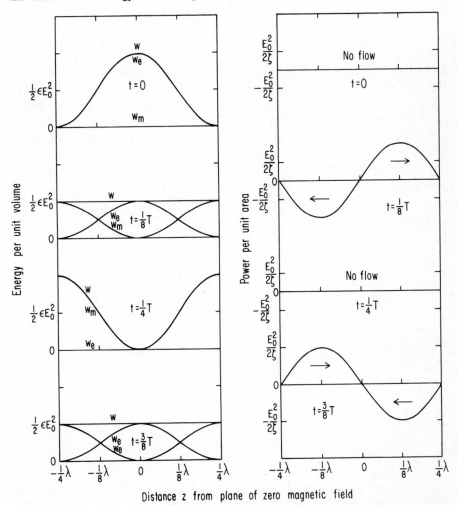

Distance z from plane of zero magnetic field

Fig. 14.2.   *Illustrating storage and flow of energy for a standing wave in homogeneous loss-free material*

Moreover, the resonant half-wavelength of line may be cut in the middle to form two resonant systems. Each section is then a quarter-wavelength of line short-circuited at one end and open-circuited at the other. The portion of the line near the open-circuited end constitutes a capacitor, and the portion near the short-circuited end constitutes an inductor. Energy oscillates back and forth between the two, as shown by the part of Fig. 14.2 between $z = 0$ and $z = \frac{1}{4}\lambda$. At location z, the circuit impedance and admittance looking in the direction z increasing are (c.f. eqns. 14.15)

$$Z = jZ_0 \cot kz, \qquad Y = -jY_0 \tan kz \qquad (14.20)$$

For a transmission line, the portion of Fig. 14.1 to the left of the vertical broken line describes the flow of energy in a wave travelling along the line in the direction z increasing. Termination of the line without reflection is then achieved by connecting the characteristic resistance Z_0 at the end of the line as shown in Fig. 14.3a. However, if the transmission line continued indefinitely to the right beyond the resistor, there would be a transmitted wave, and this would lead to a wave reflected from the resistor. To avoid this, we can short-circuit the line at a position $\frac{1}{4}\lambda$ beyond the resistor as shown in Fig. 14.3b. In accordance with eqns. 14.20 the quarter-wavelength of line between the resistor and the short-circuit then constitutes an open-circuit connected across the resistor; the rest of the line is then properly terminated.

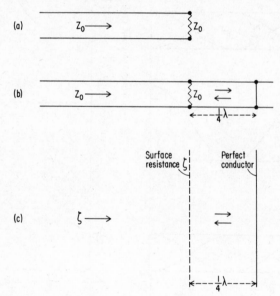

Fig. 14.3.   *Illustrating three methods for terminating a transmission system without causing reflection*

A method analogous to that illustrated in Fig. 14.3a is not available when terminating a plane wave travelling in infinite homogeneous material because it is not possible to abolish the part of space beyond the terminating plane. Use is therefore made of a method analogous to that illustrated in Fig. 14.3b; this is what is described by Fig. 14.1. As illustrated in Fig. 14.3c, one uses an infinite plane resistive sheet for which the resistance between opposite edges of a square is equal to the impedance ζ of the material, given by the first of eqns. 14.1; but at a distance $\frac{1}{4}\lambda$ beyond this resistive sheet, one places an infinite plane perfectly conducting sheet parallel to the resistive sheet. The behaviour between the sheets is then illustrated in Fig. 14.2 for $0 \leqslant z \leqslant \frac{1}{4}\lambda$. It is this behaviour that is shown in Fig. 14.1 for $0 \leqslant z \leqslant \frac{1}{4}\lambda$.

In Fig. 14.1, the resistive sheet is located at $z = 0$ in the position indicated by the vertical broken line. It will be noticed that, across the resistive sheet, the electric energy density w_e is continuous; this is because the electric field strength is continuous at the resistive sheet. On the other hand, the magnetic energy density w_m is discontinuous across the resistive sheet; this is because there is a discontinuity in the magnetic field strength at $z = 0$ equal to the surface density of current in the resistive sheet.

The method of terminating a plane electromagnetic wave described in Fig. 14.1 may be used even in free space. In accordance with the first of eqns. 14.1, the surface resistance of the terminating sheet must then be $(\mu_0/\epsilon_0)^{1/2}$. The resistance between opposite edges of a square of the sheet is then approximately 377 Ω.

14.5 Partially travelling, partially standing waves

Eqns. 14.6 describe a linearly polarised plane harmonic electromagnetic wave of angular frequency ω travelling in the positive z direction in homogeneous loss-free material of impedance ζ, admittance η and propagation constant k, given by eqns. 14.1 and 14.2. The energy behaviour of the wave is illustrated in Fig. 14.1 to the left of the vertical broken line; at no point is there any backward-flow of energy in the negative z direction. When there is a pair of such waves of equal amplitude travelling in opposite directions as described by eqns. 14.13, we have a standing wave. As illustrated in Fig. 14.2, there is as much backward-flow of energy in the negative z direction as there is forward-flow in the positive z direction. At any instant of time, the regions of forward-flow and of backward-flow occupy adjacent quarter-wavelength slabs of space. At any location in space, the intervals of forward-flow and of backward-flow occupy adjacent quarter-period blocks of time.

Let us now consider a situation in which there are waves travelling in both the positive and negative z directions, but the amplitude E_+ of the former exceeds the amplitude E_- of the latter. Let us take the plane $z = 0$ to be located where the electric vectors of the two waves are in the same phase, and let us suppose that this common phase vanishes when $t = 0$. We then obtain for the complex electromagnetic field (c.f. eqns. 14.13)

$$\tilde{E} = E_+(1, 0, 0) \exp\{j(\omega t - kz)\} + E_-(1, 0, 0) \exp\{j(\omega t + kz)\}$$
$$\tilde{H} = \eta E_+(0, 1, 0) \exp\{j(\omega t - kz)\} + \eta E_-(0, -1, 0) \exp\{j(\omega t + kz)\}$$

$$(14.21)$$

and for the actual vibrating electromagnetic field

$$\left.\begin{array}{l} E = E_+(1, 0, 0) \cos{(\omega t - kz)} + E_-(1, 0, 0) \cos{(\omega t + kz)} \\ H = \eta E_+(0, 1, 0) \cos{(\omega t - kz)} + \eta E_-(0, -1, 0) \cos{(\omega t + kz)} \end{array}\right\} \quad (14.22)$$

It follows from eqns. 14.22 that the stored electric and magnetic energy densities at location (x, y, z) at time t are

$$\left.\begin{array}{l} w_e = \tfrac{1}{2}\epsilon E_+^2 \cos^2(\omega t - kz) + \epsilon E_+ E_- \cos{(\omega t - kz)} \cos{(\omega t + kz)} \\ \qquad + \tfrac{1}{2}\epsilon E_-^2 \cos^2(\omega t + kz) \\ w_m = \tfrac{1}{2}\epsilon E_+^2 \cos^2(\omega t - kz) - \epsilon E_+ E_- \cos{(\omega t - kz)} \cos{(\omega t + kz)} \\ \qquad + \tfrac{1}{2}\epsilon E_-^2 \cos{(\omega t + kz)} \end{array}\right\}$$

$$(14.23)$$

so that the total electromagnetic energy density is

$$w = \epsilon E_+^2 \cos^2(\omega t - kz) + \epsilon E_-^2 \cos^2(\omega t + kz) \quad (14.24)$$

It also follows from eqns. 14.22 that the rate of flow of energy in the positive z direction per unit cross-sectional area is

$$f = EH = \eta E_+^2 \cos^2(\omega t - kz) - \eta E_-^2 \cos^2(\omega t + kz) \quad (14.25)$$

Using eqn. 14.24, we may rewrite eqns. 14.23 in the form

$$\left.\begin{array}{l} w_e = \tfrac{1}{2}w + \epsilon E_+ E_- \cos{(\omega t - kz)} \cos{(\omega t + kz)} \\ w_m = \tfrac{1}{2}w - \epsilon E_+ E_- \cos{(\omega t - kz)} \cos{(\omega t + kz)} \end{array}\right\}$$

Consequently, the electric and magnetic energy densities may be expressed as

$$\left.\begin{array}{l} w_e = \tfrac{1}{2}w + \tfrac{1}{2}\epsilon E_+ E_-(\cos 2\omega t + \cos 2kz) \\ w_m = \tfrac{1}{2}w - \tfrac{1}{2}\epsilon E_+ E_-(\cos 2\omega t + \cos 2kz) \end{array}\right\} \quad (14.26)$$

Curves for w, w_e and w_m are shown in the left half of Fig. 14.4, based on eqns. 14.24 and 14.26. Curves for the rate of flow of energy per unit cross-sectional area are shown in the right half of Fig. 14.4, based on eqns. 14.25. The diagrams are drawn for $E_+ = E_0$ and $E_- = \tfrac{1}{2}E_0$. Curves are shown for a single half-wavelength of space-variation and repeat in adjacent half-wavelength sections. They illustrate a single half-period of time-variation, and repeat in adjacent half-period intervals. The wavelength λ and the period T are given by eqns. (14.4).

From the right half of Fig. 14.4 we see that, at $t = 0$ and $t = \tfrac{1}{4}T$, flow of energy is exclusively in the positive z direction, that is, in the direction of propagation of the stronger wave. On the other hand, at other times such as $t = \tfrac{1}{8}T$ and $t = \tfrac{3}{8}T$, there is a section of each half-wavelength of the z co-ordinate where some back-flow of energy takes place. Such a wave is said to be partially travelling and partially standing. If E_- is increased from $\tfrac{1}{2}E_0$ to E_0, Fig. 14.4 becomes identical with Fig. 14.2, and the wave is exclusively standing. If E_- is reduced from $\tfrac{1}{2}E_0$ to zero,

Fig. 14.4 becomes identical with the portion of Fig. 14.1 where $-\frac{3}{4}\lambda \leqslant z \leqslant -\frac{1}{4}\lambda$, and the wave is exclusively travelling.

If eqn. 14.24 is averaged, either with respect to t over half a period or with respect to z over half a wavelength, we obtain

$$\bar{w} = \tfrac{1}{2}\epsilon E_+^2 + \tfrac{1}{2}\epsilon E_-^2 \tag{14.27}$$

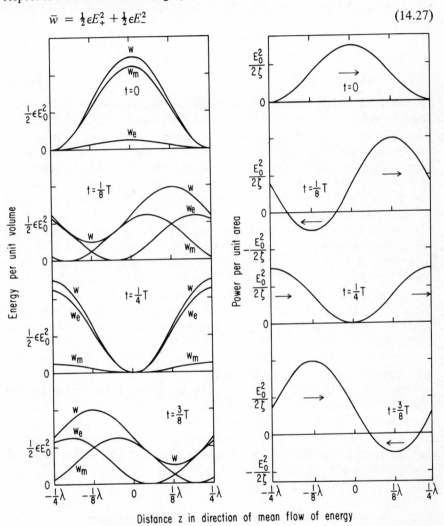

Distance z in direction of mean flow of energy

Fig. 14.4.   *Illustrating storage and flow of energy in a partially travelling, partially standing wave in a loss-free material*
$E_+ = E_0, E_- = \tfrac{1}{2}E_0$

This states, in accordance with the first of eqns. 14.12, that the mean energy density for the partially travelling, partially standing wave is the sum of the mean energy densities for the component travelling waves from which it is formed.

However, because of an oscillation of energy that is taking place, the same is not true for the electric and magnetic energy densities separately. Eqns. 14.26 show that the average values of w_e and w_m are different according as the average is taken in time or in space. The values of $\bar{w}_e$ and $\bar{w}_m$ are only equal to $\frac{1}{2}\bar{w}$ if the averages are taken both in time and in space. These features are illustrated in the left half of Fig. 14.4.

If eqn. 14.25 is averaged, either with respect to t over half a period or with respect to z over half a wavelength, we obtain

$$\bar{f} = \frac{1}{2}\eta E_+^2 - \frac{1}{2}\eta E_-^2 \tag{14.28}$$

This gives the average rate of flow of energy per unit cross-sectional area in the positive z direction; $\bar{f}$ is positive because we are assuming that $E_+ > E_-$. Eqn. 14.28 shows that the rate of flow of energy in the positive z direction is the same at all four epochs shown in the right half of Fig. 14.4 if averaged over the half-wavelength illustrated. But it is only when t is a multiple of $\frac{1}{4}T$ that this flow is unidirectional. At other times there is backflow of energy over a portion of each half-wavelength slab. This is because overall flow of energy to the right is combined with oscillation of energy parallel to the z axis.

Under purely oscillatory conditions, there is an interchange between electric energy and magnetic energy each quarter of a period as illustrated in Fig. 14.2. In the two halves of Fig. 14.4 we can see this happening in combination with an overall flow of energy to the right. In the interval $0 < t < \frac{1}{4}T$, there is a net displacement of energy to the right but, in the process, some of it is converted from magnetic to electric. Moreover, near $z = -\frac{1}{8}\lambda$, some of the energy undergoing conversion is flowing from right to left. In the interval $\frac{1}{4}T < t < \frac{1}{2}T$ there is again a net displacement of energy to the right but, in the process, some of it is converted back from electric to magnetic. Moreover, near $z = \frac{1}{8}\lambda$, some of the energy undergoing conversion is flowing from right to left.

There is a mean rate of flow of energy to the right, and this is given, per unit cross-sectional area, by eqn. 14.28. This equation states, in accordance with the second of eqns. 14.12, that the mean rate of flow of energy to the right in a partially travelling, partially standing wave is equal to the excess of the mean rate of flow to the right for the wave travelling to the right over the mean rate of flow to the left for the wave travelling to the left.

14.6   The effect of stored kinetic energy on electromagnetic waves

In the previous four sections dealing with propagation of electromagnetic waves in material, we discussed storage of electric energy and of magnetic energy; storage of kinetic energy was assumed to be negligible. In fact, vibration of bound electrons in a dielectric under the influence of an electromagnetic wave does involve oscillatory time-variation in the kinetic energy per unit volume. This is usually important only at optical and higher frequencies. In a plasma, however, vibration of free electrons under the influence of an electromagnetic wave involves, even at radio frequencies, oscillatory time-variation in the stored kinetic energy per unit volume that can be of major importance. The effect of free ions in a plasma can likewise be important.

For simplicity, let us consider a plasma in which the ions are sufficiently massive that their vibratory motion may be neglected. Let us also assume that the temperature of the plasma is not so high that the thermic velocities of particles can approach the velocity of propagation of electromagnetic waves through the plasma; such a plasma is usually called a 'cold' plasma. Let us further assume that the collisional frequency of electrons with ions and other particles is sufficiently low to be negligible. The plasma may then be modelled as a large homogeneous cloud of electrons (charge e, mass m), N per unit volume, neutralised by a smooth stationary continuum of positive charge, and it may be assumed that the electrons are at rest except in so far as they are caused to move by an applied electromagnetic field (E, B). We assume that there is no large imposed magnetic field, so that the plasma is not a magnetoplasma. The equation of motion of an electron is then given by eqn. 2.61 with the collisional frequency ν set equal to zero. It follows that, if r is the displacement of an electron caused by an applied time-varying electric field E, then

$$m\ddot{r} = eE \tag{14.29}$$

Let us first consider a slab of plasma that is used to fill the space between the plates of a parallel-plate capacitor (fringing neglected). Let a generator of angular frequency ω be applied across the plates, and let the resulting electric field strength in the plasma from the positive plate to the negative plate be

$$E = E_0 \cos \omega t \tag{14.30}$$

From eqn. 14.29, the resulting velocity of the plasma electrons in steady oscillation is

$$\dot{r} = \frac{e}{m\omega} E_0 \sin \omega t \tag{14.31}$$

and the oscillatory displacement of the plasma electrons in the direction from the positive plate to the negative plate is

$$r = -\frac{e}{m\omega^2} E_0 \cos \omega t \tag{14.32}$$

Note that positive free charges vibrate in antiphase with the applied electric field because it is their acceleration that vibrates in phase with the field.

Vibration of the electrons results in electric charges developing on the plasma surfaces adjacent to the plates. The charge per unit area existing on the plasma surface adjacent to the positive plate at time t is

$$-Ner = \frac{Ne^2}{m\omega^2} E_0 \cos \omega t = \frac{\omega_N^2}{\omega^2} \epsilon_0 E_0 \cos \omega t$$

where ω_N is the angular plasma frequency defined in eqn. 2.70. This is charge per unit area that the generator does not have to supply to the positive plate to maintain the electric field between the plates. Instead of having to supply a charge per unit area $\epsilon_0 E_0 \cos \omega t$, the generator only has to supply

$$\left(1 - \frac{\omega_N^2}{\omega^2}\right) \epsilon_0 E_0 \cos \omega t.$$

Consequently, the dielectric constant of the plasma is

$$\frac{\epsilon}{\epsilon_0} = 1 - \frac{\omega_N^2}{\omega^2} \tag{14.33}$$

as already derived in eqn. 2.69. In the special case when $\omega_N = \omega$, oscillation of the plasma electrons is supplying precisely the surface charge necessary to maintain the electric field without any action by the generator; the generator, together with the plates of the capacitor, could in these circumstances be removed, except in so far as it is necessary to overcome losses that we are neglecting. If $\omega_N > \omega$, oscillation of the plasma electrons is supplying more surface charge than is required to maintain the electric field, so that the generator then has to remove charge at the epoch when, in the absence of the plasma, it would be supplying charge.

Let us examine this behaviour in terms of energy. The magnetic energy density in the capacitor is small and may be neglected. The electric energy density is (eqn. 14.30)

$$w_e = \tfrac{1}{2}\epsilon_0 E_0^2 \cos^2 \omega t \tag{14.34}$$

The kinetic energy density is (eqn. 14.31)

$$w_k = \tfrac{1}{2}Nm\dot{r}^2 = \tfrac{1}{2}\frac{Ne^2}{m\omega^2}E_0^2 \sin^2 \omega t$$

Expressed in terms of the angular plasma frequency ω_N (eqn. 2.70), this becomes

$$w_k = \tfrac{1}{2}\epsilon_0 E_0^2 \frac{\omega_N^2}{\omega^2}\sin^2 \omega t \tag{14.35}$$

Eqns. 14.34 and 14.35 show that, in the special case when $\omega_N = \omega$, the mean electric energy density is equal to the mean kinetic energy density; there is oscillation back and forth between the electric energy of the capacitator and the kinetic energy of the vibrating plasma electrons. When $\omega_N = \omega$ no action by the generator is involved. If $\omega_N < \omega$, the kinetic energy of the vibrating plasma electrons is inadequate to provide all the electric energy needed by the capacitor, and the extra has to be supplied by the generator. If $\omega_N > \omega$, the kinetic energy of the vibrating plasma electrons supplies more electric energy than is needed by the capacitor, and the excess has to be drawn off by the generator.

It is clear that an important role is played by storage of kinetic energy in a plasma. Let us now investigate what effect this has on the propagation of electromagnetic waves in the plasma. For this purpose we employ the model of a plasma just used, but we now take the plasma to be infinite in extent. In the plasma we suppose that a plane harmonic wave exists for which the complex electromagnetic field is given by eqns. 14.6. The propagation constant is given by eqn. 14.2, and the impedance and admittance are given by eqns. 14.1, where $\mu/\mu_0 = 1$ and ϵ/ϵ_0 is given by eqn. 14.33. Hence, for the plasma,

$$k = k_0 \left(1 - \frac{\omega_N^2}{\omega^2}\right)^{1/2}, \qquad \eta = \eta_0 \left(1 - \frac{\omega_N^2}{\omega^2}\right)^{1/2} \tag{14.36}$$

where $k_0 \, [= \omega/c]$ is the propagation constant of free space, and $\eta_0 \, [=(\epsilon_0/\mu_0)^{1/2}]$ is the admittance of free space, that is, the reciprocal of 377 Ω.

If $\omega_N > \omega$, we see from eqns. 14.36 that both k and η are purely imaginary, so that the wave is evanescent (Section 13.4). If such a wave is launched from the plane $z = 0$, it is exponentially attenuated in the z direction. There is no propagation of energy away from the plane $z = 0$. There is only storage of energy in the immediate vicinity of the launching plane, and this stored energy is predominantly kinetic. But if $\omega_N < \omega$, then k and η in eqns. 14.36 are real, and we have a propagating wave. Let us examine the energy behaviour in the plasma in these circumstances.

Substituting for η from the second of eqns. 14.36 into eqns. 14.6, we obtain for the complex electric and magnetic vectors of the wave in the plasma at the point (x, y, z) at time t

$$\left.\begin{aligned} \tilde{E} &= E_0(1, 0, 0) \exp \{j(\omega t - kz)\} \\[2mm] \tilde{H} &= \eta_0 \left(1 - \frac{\omega_N^2}{\omega^2}\right)^{1/2} E_0(0, 1, 0) \exp \{j(\omega t - kz)\} \end{aligned}\right\} \tag{14.37}$$

where k is given by the first of eqns. 14.36. Remembering that $\omega_N < \omega$, it follows from eqns. 14.37 that the actual vibrating electric and magnetic vectors in the plasma at the point (x, y, z) at time t are

$$\left.\begin{aligned} E &= E_0(1, 0, 0) \cos (\omega t - kz) \\[2mm] H &= \eta_0 \left(1 - \frac{\omega_N^2}{\omega^2}\right)^{1/2} E_0(0, 1, 0) \cos (\omega t - kz) \end{aligned}\right\} \tag{14.38}$$

Hence the electric and magnetic energy densities in the wave at location (x, y, z) at time t are, respectively,

$$\left.\begin{aligned} w_e &= \tfrac{1}{2}\epsilon_0 E_0^2 \cos^2(\omega t - kz) \\[2mm] w_m &= \tfrac{1}{2}\epsilon_0 E_0^2 \left(1 - \frac{\omega_N^2}{\omega^2}\right) \cos^2(\omega t - kz) \end{aligned}\right\} \tag{14.39}$$

In consequence, the total electromagnetic energy per unit volume is

$$w_{em} = w_e + w_m = \tfrac{1}{2}\epsilon_0 E_0^2 \left(2 - \frac{\omega_N^2}{\omega^2}\right) \cos^2(\omega t - kz) \tag{14.40}$$

The flow of electromagnetic energy is in the positive z direction, and the rate of flow per unit cross-sectional area at location (x, y, z) at time t is

$$f = EH = \eta_0 E_0^2 \left(1 - \frac{\omega_N^2}{\omega^2}\right)^{1/2} \cos^2(\omega t - kz) \tag{14.41}$$

However, electromagnetic energy is not the only energy stored in the wave. There is, in addition, the kinetic energy of the vibrating plasma electrons. At location (x, y, z) at time t, these electrons are subject to an electric field obtained by replacing ωt by $\omega t - kz$ in eqn. 14.30. Their vibratory velocity is therefore obtained by replacing ωt by $\omega t - kz$ in eqn. 14.31. It follows from eqn. 14.35 that the kinetic energy density in the plasma at location (x, y, z) at time t is

$$w_k = \tfrac{1}{2}\epsilon_0 E_0^2 \frac{\omega_N^2}{\omega^2} \sin^2(\omega t - kz) \tag{14.42}$$

The total energy density in the plasma is therefore

$$w = w_{em} + w_k = \tfrac{1}{2}\epsilon_0 E_0^2 \left(2 - \frac{\omega_N^2}{\omega^2}\right) \cos^2(\omega t - kz)$$

$$+ \tfrac{1}{2}\epsilon_0 E_0^2 \frac{\omega_N^2}{\omega^2} \sin^2(\omega t - kz) \tag{14.43}$$

If we were considering a hot plasma, we would be taking the thermic motion of the electrons into account, and pressure would then be exerted in the electron-gas. A vibratory part of this pressure, combined with the vibratory motion of the electron-gas, would permit mechanical transmission of energy through the electron-gas as in a sound wave. Because we are not considering a hot plasma, mechanical transmission of energy through the plasma is negligible. The only transmission of energy through the plasma is therefore electromagnetic. The flow is in the positive z direction, and the rate of flow per unit cross-sectional area is given by eqn. 14.41. But, although there is no mechanical transmission of energy through the plasma, the fact that vibratory mechanical energy is stored in the plasma does affect the electromagnetic transmission of energy. The way in which this happens needs to be understood.

In Fig. 14.5 the solid curves illustrate the behaviour of the electromagnetic energy density w_{em} given by eqn. 14.40, and the broken curves illustrate the behaviour of the kinetic energy density w_k given by eqn. 14.42. The four diagrams illustrate the distribution of energy density in space at successive times one-eighth of a period apart. The entire pattern is moving to the right with the phase velocity ω/k. In accordance with the first of eqns. 14.36, this velocity is

$$v = c \left(1 - \frac{\omega_N^2}{\omega^2}\right)^{-1/2} \tag{14.44}$$

Hence the crests of electromagnetic energy are moving to the right with a velocity greater than the velocity of light in free space. Moreover, the crests of kinetic energy are moving to the right even though we are neglecting mechanical transmission of energy in the electron-gas. How do these phenomena occur?

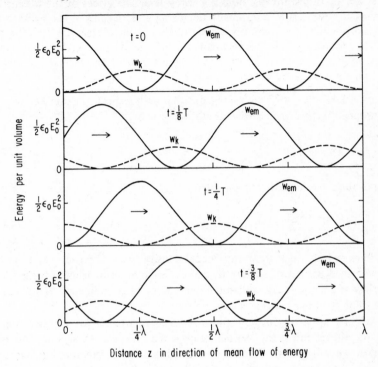

Distance z in direction of mean flow of energy

Fig. 14.5.   *Illustrating storage and flow of energy for an electromagnetic wave travelling in a homogeneous isotropic plasma, $(\omega_N/\omega)^2 = \frac{1}{2}$*

In Fig. 14.5 it can be seen that the crests of kinetic energy interleave the crests of electromagnetic energy. In the second diagram $(t = \frac{1}{8}T)$ there is a maximum of kinetic energy at $z = \frac{3}{8}\lambda$. The plasma electrons obtained this kinetic energy because they were accelerated by the electric field of the wave. They obtained it from the electromagnetic energy at $z = \frac{3}{8}\lambda$ in the previous diagram $(t = 0)$. They are to give it back to the electromagnetic energy at $z = \frac{3}{8}\lambda$ in the succeeding diagram $(t = \frac{1}{4}T)$. Hence the plasma electrons obtain their kinetic energy by detaching energy from the back of one crest of electromagnetic energy, and they give up their kinetic energy by attaching energy to the front of the succeeding crest of electromagnetic energy. Because each crest of electromagnetic energy is continually having energy detached from its back and attached to its front, its location moves forward with the fast velocity given in eqn. 14.44. But transmission of energy through the plasma is slowed down by the fact that energy is being continually handed back from each electromagnetic crest to the following electromagnetic crest via the kinetic energy of the plasma electrons.

The storage and flow of energy depicted in Fig. 14.5 for an electromagnetic wave passing through a plasma may be illustrated with the aid of a railway analogy. Let the crests of electromagnetic energy be represented by a succession of trains moving along a railway line. The effect of the plasma electrons corresponds to continually detaching a carriage from the back of each train, holding it temporarily on the line, and then attaching it to the front of the following train. Viewed from the air, the location of each moving train is speeded up by this process. But the progress of passengers towards their destination is slowed down.

If eqns. 14.41 and 14.43 are averaged, either with respect to t over half a period or with respect to z over half a wavelength, we obtain

$$\bar{f} = \tfrac{1}{2}\eta_0 E_0^2 \left(1 - \frac{\omega_N^2}{\omega^2}\right)^{1/2}, \qquad \bar{w} = \tfrac{1}{2}\epsilon_0 E_0^2 \qquad (14.45)$$

Division of these two equations then gives

$$\frac{\bar{f}}{\bar{w}} = c \left(1 - \frac{\omega_N^2}{\omega^2}\right)^{1/2} \qquad (14.46)$$

This is the speed at which the mean energy density (electromagnetic and mechanical) would need to be displaced in the z direction in order to account for the mean rate of flow of energy. It is less than the phase velocity given in eqn. 14.44, and less than the velocity of light in free space.

The velocity appearing in eqn. 14.46 is equal to what is known as the group velocity of propagation in the plasma. Suppose that the electromagnetic field under consideration is not a single plane wave travelling in the z direction but a group of plane waves travelling in this direction with angular frequencies ω that are spread over a narrow band. At the plane $z = 0$ at time zero, let the complex amplitudes of the waves have spectral density $A(\omega)$. $|A(\omega)|$ is important only over a narrow band of frequencies, and over this band there may be a slow variation of phase. The group of waves is described as a function of position and time by the Fourier synthesis

$$\frac{1}{2\pi} \int_{-\infty}^{\infty} A(\omega) \exp \{j(\omega t - kz)\} \, d\omega \qquad (14.47)$$

With suitable choice of $A(\omega)$, this represents a planar pulse travelling in the direction z increasing. In free space we would have $k = \omega/c$ and the pulse would travel without distortion. But, in a plasma, k does not depend linearly on ω, and the shape of the pulse changes somewhat as it propagates. Replacing k_0 by ω/c in the first of eqns. 14.36, we see that the dispersion relation for the plasma may be written

$$k = (\omega^2 - \omega_N^2)^{1/2}/c \qquad (14.48)$$

The maximum in the pulse occurs where the various waves in the group interfere constructively, that is, where the waves have almost the same phase. For a given location z at a given time t, this requires that a first-order variation of ω produces no first order variation of the phase $\omega t - kz$ in expr. 14.47. This means that

$$\frac{d}{d\omega}(\omega t - kz) = 0$$

or

$$t - \frac{dk}{d\omega}z = 0$$

To follow the maximum of the pulse, it is therefore necessary to move along the z axis with the velocity

$$u = \frac{z}{t} = \frac{d\omega}{dk} \tag{14.49}$$

This is the group velocity and, by differentiation of eqn. 14.48, we see that, for the plasma,

$$u = c\left(1 - \frac{\omega_N^2}{\omega^2}\right)^{1/2} \tag{14.50}$$

This is identical with the velocity appearing in eqn. 14.46.

The pulse as a whole moves with the group velocity u given in eqn. 14.50, but the individual wave-crests within the pulse move with the faster phase velocity v given in eqn. 14.44. Individual wave-crests therefore appear at the back of the pulse, move across the pulse in the direction of propagation, and disappear at the front. A crest of electromagnetic energy that disappears at the front of the pulse is used to get the plasma electrons oscillating. A crest of electromagnetic energy that appears at the back of the pulse is using what remains of the kinetic energy of vibration of the plasma electrons before they again become relatively quiescent.

14.7   Energy storage and flow in crossing plane waves

When a pair of plane waves are crossing each other at an angle, a pattern of field can be produced that is travelling in one direction but standing in a perpendicular direction. Let us revert to a homogeneous linear isotropic loss-free material for which storage of kinetic energy is unimportant. The material then has an impedance ζ and an admittance η that are real and that are given by eqns. 14.1. At angular frequency ω, the material has a propagation constant k that is real and that is given by eqn. 14.2. The wavelength λ and the periodic time T are given by eqns. 14.4.

In cartesian co-ordinates (x, y, z), let the propagation vectors of the pair of crossing waves be

where

$$k(S, O, C), \qquad k(-S, O, C) \tag{14.51}$$

$$S^2 + C^2 = 1 \tag{14.52}$$

Both waves are travelling parallel to the xz plane. If S and C are real, their directions of propagations make angles $\pm\,\theta$ with the positive z direction as shown in Fig. 14.6, and

$$S = \sin\theta, \qquad C = \cos\theta \tag{14.53}$$

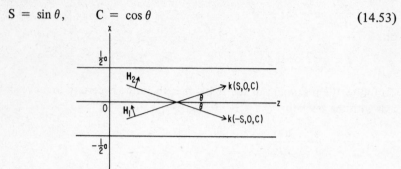

Fig. 14.6.   *Illustrating a pair of crossed plane waves forming a mode in a wave guide*

For each wave, let the electric vector vibrate parallel to the y axis, the positive direction being the positive y direction. At the origin, let the electric field in each wave have zero phase and amplitude $\frac{1}{2}E_0$. Then the complex electric and magnetic vectors for the first wave at location (x, y, z) at time t are (c.f. the right half of Table 13.1)

$$\left.\begin{aligned} \tilde{E}_1 &= \tfrac{1}{2}E_0(0, 1, 0) \exp\{j(\omega t - kSx - kCz)\} \\ \tilde{H}_1 &= \tfrac{1}{2}\eta E_0(-C, 0, +S) \exp\{j(\omega t - kSx - kCz)\} \end{aligned}\right\} \tag{14.54}$$

The complex electric and magnetic vectors for the second wave are obtained by changing the sign of S, and are therefore

$$\left.\begin{aligned} \tilde{E}_2 &= \tfrac{1}{2}E_0(0, 1, 0) \exp\{j(\omega t + kSx - kCz)\} \\ \tilde{H}_2 &= \tfrac{1}{2}\eta E_0(-C, 0, -S) \exp\{j(\omega t + kSz - kCz)\} \end{aligned}\right\} \tag{14.55}$$

By adding eqns. 14.54 and 14.55 we obtain for the complex electric and magnetic vectors of the combined waves at location (x, y, z) at time t

$$\left.\begin{aligned} \tilde{E} &= E_0(0, 1, 0) \cos kSx \exp\{j(\omega t - kCz)\} \\ \tilde{H} &= \eta E_0(-C \cos kSx, 0, -jS \sin kSx) \exp\{j(\omega t - kCz)\} \end{aligned}\right\} \tag{14.56}$$

We see that this is an electromagnetic field that constitutes a standing wave in the x direction with propagation constant

$$kS \tag{14.57}$$

and a travelling wave in the z direction with propagation constant

$$kC = k(1 - S^2)^{1/2} \tag{14.58}$$

The field impedance and admittance looking in the x direction are

$$\frac{\widetilde{E}_y}{\widetilde{H}_z} = -j\frac{1}{\eta S}\cot kSx, \qquad \frac{\widetilde{H}_z}{\widetilde{E}_y} = j\eta S \tan kSx \tag{14.59}$$

These are reactive if S is real, indicating purely oscillatory flow of energy parallel to the x axis. On the other hand, the field impedance and admittance looking in the z direction are

$$-\frac{\widetilde{E}_y}{\widetilde{H}_x} = \frac{1}{\eta C}, \qquad -\frac{\widetilde{H}_x}{\widetilde{E}_y} = \eta C \tag{14.60}$$

These are resistive if C is real, indicating progressive transmission of energy in the direction of the positive z axis.

The first of eqns. 14.56 shows that the electric field vanishes at $x = \pm\frac{1}{2}a$ if $\cos\frac{1}{2}kSa$ vanishes, that is, if

$$kSa = (2m + 1)\pi \tag{14.61}$$

where m is an integer. Concentrating on the case when $m = 0$ and regarding a as prescribed, eqn. 14.61 gives

$$S = \frac{\lambda}{2a}, \qquad C = \left\{1 - \left(\frac{\lambda}{2a}\right)^2\right\}^{1/2} \tag{14.62}$$

where $2\pi/k$ has been replaced by the wavelength λ (eqns. 14.4). With these values of S and C, perfectly conducting sheets may be inserted at $x = \pm\frac{1}{2}a$ without upsetting the field (see Fig. 14.6). Perfectly conducting sheets may also be inserted at $y = \pm\frac{1}{2}b$ because these are perpendicular to the electric field. If we retain only the field in the region $-\frac{1}{2}a \leqslant x \leqslant \frac{1}{2}a, -\frac{1}{2}b \leqslant y \leqslant \frac{1}{2}b$, we then have the fundamental mode of transmission in a perfectly conducting rectangular pipe or wave guide.

From the second of eqns. 14.62 we can see that, if $a < \frac{1}{2}\lambda$, then C is imaginary, so that the propagation constant kC along the pipe (eqn. 14.58) is also imaginary. In other words, this mode of propagation in the pipe is evanescent if $a < \frac{1}{2}\lambda$. Other modes can also exist in the pipe — for example, those corresponding to non-zero integral values of m in eqn. 14.61. Let us arrange for the fundamental mode to be the only one that can travel down the pipe. This is achieved by taking a to be a little greater than $\frac{1}{2}\lambda$, and b to be a little less than $\frac{1}{2}\lambda$. For the fundamental mode the values of both S and C are then real and less than unity in all of the equations of this section. They are simply the sine and cosine of the angle θ in Fig. 14.6. Each crossing plane wave is obtained from the other by reflection in the sides $x = \pm\frac{1}{2}a$ of the perfectly conducting pipe. The wavelength λ_g along the guide is 2π divided by the propagation constant kC along the guide appearing in eqn. 14.58. Since $2\pi/k$ is the wavelength λ of a plane wave in the material that fills the pipe, it follows that

$$\lambda_g = \lambda \sec\theta \tag{14.63}$$

Hence the wavelength along the guide exceeds λ.

By taking the real parts of the complex electric and magnetic fields in eqns. 14.56, we obtain for the actual vibrating electromagnetic vectors at location (x, y, z) at time t

$$E = E_0(0, 1, 0) \cos kSx \cos(\omega t - kCz)$$

$$H = \eta E_0 [-C \cos kSx \cos(\omega t - kCz), 0, S \sin kSx \sin(\omega t - kCz)]$$

$$(14.64)$$

It follows that the electric and magnetic energy densities at location (x, y, z) at time t are

$$w_e = \tfrac{1}{2}\epsilon E_0^2 \cos^2 kSx \cos^2(\omega t - kCz)$$

$$w_m = \tfrac{1}{2}\epsilon E_0^2 C^2 \cos^2 kSx \cos^2(\omega t - kCz)$$

$$+ \tfrac{1}{2}\epsilon E_0^2 S^2 \sin^2 kSx \sin^2(\omega t - kCz)$$

$$(14.65)$$

and the Poynting vector at location (x, y, z) at time t is

$$f = E \times H = \eta E_0^2 [\tfrac{1}{4} S \sin 2kSx \sin\{2(\omega t - kz)\}, 0,$$

$$C \cos^2 kSx \cos^2(\omega t - kCz)]$$

$$(14.66)$$

By adding the expressions for w_e and w_m in eqns. 14.65, we obtain for the total energy density

$$w = \tfrac{1}{2}\epsilon E_0^2 [(1 + C^2) \cos^2 kSx \cos^2(\omega t - kz) + S^2 \sin^2 kSx \sin^2(\omega t - kz)]$$

$$(14.67)$$

In Fig. 14.7 are shown contour maps of the total energy density w in the (x, z) plane at four intervals of time separated by one-eighth of a period. After half a period the contour maps repeat because the pattern has been displaced in the z direction through a distance $\tfrac{1}{2}\lambda_g$ given by eqn. 14.63. The velocity of this displacement is

$$v = \frac{\omega}{kC} = (\mu\epsilon)^{-1/2} \sec\theta \qquad (14.68)$$

This is the velocity of propagation of phase along the guide, and it exceeds the velocity $(\mu\epsilon)^{-1/2}$ of light in the material filling the pipe. However, it is not true that energy is being displaced along the pipe with the velocity v. Eqns. 14.59 and 14.66 show that there is oscillation of energy across the guide as well as transmission of energy along the guide.

In Fig. 14.7, crosses indicate maxima of stored energy density and circles indicate zeros of stored energy density. There is not much energy in the regions where the contour lines are shown broken. There are concentrations of energy near the centre of the guide at intervals of $\tfrac{1}{2}\lambda_g$ along the pipe. Opposite the zeros of energy density at the centre of the guide there are concentrations of energy near the edges of the guide. The latter are concentrations of magnetic energy associated with oscillation of energy between the edges and the centre, where they become concentrations of electric energy. The oscillation of energy in the x direction between the centre

t = 0

t = $\frac{1}{8}$T

t = $\frac{1}{4}$T

t = $\frac{3}{8}$T

Distance x across the guide

$-\frac{1}{4}\lambda_g$ 0 $\frac{1}{4}\lambda_g$ $\frac{1}{2}\lambda_g$ $\frac{3}{4}\lambda_g$

Distance z along the guide

Fig. 14.7.   *Illustrating storage and flow of energy for the fundamental mode of transmission in a rectangular wave guide, $\theta = 60°$*

and the edges at $z = \frac{1}{4}\lambda_g$ in Fig. 14.7 is similar to that shown for the z direction in Fig. 14.2. This sideways oscillation of energy is combined with transmission of energy along the guide in the following manner.

The energy concentrated at the edges of the guide near $z = \frac{3}{8}\lambda_g$ in the second diagram of Fig. 14.7 $(t = \frac{1}{8}T)$ was detached from the back of the concentration of energy in the previous diagram $(t = 0)$ near $z = \frac{1}{2}\lambda_g$ in the centre of the guide; this happened in the manner indicated by vertical arrows. When it is returned to the centre of the guide, it is attached to the front of the concentration of energy in the succeeding diagram $(t = \frac{1}{4}T)$ near $z = \frac{1}{4}\lambda_g$ in the centre of the guide. The effect of sideways oscillation of energy is therefore to cause continual detachment of energy from the back of a concentration in the centre of the guide, to move the detached energy to the edges, and then to attach it to the front of the following central concentration of energy. In Fig. 14.7 this speeds up the displacement along the pipe of the locations of central concentrations of energy while at the same time handing energy backwards from one central concentration to the following one.

> A railway analogy would involve a main line with side tracks on either side. The central concentrations of energy correspond to a succession of trains travelling along the main line. Carriages are being continually detached from the back of each train, placed temporarily on the sidings, and then attached to the front of the following train. Viewed from the air, the location of each moving train would be speeded up by this process, but the progress of passengers toward their destination would be slowed down.

By averaging eqns. 14.66 and 14.67 over a length $\frac{1}{2}\lambda_g$ of the pipe, both with respect to z and with respect to x, we see that the values for (i) the mean rate of flow of energy per unit cross-sectional area and (ii) the mean energy stored per unit volume are, respectively.

$$\bar{f} = \eta E_0^2(0, 0, \tfrac{1}{4}C), \qquad \bar{w} = \tfrac{1}{4}\epsilon E_0^2 \tag{14.69}$$

It follows that the mean flow of energy is down the pipe, and that the speed with which the mean energy density must be displaced in order to account for the mean rate of flow of energy is

$$\bar{f}/\bar{w} = (\mu\epsilon)^{-1/2}C = (\mu\epsilon)^{-1/2}\cos\theta \tag{14.70}$$

This is less than the phase velocity given in eqn. 14.68, and less than the velocity $(\mu\epsilon)^{-1/2}$ of light in the material filling the pipe.

As in the preceding section, a calculation can be made of the group velocity u down the pipe; it evaluates to the velocity appearing in eqns. 14.70.

14.8   Energy storage and flow for a radiating dipole

Radiation in free space from an elementary electric dipole whose moment p is a function of the time t is described by means of the Hertzian potential in the manner outlined in Section 1.11. The Hertzian potential at time t at a point distant r from the dipole is given by eqn. 1.69. Substitution of this into eqns. 1.68 then gives the electromagnetic field (E, B). Performance of the indicated differentiations leads to the formulas quoted in Problem 1.7. For a dipole of short length l carrying an oscillatory electric current of angular frequency ω given by the real part of $I_0 \exp(j\omega t)$, the complex electromagnetic field is that quoted in Problem 2.8 where, for free space, $\tilde{\epsilon}/\epsilon_0$ and $\tilde{n}$ must be replaced by unity. The field at large values of r behaves differently from that at small values of r. If $r < \lambda_0/(2\pi)$, where λ_0 is the wavelength in free space, the field is largely controlled by the r^{-3} terms and, except for the time-variation, is essentially that for an electrostatic dipole (Problem 1.1). If the dipole is regarded as a capacitor, this is its fringing field. The oscillatory generator charging the capacitor is providing energy (primarily electric energy) to the spherical volume $r < \lambda_0/(2\pi)$ in alternate quarter-periods and is reclaiming it in the intervening quarter-periods. However, the generator does not receive back quite all the energy in one quarter-period that it provided in the preceding quarter-period. This is because there is some leakage of energy across the sphere $r = \lambda_0/(2\pi)$, especially in the vicinity of the equatorial plane of the dipole ($\theta = \frac{1}{2}\pi$ in Problems 1.7 and 2.8). What leaks becomes a crest of electromagnetic energy in a wave radiated by the antenna. Successive crests of electromagnetic energy are launched into the region $r > \lambda_0/(2\pi)$ each half period, forming a wave that expands spherically. It is this radiating wave that is described by the terms of order r^{-1} in the formulas of Problems 1.7 and 2.8.

An electric dipole in free space that is short compared with $\lambda_0/(2\pi)$, being primarily a capacitor, has an input impedance that is highly reactive. When connected to the end of a transmission line having a resistive characteristic impedance, reflection occurs at the junction. One way to reduce this reflection is to use a dipole that is approximately half a wavelength long. Let us examine the storage and flow of energy in such a half-wave dipole.

In cylindrical polar co-ordinates (ρ, ψ, z) let the dipole lie along the z axis with its centre located at the origin and its ends located at $z = \pm L$, so that $L = \frac{1}{4}\lambda_0$. Let z' be a co-ordinate measured along the dipole from the centre in the direction z increasing. Let us assume that the electric current along the dipole oscillates sinusoidally with angular frequency ω and that the current is the real part of

$$I = I_0 \cos k_0 z' \exp(j\omega t) \tag{14.71}$$

where

$$k_0 = 2\pi/\lambda_0, \qquad k_0 L = \frac{1}{2}\pi \tag{14.72}$$

Eqn. 14.71 may be written

$$I = I_+ + I_- \tag{14.73}$$

where

$$I_+ = \frac{1}{2}I_0 \exp\{j(\omega t - k_0 z')\}, \qquad I_- = \frac{1}{2}I_0 \exp\{j(\omega t + k_0 z')\} \tag{14.74}$$

It is convenient to consider these two oppositely travelling current waves separately.

To calculate the complex Hertzian potential Π at location (ρ, ψ, z) in cylindrical polar co-ordinates, we apply eqn. 1.70. The integral is over the volume occupied by current, and therefore reduces to an integral along the dipole with respect to z'.

The distance r from the point $(0, \psi, z')$ on the wire to the point (ρ, ψ, z) in space is

$$r = \{\rho^2 + (z - z')^2\}^{1/2} \tag{14.75}$$

The complex electric moment of the current element $I \, dz'$ is $dz' \int I \, dt$, or $(j\omega)^{-1} I \, dz'$. Because all dipoles are oriented parallel to the z axis, there is only a z component Π of the complex Hertzian potential. For the current wave in the positive z' direction, this evaluates to

$$\Pi_+ = \frac{1}{4\pi\epsilon_0} \int_{-L}^{L} \frac{\tfrac{1}{2} I_0 \exp\{j(\omega t - k_0 z')\}}{j\omega} \frac{\exp(-jk_0 r)}{r} dz'$$

which may be written as

$$\Pi_+ = \frac{I_0}{8\pi\epsilon_0} \exp\{j(\omega t - k_0 z)\} \int_{-L}^{L} \frac{\exp\{-jk_0(r + z' - z)\}}{r} dz' \tag{14.76}$$

We introduce a new variable of integration u such that

$$u = r + z' - z, \qquad du/u = dz'/r \tag{14.77}$$

Eqn. 14.76 then becomes

$$\Pi_+ = \frac{I_0}{8\pi\epsilon_0} \exp\{j(\omega t - k_0 z)\} \int_{r_2 - L - z}^{r_1 + L - z} \frac{\exp(-jk_0 u)}{u} du \tag{14.78}$$

where r_1 and r_2 are the distances of the field point P from the ends of the dipole as shown in Fig. 14.8. If desired, the integral in eqn. 14.78 may be expressed in terms of the sine and cosine integral functions.

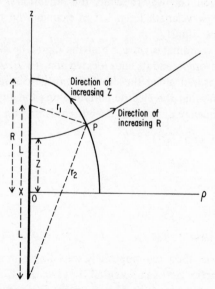

Fig. 14.8.   *Illustrating the use of confocal co-ordinates (R, Z) for calculating the electromagnetic field of a half-wave dipole of length 2L*

For the current wave in the negative z' direction, Π_+ in eqn. 14.78 is replaced by Π_-, for which the signs of z and L are reversed. The complete complex Hertzian vector in cylindrical polar co-ordinates is then

$$\Pi = (0, 0, \Pi_+ + \Pi_-) \tag{14.79}$$

This is what has to be substituted into eqns. 1.68 in order to obtain the complex electromagnetic field. The differentiation results in the final formulae for the electromagnetic field containing no integrals such as that appearing in eqn. 14.78.

It is convenient to express the electromagnetic field, not in cylindrical polar co-ordinates, but in a confocal co-ordinate system that uses the ends of the dipole as foci. This is illustrated in Fig. 14.8. An ellipse is such that the sum of r_1 and r_2 is constant; an orthogonally intersecting hyperbola is such that the difference between r_1 and r_2 is constant. An ellipse is identified by the length R of its semi-major axis. One branch of a hyperbola is identified by the length Z of its semi-major axis; the other branch is obtained by changing the sign of Z. It follows that

$$R = \tfrac{1}{2}(r_2 + r_1), \qquad Z = \tfrac{1}{2}(r_2 - r_1) \tag{14.80}$$

so that

$$r_1 = R - Z, \qquad r_2 = R + Z \tag{14.81}$$

A point P that is identified in cylindrical polar co-ordinates as (ρ, ψ, z) is identified in confocal co-ordinates as (R, Z, ψ). From the geometry of the ellipses and hyperbolae it follows that

$$\rho = (R^2 - L^2)^{1/2}(L^2 - Z^2)^{1/2}/L, \qquad z = RZ/L \tag{14.82}$$

The dipole is the degenerate prolate ellipsoid $R = L$ and, on the dipole, Z is distance measured from the centre in the direction z increasing. At large distances from the dipole we have approximately

$$R = r, \qquad Z = L \cos \theta \tag{14.83}$$

where (r, θ, ψ) are spherical polar co-ordinates. Expressed in confocal co-ordinates, eqn. 14.78 becomes

$$\Pi_+ = \frac{I_0}{8\pi\epsilon_0} \exp\{j(\omega t - k_0 RZ/L)\} \int_{R+Z-L-RZ/L}^{R-Z+L-RZ/L} \frac{\exp(-jk_0 u)}{u} \, du \tag{14.84}$$

and Π_- is obtained by changing the sign of L.

At the point P in Fig. 14.8, the direction R increasing is along a hyperbola, the direction Z increasing is along an ellipse, and the direction ψ increasing is perpendicular to the paper round a circle coaxial with the dipole. Actual distances in these directions corresponding to increments dR, dZ and $d\psi$ in the confocal co-ordinates are $h_1 dR$, $h_2 dZ$ and $h_3 d\psi$ where

$$h_1 = \left(\frac{R^2 - Z^2}{R^2 - L^2}\right)^{1/2}, \qquad h_2 = \left(\frac{R^2 - Z^2}{L^2 - Z^2}\right)^{1/2},$$

$$h_3 = \frac{(R^2 - L^2)(L^2 - Z^2)^{1/2}}{L} \left.\begin{array}{c}\\\\\\\\\end{array}\right\} \tag{14.85}$$

In terms of these quantities, the expressions for gradient, divergence and curl in confocal coordinates are as given in Appendix A. Use of these expressions in eqns. 1.68 then gives the components of the complex electromagnetic field in the directions R, Z and ψ increasing.

There is no azimuthal component of the electric field, and no component of the magnetic field in an azimuthal plane. It is the R and Z components of the electric vector, and the ψ component of the magnetic vector, that are non-zero. The complex electromagnetic field evaluates to

$$\tilde{E}_R = \zeta_0 \frac{I_0 L}{2\pi} \frac{\sin k_0 Z \exp\{j(\omega t - k_0 R)\}}{(R^2 - Z^2)^{1/2}(R^2 - L^2)^{1/2}}$$

$$\tilde{E}_Z = -j\zeta_0 \frac{I_0 L}{2\pi} \frac{\cos k_0 Z \exp\{j(\omega t - k_0 R)\}}{(R^2 - Z^2)^{1/2}(L^2 - Z^2)^{1/2}} \left.\begin{array}{c}\\\\\\\\\\\end{array}\right\} \tag{14.86}$$

$$\tilde{H}_\psi = j\frac{I_0 L}{2\pi} \frac{\cos k_0 Z \exp\{j(\omega t - k_0 R)\}}{(R^2 - L^2)^{1/2}(L^2 - Z^2)^{1/2}}$$

where ζ_0 is the impedance of free space (377 Ω). The exponential behaviour in the R direction connotes travelling-wave behaviour along the hyperbolae, the direction of propagation being away from the dipole. The sinusoidal behaviour in the Z direction connotes standing-wave behaviour round the ellipses. There is outward transmission of energy along the hyperbolae combined with oscillation of energy in the perpendicular direction round the ellipses. This is reminiscent of the behaviour encountered in the preceding section, but now the coordinates are curvilinear. At the point (R, Z, ψ), the field impedances looking in the directions R increasing and Z increasing are, respectively,

$$-\frac{\tilde{E}_Z}{\tilde{H}_\psi} = \zeta_0 \left(\frac{R^2 - L^2}{R^2 - Z^2}\right)^{1/2}, \qquad \frac{\tilde{E}_R}{\tilde{H}_\psi} = -j\zeta_0 \left(\frac{L^2 - Z^2}{R^2 - Z^2}\right)^{1/2} \tan k_0 Z$$

$$\tag{14.87}$$

When $R \gg L$, these become approximately ζ_0 and zero, while the complex electromagnetic field in eqns. 14.86 becomes

$$\tilde{E}_R = 0$$

$$\tilde{E}_Z = -j\frac{\zeta_0 I_0}{2\pi} \frac{\cos(\tfrac{1}{2}\pi \cos\theta)}{\sin\theta} \frac{\exp\{j(\omega t - k_0 r)\}}{r} \left.\begin{array}{c}\\\\\\\\\end{array}\right\} \tag{14.88}$$

$$\tilde{H}_\psi = j\frac{I_0}{2\pi} \frac{\cos(\tfrac{1}{2}\pi \cos\theta)}{\sin\theta} \frac{\exp\{j(\omega t - k_0 r)\}}{r}$$

The θ dependence in these equations describes the radiation pattern of a half-wave dipole.

By taking the real parts of eqns. 14.86, we obtain for the actual vibrating electromagnetic field at any location (R, Z, ψ) at time t

$$
\left.
\begin{aligned}
E_R &= \zeta_0 \frac{I_0 L}{2\pi} \frac{\sin k_0 Z \cos(\omega t - k_0 R)}{(R^2 - Z^2)^{1/2}(R^2 - L^2)^{1/2}} \\[2mm]
E_Z &= \zeta_0 \frac{I_0 L}{2\pi} \frac{\cos k_0 Z \sin(\omega t - k_0 R)}{(R^2 - Z^2)^{1/2}(L^2 - Z^2)^{1/2}} \\[2mm]
H_\psi &= -\frac{I_0 L}{2\pi} \frac{\cos k_0 Z \sin(\omega t - k_0 R)}{(R^2 - L^2)^{1/2}(L^2 - Z^2)^{1/2}}
\end{aligned}
\right\}
\tag{14.89}
$$

It follows that the electric and magnetic energy densities are

$$
\left.
\begin{aligned}
w_e &= \mu_0 \frac{I_0^2 L^2}{8\pi^2} \frac{\sin^2 k_0 Z \cos^2(\omega t - k_0 R)}{(R^2 - Z^2)(R^2 - L^2)} \\[2mm]
&+ \mu_0 \frac{I_0^2 L^2}{8\pi^2} \frac{\cos^2 k_0 Z \sin^2(\omega t - k_0 R)}{(R^2 - Z^2)(L^2 - Z^2)} \\[2mm]
w_m &= \mu_0 \frac{I_0^2 L^2}{8\pi^2} \frac{\cos^2 k_0 Z \sin^2(\omega t - k_0 R)}{(R^2 - L^2)(L^2 - Z^2)}
\end{aligned}
\right\}
\tag{14.90}
$$

and that the R, Z and ψ components of the Poynting vector at location (R, Z, ψ) at time t are given by

$$
f = E \times H = \zeta_0 \frac{I_0^2 L^2}{4\pi^2} \left[\frac{\cos^2 k_0 Z \sin^2(\omega t - k_0 R)}{(R^2 - Z^2)^{1/2}(R^2 - L^2)^{1/2}(L^2 - Z^2)}, \right.
$$
$$
\left. -\frac{\tfrac{1}{4} \sin 2k_0 Z \sin(2\omega t - 2k_0 R)}{(R^2 - Z^2)^{1/2}(L^2 - Z^2)^{1/2}(R^2 - L^2)}, \; 0 \right]
\tag{14.91}
$$

The total energy density w at location (R, Z, ψ) at time t is obtained by adding the expressions for w_e and w_m given in eqns. 14.90. Contour maps for w in an azimuthal plane $\psi = $ constant are shown in Fig. 14.9. Each diagram should be reflected in the horizontal axis (the equatorial plane of the dipole), and in addition rotated round the axis of the dipole. Contour maps are shown at intervals of one-eighth of a period over half a period, after which time they repeat. Crosses indicate maxima of energy density and circles indicate zeros. Contours are marked in decibels above $\mu_0 I_0^2/(8\pi^2 L^2)$.

Let us assume that the surface of the dipole is one of the thin confocal spheroids. The surface then has a coordinate R that is slightly greater than L. If a is the small

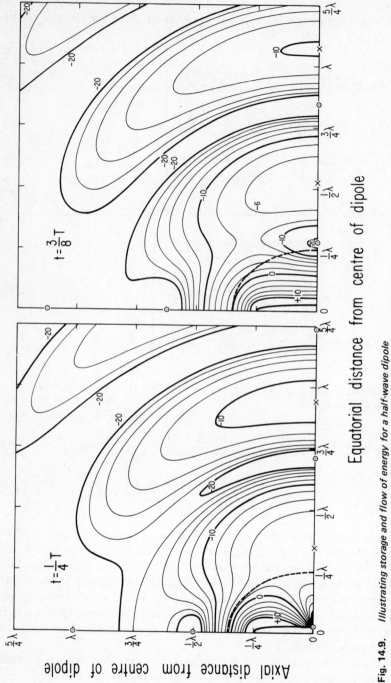

Equatorial distance from centre of dipole

Axial distance from centre of dipole

Fig. 14.9.   *Illustrating storage and flow of energy for a half-wave dipole*
The region between the dipole and the broken-line ellipsoid acts as a coaxial
transmission line, leakage from which launches the radiated wave

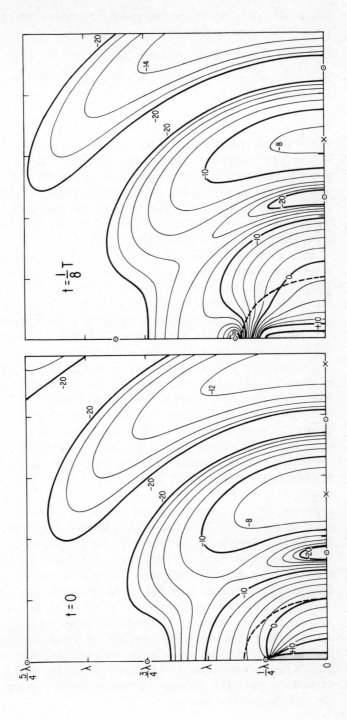

semi-minor axis of the dipole, then it follows from the first of eqns. 14.80 that the surface of the dipole is given by

$$R = (L^2 + a^2)^{1/2} \doteq L + \tfrac{1}{2}a^2/L \qquad (14.92)$$

Also indicated in Fig. 14.9 by a broken curve is the coaxial prolate spheroid for which

$$R = 2^{1/2}L. \qquad (14.93)$$

This is the one whose radius in the equatorial plane of the dipole is equal to L. Between the broken-line spheroid and the dipole, the flow of energy consists primarily of an oscillation back and forth along the ellipses $R = $ constant in a manner similar to that occuring in a half-wavelength of coaxial transmission line open-circuited at its ends (c.f. Section 14.4). This oscillation is described by the Z component of f in eqn. 14.91. It is also described by the reactive field impedance appearing in the second of eqns. 14.87.

Outside the broken-line spheroid in Fig. 14.9, the dominant behaviour is that of an electromagnetic wave travelling outwards along the hyperbolae $Z = $ constant. This is described by the R component of f in eqn. 14.91. It is also described by the resistive field impedance appearing in the first of eqns. 14.87. The broken-line spheroid in Fig. 14.9 may therefore be thought of as constituting the outer boundary of a coaxial transmission line from which leakage of electromagnetic energy is taking place, especially in the vicinity of the equatorial plane of the dipole.

The sequence of contour maps shown in Fig. 14.9 illustrates how oscillation of energy in the coaxial transmission line fits together with outward propagation of energy outside the transmission line. Let us assume that the half-wave dipole is energised by means of a generator at its centre delivering an oscillatory electric current $I_0 \cos \omega t$. In each of the diagrams in Fig. 14.9 high densities of energy occur between the dipole and the 10 decibel contours. At $t = 0$ as shown in the top left-hand diagram, storage of energy is strong in the coaxial transmission line near the centre of the dipole. This is magnetic energy which then flows approximately along the ellipses towards the ends of the coaxial transmission line and is at the same time converted into electric energy. A quarter of a period later, at $t = \tfrac{1}{4}T$ in the bottom left hand diagram, we have concentrations of energy near the ends of the dipole. In the ensuing quarter of a period, energy flows back along the coaxial transmission line from the ends towards the centre, and is largely converted back into magnetic energy. However, less energy returns to the centre then started out because some has leaked across the outer surface of the coaxial transmission line. The leakage forms a crest of electromagnetic energy propagating outwards along the hyperbolae. Moreover, although less energy has returned to the centre of the coaxial transmission line, this loss is made up from the generator. The flow of energy from the centre to the ends and back then repeats, and the next crest of electromagnetic energy is launched through the outer surface of the coaxial transmission line. The sequence of contour maps of energy density shown in Fig. 14.9 illustrates the process.

Leakage of energy through the outer surface of the coaxial transmission line is much greater than would occur for a metallic outer boundary. Substitution from eqn. 14.93 into the first of eqns. 14.87 shows that, in Fig. 14.9, the field impedance looking across the broken-line spheroid in the direction R increasing is

$$\zeta_0 \frac{L}{(2L^2 - Z^2)^{1/2}} \tag{14.94}$$

This is of the order of the impedance of free space, and varies from ζ_0 on the axis of the dipole ($Z = L$) to $2^{-1/2}\zeta_0$ on the equatorial plane ($Z = 0$). Even so, for a thin dipole, the energy radiated per half-period is a small fraction of the energy stored in the coaxial transmission line (see Problem 14.2). This is because of the large amount of energy stored close to the thin wire.

Let us calculate the rate at which energy is crossing the confocal prolate spheroidal surface whose co-ordinate is R. Per unit area, this is given by the R component of f in eqn. 14.91. The area on the ellipsoid between the confocal hyperbolae Z and $Z + dZ$ is $2\pi\rho h_2 dZ$, where ρ is given by the first of eqns. 14.82 and h_2 by the second of eqns. 14.85. Hence the rate at which energy is crossing the R ellipsoid at time t is

$$\int_{-L}^{L} f_R 2\pi\rho h_2 dZ = I_0^2 \sin^2(\omega t - k_0 R) \frac{\zeta_0}{2\pi} \int_{-1}^{1} \frac{\cos^2 \frac{1}{2}\pi v}{1 - v^2} dv \tag{14.95}$$

where $v = Z/L$. On evaluating the integral and replacing ζ_0 by $377\,\Omega$, we obtain approximately

$$\frac{\zeta_0}{2\pi}\int_{-1}^{1} \frac{\cos^2 \frac{1}{2}\pi v}{1 - v^2} dv = 73\,\Omega \tag{14.96}$$

Hence the rate in watts at which energy is crossing the R ellipsoid at time t is

$$P = 73I_0^2 \sin^2(\omega t - k_0 R) \tag{14.97}$$

This result applies to any of the confocal ellipsoids. In particular, it applies to the outer surface of the coaxial transmission line given by eqn. 14.93. But it also applies to a confocal ellipsoid consisting of a large sphere of radius r concentric with the dipole; for this surface, in accordance with the first of eqns. 14.83, eqn. 14.97 becomes

$$P = 73I_0^2 \sin^2(\omega t - k_0 r) \tag{14.98}$$

Eqn. 14.97 also applies to the surface of the dipole; here $R \doteq L = \frac{1}{4}\lambda_0$, so that the rate in watts at which energy is leaving the dipole at time t is

$$P = 73I_0^2 \cos^2 \omega t \tag{14.99}$$

If the dipole is fed from a transmission line connected at the centre, behaviour on the feeder line is the same as if a $73\,\Omega$ resistor were connected at its output terminals. This is known as the radiation resistance of the dipole. If the dipole does not have precisely the form of a prolate spheroid, there is also a reactive component to the input impedance of the dipole. This can, however, be eliminated by a minor adjustment in the length of the dipole.

For a centre-fed half-wave dipole, the assumption in eqn. 14.71 of a strictly sinusoidal distribution of current is an approximation. For a centre-fed half-wave dipole, all energy must come from the feeding generator at the centre. But eqn. 14.91 shows that, for a strictly sinusoidal distribution of current, the mean flow of energy is along the hyperbolae $Z = $ constant. These do not run from the centre of the dipole. Eqn. 14.95, with the integration with respect to Z removed, gives the energy per unit length per unit time leaving the dipole at location Z at time t. To produce a strictly sinusoidal distribution of current, the dipole must be fed, not just at its centre, but along its whole length. It must be fed in such a way as to create, at the surface of the dipole, the parallel component (Z component) of electric field indicated in eqns. 14.86 and 14.89. But, for a perfectly conducting centre-fed dipole, the parallel component of electric field at the surface of the dipole is zero except at the feeding generator. This modifies the distribution of current along the dipole slightly. It also modifies the mean flow of energy close to the dipole so as to avoid outward flow of energy through the surface of the wire, and make all energy emanate from the generator at the centre.

A practical dipole is made of wire that in fact has a little resistance. In this case, flow of energy through the surface of the wire does occur. But it is inward flow to overcome the metal loss, not outward flow to create radiation to a large distance from the dipole. Such considerations affect, especially near the wire, details of the oscillation of energy in the coaxial transmission line bounded by the broken-line spheroid in Fig. 14.9. But outside this surface, the outward transmission of energy along the hyperbolae $Z = $ constant is virtually unaffected.

Problems

14.1   In a system of cartesian co-ordinates (x, y, z), the regions $z < 0$ and $z > 0$ are occupied by homogeneous non-conducting materials of impedances ζ_1 and ζ_2, respectively. The two materials are moved apart sufficiently to insert between them, in contact with both, a slab of homogeneous non-conducting material of impedance $(\zeta_1\zeta_2)^{1/2}$. A plane electromagnetic wave of angular frequency ω, travelling in material 1 in the direction z increasing, is incident normally upon the slab. The frequency is such that the slab is a quarter of a wavelength thick, so that it constitutes a quarter-wave transformer matching materials 1 and 2 at normal incident (Problem 13.5). If $\zeta_1 = 9\zeta_2$ show that the storage and flow of energy in the slab is as shown in Fig. 14.4 for $-\frac{1}{4}\lambda \leqslant z \leqslant 0$. If $\zeta_2 = 9\zeta_1$ show that the storage and flow of energy in the slab is as shown in Fig. 14.4 for $0 \leqslant z \leqslant \frac{1}{4}\lambda$.

14.2   A half-wave dipole of length $2L$ in free space has the form of a thin prolate spheroid of minor axis a as illustrated in Fig. 14.9. The current in the dipole oscillates sinusoidally with angular frequency ω, and its value at the centre at time t is $I_0 \cos \omega t$. Show that the electromagnetic energy stored inside the confocal spheroid given by eqn. 14.93 and indicated in Fig. 14.9 by a broken curve is approximately

$$\frac{\mu_0 I_0^2 L}{4\pi} \ln \frac{L}{a}$$

Show also that, if one includes the small amount of energy oscillating round the ellipses outside the broken-line spheroid, then the stored electromagnetic energy is increased to approximately

$$\frac{\mu_0 I_0^2 L}{4\pi} \ln \frac{2L}{a}$$

Deduce that, if this is expressed as a ratio to the mean energy radiated by the dipole during a time-interval equal to a period divided by 2π, one obtains

$$Q = 1 \cdot 3 \ln (2L/a).$$

Appendices

Appendix A   Orthogonal Co-ordinate Systems

Cartesian co-ordinates (x, y, z)

Element of length $ds = (dx, dy, dz)$

$$\nabla\phi = \text{grad}\,\phi = \left(\frac{\partial\phi}{\partial x}, \frac{\partial\phi}{\partial y}, \frac{\partial\phi}{\partial z}\right)$$

$$\nabla\cdot A = \text{div}\,A = \frac{\partial A_x}{\partial x} + \frac{\partial A_y}{\partial y} + \frac{\partial A_z}{\partial z}$$

$$\nabla \times A = \text{curl}\,A = \begin{vmatrix} \hat{x} & \hat{y} & \hat{z} \\ \dfrac{\partial}{\partial x} & \dfrac{\partial}{\partial y} & \dfrac{\partial}{\partial z} \\ A_x & A_y & A_z \end{vmatrix}$$

$$\nabla^2\phi = \text{div}\,\text{grad}\,\phi = \frac{\partial^2\phi}{\partial x^2} + \frac{\partial^2\phi}{\partial y^2} + \frac{\partial^2\phi}{\partial z^2}$$

Orthogonal curvilinear co-ordinates (q_1, q_2, q_3)

Element of length $ds = (h_1 dq_1, h_2 dq_2, h_3 dq_3)$

$$\text{grad}\,\phi = \left(\frac{1}{h_1}\frac{\partial\phi}{\partial q_1}, \frac{1}{h_2}\frac{\partial\phi}{\partial q_2}, \frac{1}{h_3}\frac{\partial\phi}{\partial q_3}\right)$$

$$\text{div}\,A = \frac{1}{h_1 h_2 h_3}\left[\frac{\partial}{\partial q_1}(h_2 h_3 A_1) + \frac{\partial}{\partial q_2}(h_3 h_1 A_2) + \frac{\partial}{\partial q_3}(h_1 h_2 A_3)\right]$$

$$\text{curl}\,A = \frac{1}{h_1 h_2 h_3}\begin{vmatrix} h_1\hat{q}_1 & h_2\hat{q}_2 & h_3\hat{q}_3 \\ \dfrac{\partial}{\partial q_1} & \dfrac{\partial}{\partial q_2} & \dfrac{\partial}{\partial q_3} \\ h_1 A_1 & h_2 A_2 & h_3 A_3 \end{vmatrix}$$

$$\nabla^2\phi = \frac{1}{h_1 h_2 h_3}\left[\frac{\partial}{\partial q_1}\left(\frac{h_2 h_3}{h_1}\frac{\partial\phi}{\partial q_1}\right) + \frac{\partial}{\partial q_2}\left(\frac{h_3 h_1}{h_2}\frac{\partial\phi}{\partial q_2}\right) + \frac{\partial}{\partial q_3}\left(\frac{h_1 h_2}{h_3}\frac{\partial\phi}{\partial q_3}\right)\right]$$

Cylindrical polar Co-ordinates (r, θ, z)

and
$$q_1 = r, \quad q_2 = \theta, \quad q_3 = z$$
so that
$$ds = (dr, rd\theta, dz)$$
$$h_1 = 1, \quad h_2 = r, \quad h_3 = 1.$$

Spherical polar co-ordinates (r, θ, ϕ)

and
$$q_1 = r, \quad q_2 = \theta, \quad q_3 = \phi$$
so that
$$ds = (dr, rd\theta, r\sin\theta\, d\phi)$$
$$h_1 = 1, \quad h_2 = r, \quad h_3 = r\sin\theta.$$

Appendix B Vector Identities

1. $A \cdot B = A_x B_x + A_y B_y + A_z B_z = AB\cos\angle(A, B)$

2. $A \times B = \begin{vmatrix} \hat{x} & \hat{y} & \hat{z} \\ A_x & A_y & A_z \\ B_x & B_y & B_z \end{vmatrix} = -B \times A$

   $|A \times B| = AB\sin|\angle(A, B)| = $ area of parallelogram defined by A and B

3. $A \cdot B \times C = B \cdot C \times A = C \cdot A \times B$

   $= A \times B \cdot C = B \times C \cdot A = C \times A \cdot B$

   $= \begin{vmatrix} A_x & A_y & A_z \\ B_x & B_y & B_y \\ C_x & C_y & C_z \end{vmatrix} = \pm$ volume of parallelopiped defined by A, B, and C.

4. $\nabla \times (\nabla\phi) = 0$

5. $\nabla \cdot (\nabla \times A) = 0$

6. $\nabla(\phi\psi) = \phi\nabla\psi + \psi\nabla\phi$

7. $\nabla \cdot (\phi A) = \phi\nabla \cdot A + (\nabla\phi)\cdot A$

8. $\nabla \times (\phi A) = \phi\nabla \times A + (\nabla\phi) \times A$

9. $\nabla \cdot (A \times B) = -A \cdot (\nabla \times B) + B \cdot (\nabla \times A)$

10. $A \times (B \times C) = (A \cdot C)B - (A \cdot B)C$

11. $\nabla \times (\nabla \times A) = \nabla(\nabla \cdot A) - \nabla^2 A$

12. $\nabla^2 A = (\nabla^2 A_x)\hat{x} + (\nabla^2 A_y)\hat{y} + (\nabla^2 A_z)\hat{z}$

$\neq (\nabla^2 A_1)\hat{q}_1 + (\nabla^2 A_2)\hat{q}_2 + (\nabla^2 A_3)\hat{q}_3$

13. $(A \cdot \nabla)B = \left(A_x \dfrac{\partial B_x}{\partial x} + A_y \dfrac{\partial B_x}{\partial y} + A_z \dfrac{\partial B_x}{\partial z}\right)\hat{x} + \left(A_x \dfrac{\partial B_y}{\partial x} + A_y \dfrac{\partial B_y}{\partial y} + A_z \dfrac{\partial B_y}{\partial z}\right)\hat{y}$

$+ \left(A_x \dfrac{\partial B_z}{\partial x} + A_y \dfrac{\partial B_z}{\partial y} + A_z \dfrac{\partial B_z}{\partial z}\right)\hat{z}$

14. $\nabla(A \cdot B) = \{A \times (\nabla \times B) + (A \cdot \nabla)B\} + \{B \times (\nabla \times A) + (B \cdot \nabla)A\}$

15. $\nabla \times (A \times B) = \{(B \cdot \nabla)A - (A \cdot \nabla)B\} - \{B(\nabla \cdot A) - A(\nabla \cdot B)\}$

16. At a point moving with velocity v in a vector field A

$$\frac{dA}{dt} = \frac{\partial A}{\partial t} + (v \cdot \nabla)A$$

17. The divergence theorem: if V is a volume (having element of volume $d\tau$) bounded by a closed surface Σ (having outward directed vector element of area dS), then

$$\int_{\Sigma} A \cdot dS = \int_V \nabla \cdot A \, d\tau.$$

18. The curl theorem: if S is a surface (having vector element of area dS) bounded by a rim C (having right-hand-related vector element of length ds), then

$$\int_C A \cdot ds = \int_S (\nabla \times A) \cdot dS.$$

Appendic C Relativistic mechanics

Suppose that an observer O sees a particle of mass m, velocity u, relativistic momentum p and relativistic energy E at position r at time t, and that the particle is subject to an acceleration a due to a force F. Another observer $\vec{O}$ moving relative to O with steady velocity v perceives the same particle to have mass $\vec{m}$, velocity $\vec{u}$, momentum $\vec{p}$, and energy $\vec{E}$ at position $\vec{r}$ at time $\vec{t}$; he regards the particle as having an acceleration $\vec{a}$ due to a force $\vec{F}$. If the cartesian frames of reference (x, y, z) and $(\vec{x}, \vec{y}, \vec{z})$ for the two observers coincide at time zero ($t = \vec{t} = 0$), and the origin for observer $\vec{O}$ is moving along the z axis of observer O in the positive direction, then the relations between corresponding mechanical quantities are as shown in Table C, where (c.f. eqn. 1.109)

$$\gamma = \{1 - (v/c)^2\}^{-1/2} \tag{C.1}$$

The relativistic momentum of the particle is mu for observer O and $\vec{m}\vec{u}$ for observer

Table C   *The relativistic Lorentz transformation in mechanics*

Space, time	$\vec{x} = x,\quad \vec{y} = y,\quad \vec{z} = \gamma(z - vt),\quad \vec{t} = \gamma(t - c^{-2}vz)$	$x = \vec{x},\quad y = \vec{y},\quad z = \gamma(\vec{z} + v\vec{t}),\quad t = \gamma(\vec{t} + c^{-2}v\vec{z})$
Velocity	$[\vec{u}_x, \vec{u}_y, \vec{u}_z] = \dfrac{[u_x, u_y, \gamma(u_z - v)]}{\gamma(1 - c^{-2}vu_z)}$	$[u_x, u_y, u_z] = \dfrac{[\vec{u}_x, \vec{u}_y, \gamma(\vec{u}_z + v)]}{\gamma(1 + c^{-2}v\vec{u}_z)}$
Acceleration	$[\vec{a}_x, \vec{a}_y, \vec{a}_z] = \dfrac{[a_x, a_y, 0]}{\gamma^2(1 - c^{-2}vu_z)^2} + \dfrac{[\gamma c^{-2}vu_x a_y, \gamma c^{-2}vu_y a_x, a_z]}{\gamma^3(1 - c^{-2}vu_z)^3}$	$[a_x, a_y, a_z] = \dfrac{[\vec{a}_x, \vec{a}_y, 0]}{\gamma^2(1 + c^{-2}v\vec{u}_z)^2} + \dfrac{[-\gamma c^{-2}v\vec{u}_x \vec{a}_y, -\gamma c^{-2}v\vec{u}_y \vec{a}_x, \vec{a}_z]}{\gamma^3(1 + c^{-2}v\vec{u}_z)^3}$
Mass	$\vec{m} = \gamma(1 - c^{-2}vu_z)m$	$m = \gamma(1 + c^{-2}v\vec{u}_z)\vec{m}$
Momentum, energy	$\vec{p}_x = p_x,\quad \vec{p}_y = p_y,\quad \vec{p}_z = \gamma(p_z - c^{-2}vE),\quad \vec{E} = \gamma(E - vp_z)$	$p_x = \vec{p}_x,\quad p_y = \vec{p}_y,\quad p_z = \gamma(\vec{p}_z + c^{-2}v\vec{E}),\quad E = \gamma(\vec{E} + v\vec{p}_z)$
Force	$[\vec{F}_x, \vec{F}_y, \vec{F}_z] = \dfrac{[F_x, F_y, \gamma(1 - c^{-2}vu_z)F_z - \gamma c^{-2}v(u_x F_x + u_y F_y)]}{\gamma(1 - c^{-2}vu_z)}$	$[F_x, F_y, F_z] = \dfrac{[\vec{F}_x, \vec{F}_y, \gamma(1 + c^{-2}v\vec{u}_z)\vec{F}_z + \gamma c^{-2}v(\vec{u}_x \vec{F}_x + \vec{u}_y \vec{F}_y)]}{\gamma(1 + c^{-2}v\vec{u}_z)}$

A superscript arrow indicates a quantity appropriate to the 'moving' observer.

$\vec{O}$. The relativistic energy of the particle is mc^2 for observer O and $\vec{m}c^2$ for observer $\vec{O}$. Relativistic energy includes the rest energy m_0c^2 applicable when the particle is at rest relative to the observer and possesses its rest-mass m_0. The relativistic kinetic energy of the particle is $mc^2 - m_0c^2$ for observer O and $\vec{m}c^2 - m_0c^2$ for observer $\vec{O}$.

Consider a situation in which the particle is at rest relative to observer $\vec{O}$. Then

$$\vec{u} = 0, \quad \vec{m} = m_0, \quad \vec{p} = 0, \quad \vec{E} = m_0c^2 \tag{C.2}$$

It follows from Table C that

$$u = (0, 0, v), \quad m = \gamma m_0, \quad p = (0, 0, \gamma m_0 v), \quad E = \gamma m_0 c^2 \tag{C.3}$$

From the second of eqns. C3 the particle is seen by observer O to possess a mass

$$m = m_0\{1 - (v/c)^2\}^{-1/2} \tag{C.4}$$

We notice that $m \to m_0$ as $v \to 0$, but that $m \to \infty$ as $v \to c$. From the last of eqns. C3 the relativistic energy of the particle as seen by observer O is

$$E = m_0c^2 \{1 - (v/c)^2\}^{-1/2} \tag{C.5}$$

We notice that $E \to m_0c^2$ as $v \to 0$, but that $E \to \infty$ as $v \to c$.

For $v \ll c$ we can expand eqn. C.1 by the binomial theorem and obtain

$$\gamma = 1 + \tfrac{1}{2}(v/c)^2 \tag{C.6}$$

The last of eqns. C.3 then becomes

$$E = m_0c^2 + \tfrac{1}{2}m_0v^2 \tag{C.7}$$

For a slow particle, therefore, the relativistic energy is the sum of the rest-energy m_0c^2 and the Newtonian kinetic energy $\tfrac{1}{2}m_0v^2$.

We may recover from Table C the results that are familiar in Newtonian mechanics by letting $c \to \infty$ and $\gamma \to 1$. We obtain

$$\vec{t} = t, \; \vec{r} = r - vt, \; \vec{u} = u - v, \; \vec{a} = a, \; \vec{m} = m = m_0, \; \vec{F} = F \tag{C.8}$$

But care is necessary with momentum and energy because relativistic energy includes the rest energy in accordance with eqn. C.7. Hence, for non-relativistic velocities, we must write

$$\vec{E} = m_0c^2 + \tfrac{1}{2}m_0\vec{u}^2, \quad E = m_0c^2 + \tfrac{1}{2}m_0u^2 \tag{C.9}$$

The momentum relations in Table C then become, as $c \to \infty$,

$$\vec{p} = p - m_0v, \quad p = \vec{p} + m_0v \tag{C.10}$$

as they should. In the energy relations in Table C, however, it is also necessary to use eqn. C.6. For slow velocities the energy relations in Table C then reduce to

$$\tfrac{1}{2}m_0\vec{u}^2 = \tfrac{1}{2}m_0(u - v)^2, \quad \tfrac{1}{2}m_0u^2 = \tfrac{1}{2}m_0(\vec{u} + v)^2 \tag{C.11}$$

Appendix D Numerical values

Velocity of light in free space $(\mu_0\epsilon_0)^{-1/2}$	$2{\cdot}998 \times 10^8 \, \text{m/s}$
Inductivity of free space μ_0	$4\pi \times 10^{-7} \, \text{H/m}$
Capacitivity of free space ϵ_0	$8{\cdot}854 \times 10^{-12} \, \text{F/m}$
Constant of electrostatics $(4\pi\epsilon_0)^{-1}$	$8{\cdot}988 \times 10^9 \, \text{m/F}$
Impedance of free space $(\mu_0/\epsilon_0)^{1/2}$	$376{\cdot}7 \, \Omega$
Mass of proton	$1{\cdot}67 \times 10^{-27} \, \text{kg}$
Mass of electron m	$9{\cdot}11 \times 10^{-31} \, \text{kg}$
Mass of proton/mass of electron	1836
Charge of proton	$1{\cdot}60 \times 10^{-19} \, \text{C}$
Charge of electron e	$-1{\cdot}60 \times 10^{-19} \, \text{C}$
Charge of proton/mass of proton	$9{\cdot}58 \times 10^7 \, \text{C/kg}$
Charge of electron/mass of electron	$-1{\cdot}76 \times 10^{11} \, \text{C/kg}$
Magnetic moment of electron	$9{\cdot}28 \times 10^{-24} \, \text{A m}^2$
Classical size of electron $\mu_0 e^2/(4\pi m)$	$2{\cdot}82 \times 10^{-15} \, \text{m}$
Gauss	$10^{-4} \, \text{weber/m}^2$
Boltzmann's constant	$1{\cdot}380 \times 10^{-23} \, \text{J/deg}$
Neper	$8{\cdot}686 \, \text{dB}$

Index